Be prepared...
To pass...
To succeed...

Get **REA**dy. It all starts here.
REA's PSSA Math test prep is **fully aligned** with Pennsylvania's core curriculum standards.

Visit us online at
www.rea.com

The Best Test Preparation for the

8th Grade Mathematics

Stephen Hearne, Ph.D.

Penny Luczak, M.A.

and the Staff of
Research & Education Association

Research & Education Association
61 Ethel Road West
Piscataway, New Jersey 08854

The Best Test Preparation for the
Pennsylvania PSSA–Eighth Grade Mathematics

Printed in the United States of America

Library of Congress Control Number 2005901829

International Standard Book Number 0-7386-0029-6

B05

About Our Editors

Stephen Hearne is a professor at Skyline College in San Bruno, California, where he teaches Quantitative Reasoning. He earned his Ph.D. degree from the University of Mississippi in 1998, specializing in Quantitative Psychology. For the past twenty years, Dr. Hearne has tutored students of all ages in math, algebra, statistics, and test preparation. He prides himself in being able to make the complex simple.

Penny Luczak received her B.A. in Mathematics from Rutgers University and her M.A. in Mathematics from Villanova University. She is currently a full-time faculty member at Camden County College in New Jersey. She previously worked as an adjunct instructor at Camden County College as well as Burlington County College and Rutgers University.

About Research & Education Association

Founded in 1959, Research & Education Association is dedicated to publishing the finest and most effective educational materials—including software, study guides, and test preps—for students in middle school, high school, college, graduate school, and beyond. Today, REA's wide-ranging catalog is a leading resource for teachers, students, and professionals. We invite you to visit us at *www.REA.com* to find out how "REA is making the world smarter."

Acknowledgments

In addition to Dr. Hearne and Penny Luczak, we would like to thank Larry B. Kling, Vice President, Editorial, for his editorial direction; Pam Weston, Vice President, Publishing, for ensuring press readiness; Gianfranco Origliato, Project Manager, for coordinating development; Jeanne Audino, Senior Editor, for preflight editorial review; Diane Goldschmidt, Associate Editor, for post-production quality assurance; and Jeremy Rech, Graphic Designer, for graphic design support.

TABLE OF CONTENTS

PASSING THE PSSA ASSESSMENT IN MATHEMATICS

ARITHMETIC REVIEW

ALGEBRA REVIEW

GEOMETRY REVIEW

WORD PROBLEMS REVIEW

QUANTITATIVE ABILITY

SPATIAL SENSE/RELATIONSHIPS

STATISTICS

DIAGNOSTIC TESTS FOR CLASS AND HOMEWORK ASSIGNMENTS

PASSING THE PENNSYLVANIA SYSTEM OF SCHOOL ASSESSMENT (PSSA)–MATHEMATICS

ABOUT THIS BOOK

Our book provides excellent preparation for the Pennsylvania System of School Assessment (PSSA) in Mathematics. Inside you will find reviews that are designed to provide you with the information and strategies needed to do well on this exam. We also provide a full-length practice test, so you can get a good idea of what you'll be facing on test day. Detailed explanations follow the practice test, so if you are having a problem with a particular question, we'll tell you how to solve it.

Our **Teacher's Answer Guide** contains full explanations to the "Class and Homework Assignment" questions in the diagnostic tests at the back of this book. Teachers may obtain the answer guide by contacting REA.

ABOUT THE TEST

Since 1999, the Pennsylvania Department of Education has administered an eighth grade assessment to determine how well a student is advancing. The annual Pennsylvania System of School Assessment (PSSA) is a standards-based assessment used to measure a student's attainment of the academic standards while also determining the degree to which school programs enable students to attain proficiency of the standards. Every Pennsylvania student in 5th, 8th and 11th grade is assessed in reading, math, and writing. It is one of the key tools used to identify students who need additional instruction to master the knowledge and skills detailed in the Pennsylvania School Code and Pennsylvania Academic Standards, the standards which guide education in Pennsylvania.

Students are given 2 hours and 30 minutes to complete all three of the PSSA's sessions. The test comprises both multiple-choice and open-ended questions. A correct answer on a multiple-choice question earns one point, while responses to the open-ended questions can earn from zero to four points. The use of calculators is permitted during this test, but not for the first several items. Students are provided with a Reference Sheet that supplies common formulas and conversion tables. This sheet may be used to help in solving any of the test questions. The test questions are designed to test student mastery of the five content areas:

1. Numbers and Operations
2. Measurement
3. Geometry
4. Algebraic Concepts
5. Data Analysis and Probability

HOW TO USE THIS BOOK

What do I study first?

Read through the review and our suggestions for test-taking. Studying the review thoroughly will reinforce the basic skills you will need to do well on the test. Our practice drills and diagnostic tests feature five answer choices, whereas the actual exam has only four choices. This results in a greater challenge and more rigorous preparation.

The PSSA's four-choice format is accurately reflected in our practice test, which you'll find in the back of this book. Our practice test is designed to capture the spirit of the PSSA, providing you with an experience that mimics the administration of the exam.

When should I start studying?

It is never too early to start studying for the exam. The earlier you begin, the more time you will have to sharpen your skills. Do not procrastinate! Cramming is *not* an effective way to study, since it does not allow you the time needed to learn the test material. The sooner you learn the format of the exam, the more time you will have to familiarize yourself with the exam content.

ABOUT THE REVIEW SECTIONS

The review sections in this book are designed to help you sharpen the basic skills needed to approach the exam, as well as to provide strategies for attacking each type of question. You will also find exercises to reinforce what you have learned. By using the reviews in conjunction with the drills and practice test, you will put yourself in a position to succeed on the exam.

TEST-TAKING TIPS

There are many ways to acquaint yourself with this type of examination and help alleviate your test-taking anxieties. Listed below are ways to help yourself.

Become comfortable with the format. When you are practicing, simulate the conditions under which you will be taking the actual test. Take the practice test in a quiet room, free of distractions. Stay calm and pace yourself. After simulating the test only a couple of times, you will boost your chances of doing well, and you will be able to sit down for the actual exam with much more confidence.

Read all of the possible answers. Just because you think you have found the correct response, do not automatically assume that it is the best answer. Read through each choice to be sure that you are not making a mistake by jumping to conclusions.

Use the process of elimination. Go through each answer to a question and eliminate those that are obviously incorrect. By eliminating two answer choices, you can vastly improve your chances of getting the item correct, since there will only be two choices left from which to make your guess. It is recommended that you attempt to answer each question, since your score is calculated based on how many questions you get right, and unanswered or incorrectly answered questions receive no credit.

Work quickly and steadily. Avoid focusing on any one problem for too long. Even so, you should never rush. Rushing leads to careless errors. Taking the practice test in this book will help you learn to budget your time.

Learn the directions and format for the test. Familiarizing yourself with the directions and format of the test will not only save time, but will also help you avoid anxiety (and the mistakes caused by getting anxious).

Work on the easier questions first. If you find yourself working too long on one question, make a mark next to it on your test booklet and continue. After you have answered all of the questions that you can, go back to the ones you have skipped.

Avoid errors when indicating your answers on the answer sheet. Marking one answer out of sequence can throw off your answer key and thus your score. Be extremely careful.

Eliminate obvious wrong answers. This ties in with using the process of elimination. Sometimes a question will have one or two answer choices that are a little odd. These answers will be obviously wrong for one of several reasons: they may be impossible given the conditions of the problem, they may violate mathematical rules or principles, or they may be illogical. Being able to spot obvious wrong answers before you finish a problem gives you an advantage because you will be able to make a better educated guess from the remaining choices even if you are unable to fully solve the problem.

Work from answer choices. One of the ways you can use a multiple-choice format to your advantage is to work backwards from the answer choices to solve a problem. This is not a strategy you can use all the time, but it can be helpful if you can just plug the choices into a given statement or equation. The answer choices can often narrow the scope of responses. You may be able to make an educated guess based on eliminating choices that you know do not fit into the problem.

THE DAY OF THE TEST

Before the Test

On the day of the test, you should wake up early (hopefully, after a decent night's rest) and have a good breakfast. Make sure to dress comfortably, so that you are not distracted by being too hot or too cold while taking the test. Also plan on arriving at school early. This will allow you to collect your thoughts and relax before the test, and will also spare you the anguish that comes with being late.

During the Test

Follow all of the rules and instructions given by your teacher or test supervisor.

When all of the test materials have been passed out, you will receive directions for filling out your answer sheet. You must fill out this sheet carefully since this information will be printed on your score report. Fill out your name exactly as it appears on your identification documents, unless otherwise instructed.

You can write in your test booklet or on scratch paper, which will be provided. However, you must be sure to mark your answers in the appropriate spaces in the answer folder. Each numbered row will contain four ovals corresponding to each answer choice for that question. Fill in the oval that corresponds to your answer darkly, completely, and neatly. You can change your answer, but be sure to completely erase your old answer. Only one answer should be marked. This is very important, as your answer sheet will be machine-scored and stray lines or unnecessary marks may cause the machine to score your answers incorrectly.

ARITHMETIC REVIEW

1. INTEGERS AND REAL NUMBERS

Most of the numbers used in algebra belong to a set called the **real numbers** or **reals**. This set can be represented graphically by the real number line.

Given the number line below, we arbitrarily fix a point and label it with the number 0. In a similar manner, we can label any point on the line with one of the real numbers, depending on its position relative to 0. Numbers to the right of zero are positive, while those to the left are negative. Value increases from left to right, so that if a is to the right of b, it is said to be greater than b.

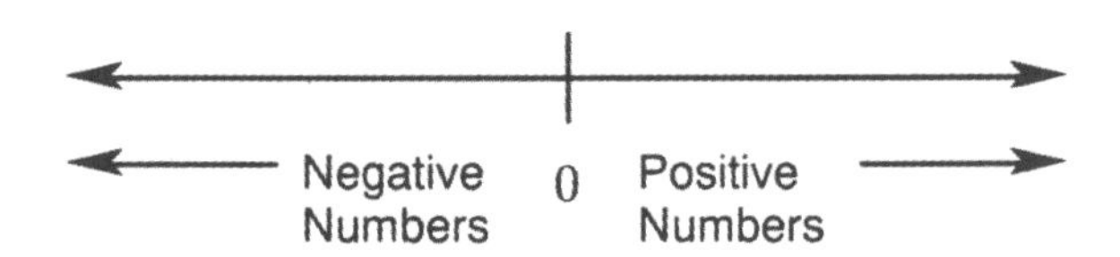

If we now divide the number line into equal segments, we can label the points on this line with real numbers. For example, the point 2 lengths to the left of zero is – 2, while the point 3 lengths to the right of zero is + 3 (the + sign is usually assumed, so + 3 is written simply as 3). The number line now looks like this:

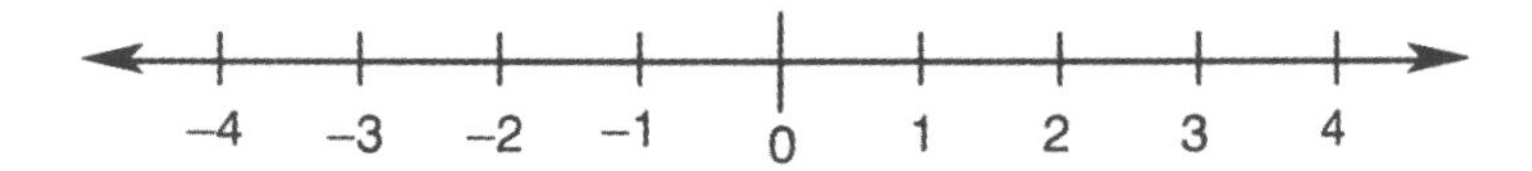

These boundary points represent the subset of the reals known as the **integers**. The set of integers is made up of both the positive and negative whole numbers: {... – 4, – 3, – 2, – 1, 0, 1, 2, 3, 4, ...}. Some subsets of integers are:

Natural Numbers or Positive Numbers—the set of integers starting with 1 and increasing: $\mathcal{N}$= {1, 2, 3, 4, ...}.

Whole Numbers—the set of integers starting with 0 and increasing: $\mathcal{W}$= {0, 1, 2, 3, ...}.

Negative Numbers—the set of integers starting with – 1 and decreasing: $\mathcal{Z}$= { – 1, – 2, – 3 ...}.

Prime Numbers—the set of positive integers greater than 1 that are divisible only by 1 and themselves: {2, 3, 5, 7, 11, ...}.

Even Integers—the set of integers divisible by 2: {..., – 4, – 2, 0, 2, 4, 6, ...}.

Odd Integers—the set of integers not divisible by 2: { ..., – 3, – 1, 1, 3, 5, 7, ...}.

RATIONAL AND IRRATIONAL NUMBERS

A rational number is any number that can be written in the form $\frac{a}{b}$ where a is any integer and b is any integer except zero. An irrational number is a number that cannot be written as a simple fraction. It is an infinite and non-repeating decimal.

The tree diagram below shows you the relationships between the different types of numbers.

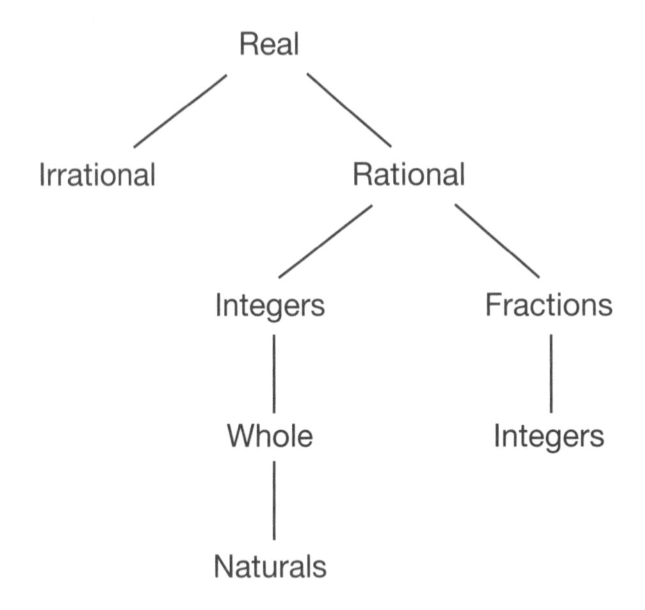

EXAMPLE

Here are some examples of some rational numbers.

2 3 5 10 32 –2 –4 –18 –25

$\frac{1}{4}$ $\frac{1}{2}$ $\frac{2}{3}$ $-\frac{1}{4}$ $-\frac{4}{7}$ $-\frac{10}{55}$ $\frac{21}{9}$ $\frac{101}{635}$

EXAMPLE

Here are some examples of irrational numbers.

π — approximately equal to 3.14159

e — approximately equal to 2.71828

$\sqrt{2}$ — approximately equal to 1.41421

$\sqrt{3}$ — approximately equal to 1.73205

$\sqrt{5}$ — approximately equal to 2.23607

PROBLEM

List the numbers shown below from least to greatest.

$$\frac{1}{3}, \quad \sqrt{3}, \quad 3, \quad 0.3$$

SOLUTION

$\frac{1}{3} \approx 0.33333$

$\sqrt{3} \approx 1.73205$

Therefore, the numbers from least to greatest are: $0.3, \frac{1}{3}, \sqrt{3}, 3$

PROBLEM

List the numbers shown below from greatest to least

$$2^2, \quad \sqrt{9}, \quad e, \quad \pi$$

SOLUTION

$2^2 = 4$; $\sqrt{9} = 3$; $e \approx 2.71828$; $\pi \approx 3.14159$

Therefore, the numbers from greatest to least are: $2^2, \pi, \sqrt{9}, e$

PROBLEM

List the numbers shown below from least to greatest.

$$-5, \quad \sqrt{5}, \quad -8, \quad \frac{16}{7}$$

SOLUTION

$\sqrt{5} \approx 2.23607$

$\frac{16}{7} \approx 2.28571$

Therefore, the numbers from least to greatest are: $-8, -5, \sqrt{5}, \frac{16}{7}$

PROBLEM

Classify each of the following numbers into as many different sets as possible. Example: real, integer …

(1) 0 (2) 9 (3) $\sqrt{6}$

(4) $^1/_2$ (5) $^2/_3$ (6) 1.5

SOLUTION

(1) Zero is a real number and an integer.

(2) 9 is a real, natural number, and an integer.

(3) $\sqrt{6}$ is a real number.

(4) $^1/_2$ is a real number.

(5) $^2/_3$ is a real number.

(6) 1.5 is a real number.

PROBLEM

Write each integer below as a product of its primes.

2 12 5 22 18 36

SOLUTION

A prime number is a number that has no factors other than itself and 1. For example, the numbers 1, 3, 5, and 7 are prime numbers. Write each integer as a product of its primes.

$$2 = 2 \times 1$$
$$12 = 4 \times 3 = 2 \times 2 \times 3$$
$$5 = 5 \times 1$$
$$22 = 11 \times 2$$
$$18 = 6 \times 3 = 2 \times 3 \times 3$$
$$36 = 6 \times 6 = 3 \times 2 \times 3 \times 2$$
$$\text{or } 18 \times 2 = 9 \times 2 \times 2 = 3 \times 3 \times 2 \times 2$$

PROBLEM

Find the greatest common divisor (GCD) for the following numbers: 12 and 24.

SOLUTION

Step 1 is to write out each number as a product of its primes.

$$12 = 6 \times 2 = 3 \times 2 \times 2$$
$$24 = 12 \times 2 = 6 \times 2 \times 2 = 3 \times 2 \times 2 \times 2$$

$3 \times 2 \times 2$ is the common factor in both sets of prime factors.

Step 2 is to multiply the common factors together.

$$3 \times 2 \times 2 = 12$$

Therefore, 12 is the GCD.

ABSOLUTE VALUE

The **absolute value** of a number is represented by two vertical lines around the number, and is equal to the given number, regardless of sign.

The absolute value of a real number A is defined as follows:

$$|A| = \begin{cases} A \text{ if } A \geq 0 \\ -A \text{ if } A < 0 \end{cases}$$

EXAMPLE

$|5| = 5, |-8| = -(-8) = 8.$

Absolute values follow the given rules:

(A) $|-A| = |A|$

(B) $|A| \geq 0$, equality holding only if $A = 0$

(C) $\left|\frac{A}{B}\right| = \frac{|A|}{|B|}, B \neq 0$

(D) $|AB| = |A| \times |B|$

(E) $|A|^2 = A^2$

Absolute value can also be expressed on the real number line as the distance of the point represented by the real number from the point labeled 0.

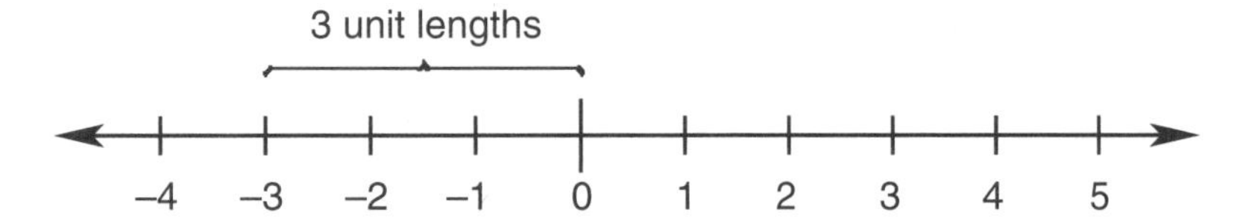

So $|-3| = 3$ because -3 is 3 units to the left of 0.

PROBLEM

Classify each of the following statements as true or false. If it is false, explain why.

(1) $|-120| > 1$

(2) $|4 - 12| = |4| - |12|$

(3) $|4 - 9| = 9 - 4$

(4) $|12 - 3| = 12 - 3$

(5) $|-12a| = 12|a|$

SOLUTION

(1) True

(2) False, $|4 - 12| = |4| - |12|$

$$|-8| = 4 - 12$$

$$8 \neq -8$$

In general, $|a + b| \neq |a| + |b|$

(3) True

(4) True

(5) True

PROBLEM

Calculate the value of each of the following expressions:

(1) $||2 - 5| + 6 - 14|$

(2) $\dfrac{11}{6} + \dfrac{5}{16} = \dfrac{88}{48} + \dfrac{15}{48} = \dfrac{103}{48}$

SOLUTION

Before solving this problem, one must remember the order of operations: parenthesis, multiplication and division, addition and subtraction.

(1) $||-3| + 6 - 14| = |3 + 6 - 14| = |9 - 14| = |-5| = 5$

(2) $(5 \times 4) + {}^{12}/_{4} = 20 + 3 = 23$

PROBLEM

Find the absolute value for each of the following:

(1) zero

(2) 4

(3) $-\pi$

(4) a, where a is a real number

SOLUTION

(1) $|0| = 0$

(2) $|4| = 4$

(3) $|-\pi| = \pi$

(4) for $a > 0$, $|a| = a$

for $a = 0$, $|a| = 0$

for $a < 0$, $|a| = -a$

i.e., $|a| = \begin{cases} a \text{ if } a > 0 \\ 0 \text{ if } a = 0 \\ -a \text{ if } a < 0 \end{cases}$

PROBLEM

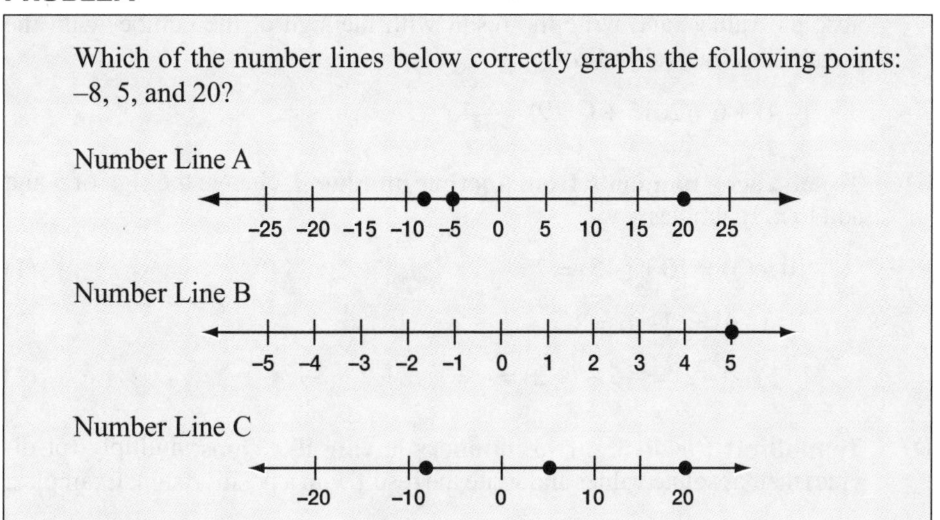

SOLUTION

The correct answer is Number Line C. The numbers in Number Line A are incremented correctly, but –5 is graphed instead of 5. Number Line B is incorrect because only 5 is graphed. Number Line C is correct because the numbers are incremented correctly and the points are correctly graphed.

PROBLEM

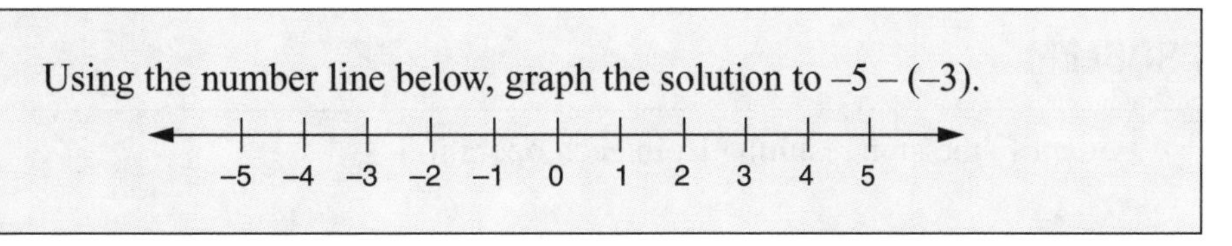

SOLUTION

Step 1 is to graph point –5 on the number line.

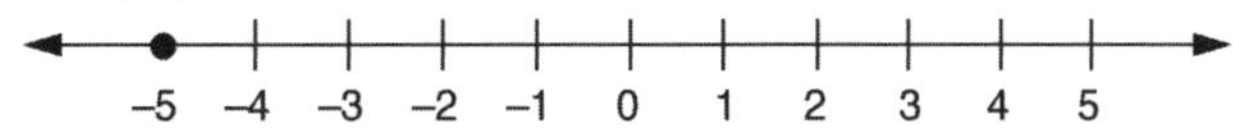

Step 2 is to move 3 units to the *right* of –5. In this problem we move to the right of –5 because a negative number is being subtracted. Since –2 is 3 units to the right of –5, graph –2 on the number line.

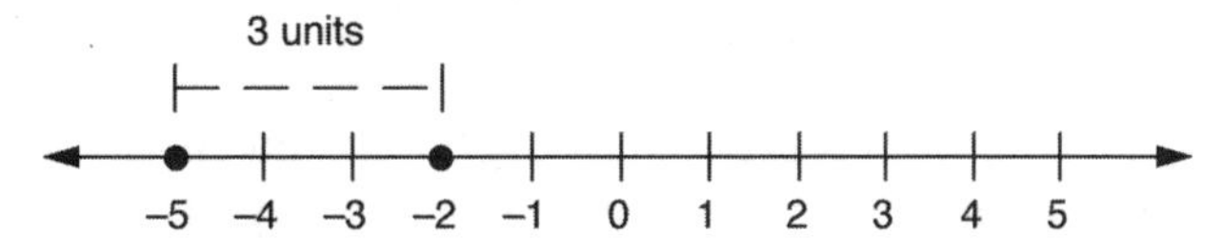

POSITIVE AND NEGATIVE NUMBERS

A) **To add two numbers with like signs,** add their absolute values and write the sum with the common sign. So,

$$6 + 2 = 8, (-6) + (-2) = -8$$

B) **To add two numbers with unlike signs,** find the difference between their absolute values, and write the result with the sign of the number with the greater absolute value. So,

$(-4) + 6 = 2$, $15 + (-19) = -4$

C) **To subtract a number *b* from another number *a*,** change the sign of b and add to a. Examples:

$10 - (3) = 10 + (-3) = 7$ (1)

$2 - (-6) = 2 + 6 = 8$ (2)

$(-5) - (-2) = -5 + (+2) = -3$ (3)

D) **To multiply (or divide) two numbers having like signs,** multiply (or divide) their absolute values and write the result with a positive sign. Examples:

$(5)(3) = 15$ (1)

$-6 / -3 = 2$ (2)

E) **To multiply (or divide) two numbers having unlike signs,** multiply (or divide) their absolute values and write the result with a negative sign. Examples:

$(-2)(8) = -16$ (1)

$9 / -3 = -3$ (2)

PROBLEM

Identify the sign resulting from each operation.

a) (+)(+) c) (+)(–)

b) (–)(–) d) (–)(+)

SOLUTION

The correct answer for problem a is a positive number. When multiplying two positive numbers, the product is always positive.

The correct answer for problem b is a positive number. When multiplying two negative numbers, the product is always positive.

The correct answer for problem c is a negative number. When multiplying a positive and a negative number, the product is always negative.

The correct answer for problem d is a negative number. When multiplying a negative and a positive number, the product is always negative.

According to the law of signs for real numbers, the square of a positive or negative number is always positive. This means that it is impossible to take the square root of a negative number in the real number system.

ORDER OF OPERATIONS

When a series of operations involving addition, subtraction, multiplication, or division is indicated, first resolve any operations in parentheses, then resolve exponents, then resolve multiplication and/or division, and finally perform addition and/or subtraction. One way to remember this is to recite "Please excuse my dear Aunt Sally." The **P** stands for parentheses, the **E** for exponents, the **M** for multiplication, the **D** for division, the **A** for addition and the **S** for subtraction. Now let's try using the order of operations.

Consider

$$60 - 25 \div 5 + 15 - 100 + 4 \times 10$$

$$= 60 - 25 \div 5 + 15 - 100 + 40$$

$$= 60 - 5 + 15 - 100 + 40$$

$$= 115 - 105$$

$$= 10$$

Notice that $25 \div 5$ could be evaluated at the same time that 4×10 is evaluated, since they are both part of the multiplication/division step.

DRILL: INTEGERS AND REAL NUMBERS

Addition

1. Simplify $4 + (-7) + 2 + (-5)$.

(A) -6 (B) -4 (C) 0 (D) 6 (E) 18

2. Simplify $144 + (-317) + 213$.

(A) -357 (B) -40 (C) 40 (D) 357 (E) 674

3. Simplify $|4 + (-3)| + |-2|$.

(A) -2 (B) -1 (C) 1 (D) 3 (E) 9

4. What integer makes the equation $-13 + 12 + 7 + ? = 10$ a true statement?

(A) -22 (B) -10 (C) 4 (D) 6 (E) 10

5. Simplify $4 + 17 + (-29) + 13 + (-22) + (-3)$.

(A) -44 (B) -20 (C) 23 (D) 34 (E) 78

Subtraction

6. Simplify $319 - 428$.

(A) -111 (B) -109 (C) -99 (D) 109 (E) 747

7. Simplify 91,203 – 37,904 + 1,073.

(A) 54,372 (B) 64,701 (C) 128,034 (D) 129,107 (E) 130,180

8. Simplify $| 43 - 62 | - | - 17 - 3 |$.

(A) – 39 (B) – 19 (C) – 1 (D) 1 (E) 39

9. Simplify $-(-4-7)+(-2)$.

(A) – 22 (B) – 13 (C) – 9 (D) 7 (E) 9

10. In Great Smoky Mountains National Park, Mt. LeConte rises from 1,292 feet above sea level to 6,593 feet above sea level. How tall is Mt. LeConte?

(A) 4,009 ft (B) 5,301 ft (C) 5,699 ft (D) 6,464 ft (E) 7,885 ft

Multiplication

11. Simplify $-3(-18)(-1)$.

(A) – 108 (B) – 54 (C) – 48 (D) 48 (E) 54

12. Simplify $| -42 | \times | 7 |$.

(A) – 294 (B) – 49 (C) – 35 (D) 284 (E) 294

13. Simplify $-6 \times 5(-10)(-4) 0 \times 2$.

(A) – 2,400 (B) – 240 (C) 0 (D) 280 (E) 2,700

14. Simplify $-| -6 \times 8 |$.

(A) – 48 (B) – 42 (C) 2 (D) 42 (E) 48

15. A city in Georgia had a record low temperature of –3°F one winter. During the same year, a city in Michigan experienced a record low that was nine times the record low set in Georgia. What was the record low in Michigan that year?

(A) – 31°F (B) – 27°F (C) – 21°F (D) – 12°F (E) – 6°F

Division

16. Simplify $-24 \div 8$.

(A) – 4 (B) – 3 (C) – 2 (D) 3 (E) 4

17. Simplify $(-180) \div (-12)$.

(A) – 30 (B) – 15 (C) 1.5 (D) 15 (E) 216

18. Simplify $| -76 | \div | -4 |$.

(A) – 21 (B) – 19 (C) 13 (D) 19 (E) 21.5

19. Simplify $|216 \div (-6)|$.

(A) -36 (B) -12 (C) 36 (D) 38 (E) 43

20. At the end of the year, a small firm has $2,996 in its account for bonuses. If the entire amount is equally divided among the 14 employees, how much does each one receive?

(A) $107 (B) $114 (C) $170 (D) $210 (E) $214

Order of Operations

21. Simplify $\frac{4+8\times 2}{5-1}$

(A) 4 (B) 5 (C) 6 (D) 8 (E) 12

22. $96 \div 3 \div 4 \div 2 =$

(A) 65 (B) 64 (C) 16 (D) 8 (E) 4

23. $3 + 4 \times 2 - 6 \div 3 =$

(A) -1 (B) 5/3 (C) 8/3 (D) 9 (E) 12

24. $[(4 + 8) \times 3] \div 9 =$

(A) 4 (B) 8 (C) 12 (D) 24 (E) 36

25. $18 + 3 \times 4 \div 3 =$

(A) 3 (B) 5 (C) 10 (D) 22 (E) 28

26. $(29 - 17 + 4) \div 4 + |-2| =$

(A) $2^2/_3$ (B) 4 (C) $4^2/_3$ (D) 6 (E) 15

27. $(-3) \times 5 - 20 \div 4 =$

(A) -75 (B) -20 (C) -10 (D) $-8^3/_4$ (E) 20

28. $\frac{11\times 2+2}{16-2\times 2} =$

(A) 11/16 (B) 1 (C) 2 (D) 3 2/3 (E) 4

29. $|-8-4| \div 3 \times 6 + (-4) =$

(A) 20 (B) 26 (C) 32 (D) 62 (E) 212

30. $32 \div 2 + 4 - 15 \div 3 =$

(A) 0 (B) 7 (C) 15 (D) 23 (E) 63

2. FRACTIONS

The fraction, a/b, where the **numerator** is a and the **denominator** is b, implies that a is being divided by b. The denominator of a fraction can never be zero since a number divided by zero is not defined. If the numerator is greater than the denominator, the fraction is called an **improper fraction**. A **mixed number** is the sum of a whole number and a fraction, i.e., $4^3/_8 = 4 + {}^3/_8$.

OPERATIONS WITH FRACTIONS

A) **To change a mixed number to an improper fraction,** simply multiply the whole number by the denominator of the fraction and add the numerator. This product becomes the numerator of the result and the denominator remains the same. E.g.,

$$5\frac{2}{3} = \frac{(5 \times 3) + 2}{3} = \frac{15 + 2}{3} = \frac{17}{3}$$

PROBLEM

Simplify the following fraction: $2\frac{12}{8}$.

SOLUTION

Step 1 is to convert $2\frac{12}{8}$ into an improper fraction.

The whole number "2" gets multiplied by the denominator "8."

$$2 \times 8 = 16$$

Next, add the numerator to the previous result.

$$16 + 12 = 28$$

The improper fraction is $\frac{28}{8}$.

Step 2 is to simplify $\frac{28}{8}$.

Find the greatest common divisor (GCD) of 28 and 8. This is done by writing the numbers as products of their primes.

$$28 = 14 \times 2 \qquad 8 = 4 \times 2$$

$$= 7 \times 2 \times 2 \qquad = 2 \times 2 \times 2$$

The GCD of 28 and 8 is 4.

Next, divide the numerator and denominator by the GCD.

$$\frac{28}{4} = 7 \qquad \frac{8}{4} = 2$$

Therefore, $2\frac{12}{8}$ simplified is $\frac{7}{2}$. $\frac{7}{2}$ can also be written as a mixed number, $3\frac{1}{2}$.

To change an improper fraction to a mixed number, simply divide the numerator by the denominator. The remainder becomes the numerator of the fractional part of the mixed number, and the denominator remains the same. E.g.,

$$\frac{35}{4} = 35 \div 4 = 8\frac{3}{4}$$

To check your work, change your result back to an improper fraction to see if it matches the original fraction.

B) **To find the sum of two fractions having a common denominator**, simply add together the numerators of the given fractions and put this sum over the common denominator.

$$\frac{11}{3} + \frac{5}{3} = \frac{11+5}{3} = \frac{16}{3}$$

Similarly for subtraction,

$$\frac{11}{3} - \frac{5}{3} = \frac{11-5}{3} = \frac{6}{3} = 2$$

PROBLEM

Find the solution to the following problem. Simplify the answer, if possible.

$$-\frac{8}{16} - \left(-\frac{4}{16}\right)$$

SOLUTION

Step 1 is to subtract the numerators.

$-8 - (-4) = -4$

Since the denominators in $-\frac{8}{16}$ and $-\frac{4}{16}$ are equal, keep the common denominator.

The correct answer is $-\frac{4}{16}$. To simplify the answer, write out the numerator and denominator as products of their primes to find the greatest common divisor (GCD).

$4 = 2 \times 2$

$16 = 4 \times 4$
$= 2 \times 2 \times 2 \times 2$

The GCD of 4 and 16 is 4. The next step is to divide the numerator and denominator by the GCD.

$4 \div 4 = 1 \qquad 16 \div 4 = 4$

Simplified, $-\frac{4}{16}$ becomes $-\frac{1}{4}$.

The correct answer is $-\frac{1}{4}$.

C) **To find the sum of the two fractions having different denominators**, it is necessary to find the **lowest common denominator**, (**LCD**) of the different denominators using a process called **factoring**.

To **factor** a number means to find two numbers that when multiplied together have a product equal to the original number. These two numbers are then said to be **factors** of the original number. E.g., the factors of 6 are

(1) 1 and 6 since $1 \times 6 = 6$.

(2) 2 and 3 since $2 \times 3 = 6$.

Every number is the product of itself and 1. A **prime factor** is a number that does not have any factors besides itself and 1. This is important when finding the LCD of two fractions having different denominators.

To find the LCD of $^{11}/_{6}$ and $^{5}/_{16}$, we must first find the prime factors of each of the two denominators.

$6 = 2 \times 3$

$16 = 2 \times 2 \times 2 \times 2$

$\text{LCD} = 2 \times 2 \times 2 \times 2 \times 3 = 48$

Note that we do not need to repeat the 2 that appears in both the factors of 6 and 16.

Once we have determined the LCD of the denominators, each of the fractions must be converted into equivalent fractions having the LCD as a denominator.

Rewrite 11/6 and 5/16 to have 48 as their denominators.

$6 \times ? = 48 \qquad 16 \times ? = 48$

$6 \times 8 = 48 \qquad 16 \times 3 = 48$

If the numerator and denominator of each fraction is multiplied (or divided) by the same number, the value of the fraction will not change. This is because a fraction *b/b*, *b* being any number, is equal to the multiplicative identity, 1.

Therefore,

$\frac{11}{6} \times \frac{8}{8} = \frac{88}{48} \qquad \frac{5}{16} \times \frac{3}{3} = \frac{15}{48}$

We may now find

$$\frac{11}{6}+\frac{5}{16}=\frac{88}{48}+\frac{15}{48}=\frac{103}{48}$$

Similarly for subtraction,

$$\frac{11}{6}-\frac{5}{16}=\frac{88}{48}-\frac{15}{48}=\frac{73}{48}$$

PROBLEM

Find the lowest common denominator (LCD) for $\frac{5}{6}$, $\frac{4}{21}$, and $\frac{1}{7}$.

SOLUTION

Step 1 is to list multiples of each denominator.

6 = 6, 12, 18, 24, 30, 36, 42, 48, 54, 60, 66, 72

21 = 21, 42, 63, 84, 105

7 = 7, 14, 21, 28, 35, 42, 49, 56, 63, 70

The lowest common multiple of 6, 21, and 7 is 42. Since 42 is the lowest common multiple, it is also the LCD.

PROBLEM

Find the solution to the problem below.

$$\frac{108}{109}+\frac{0}{144}$$

SOLUTION

It is unnecessary to find the lowest common denominator (LCD) in this problem.

Step 1 is to rewrite the problem substituting 0 for $\frac{0}{144}$.

$$\frac{108}{109}+0$$

Step 2 is to perform the addition.

$$\frac{108}{109}+0=\frac{108}{109}$$

The correct answer is $\frac{108}{109}$.

D) **To find the product of two or more fractions,** simply multiply the numerators of the given fractions to find the numerator of the product and multiply the denominators of the given fractions to find the denominator of the product. E.g.,

$$\frac{2}{3} \times \frac{1}{5} \times \frac{4}{7} = \frac{2 \times 1 \times 4}{3 \times 5 \times 7} = \frac{8}{105}$$

E) **To find the quotient of two fractions**, simply invert the divisor and multiply. E.g.,

$$\frac{8}{9} \div \frac{1}{3} = \frac{8}{9} \times \frac{3}{1} = \frac{24}{9} = \frac{8}{3}$$

PROBLEM

Find the solution to the following problem. Simplify the answer.

$$\frac{1}{4} \times \frac{6}{8} \div \frac{2}{3}$$

SOLUTION

In mathematics, multiplication precedes division in the order of operations. Therefore, the first step is to multiply $\frac{1}{4}$ and $\frac{6}{8}$.

$$\frac{1}{4} \times \frac{6}{8}$$

The answer is $\frac{6}{32}$.

Next, perform the division.

$$\frac{6}{32} \div \frac{2}{3}$$

Invert the divisor.

$$\frac{6}{32} \div \frac{3}{2}$$

Change the operation to multiplication.

$$\frac{6}{32} \times \frac{3}{2}$$

Next, multiply the numerators.

$$6 \times 3 = 18$$

Next, multiply the denominators.

$$32 \times 2 = 64$$

The correct answer is $\frac{18}{64}$, or $\frac{9}{32}$.

F) **To simplify a fraction** is to convert it into a form in which the numerator and denominator have no common factor other than 1. E.g.,

$$\frac{12}{18} = \frac{12 \div 6}{18 \div 6} = \frac{2}{3}$$

G) A **complex fraction** is a fraction whose numerator and/or denominator is made up of fractions. To simplify the fraction, find the LCD of all the fractions. Multiply both the numerator and denominator by this number and simplify.

PROBLEM

If $a = 4$ and $b = 7$, find the value of $\dfrac{a + \frac{a}{b}}{a - \frac{a}{b}}$

SOLUTION

By substitution,

$$\frac{a + \frac{a}{b}}{a - \frac{a}{b}} = \frac{4 + \frac{4}{7}}{4 - \frac{4}{7}}$$

In order to combine the terms, we must find the LCD of 1 and 7. Since both are prime factors, the LCD = $1 \times 7 = 7$.

Multiplying both numerator and denominator by 7, we get:

$$\frac{7(4 + \frac{4}{7})}{7(4 - \frac{4}{7})} = \frac{28 + 4}{28 - 4} = \frac{32}{24}$$

By dividing both numerator and denominator by 8, 32/24 can be reduced to 4/3.

PROBLEM

Simplify the following fraction: $\dfrac{244}{12}$.

SOLUTION

Step 1 is to find the greatest common divisor (GCD) of 244 and 12. To do this, write out each number as products of its primes.

$$244 = 2 \times 2 \times 61 \qquad 12 = 4 \times 3$$
$$= 2 \times 2 \times 3$$

The GCD of 244 and 12 is 4.

Step 2 is to divide the numerator and denominator by the GCD.

$$\frac{244}{4} = 61 \qquad \frac{12}{4} = 3$$

The correct answer is $\dfrac{61}{3}$ or $20\dfrac{1}{3}$.

DRILL: FRACTIONS

Fractions

DIRECTIONS: Add and write the answer in simplest form.

1. 5/12 + 3/12 =

(A) 5/24 (B) 1/3 (C) 8/12 (D) 2/3 (E) 1 1/3

2. 5/8 + 7/8 + 3/8 =

(A) 15/24 (B) 3/4 (C) 5/6 (D) 7/8 (E) 1 7/8

3. 131 2/15 + 28 3/15 =

(A) 159 1/6 (B) 159 1/5 (C) 159 1/3 (D) 159 1/2 (E) 159 3/5

4. 3 5/18 + 2 1/18 + 8 7/18 =

(A) 13 13/18 (B) 13 3/4 (C) 13 7/9 (D) 14 1/6 (E) 14 2/9

5. 17 9/20 + 4 3/20 + 8 11/20 =

(A) 29 23/60 (B) 29 23/20 (C) 30 3/20
(D) 30 1/5 (E) 30 3/5

Subtract Fractions with the Same Denominator

DIRECTIONS: Subtract and write the answer in simplest form.

6. 4 7/8 – 3 1/8 =

(A) 1 1/4 (B) 1 3/4 (C) 1 12/16 (D) 1 7/8 (E) 2

7. 132 5/12 – 37 3/12 =

(A) 94 1/6 (B) 95 1/12 (C) 95 1/6 (D) 105 1/6 (E) 169 2/3

8. 19 1/3 – 2 2/3 =

(A) 16 2/3 (B) 16 5/6 (C) 17 1/3 (D) 17 2/3 (E) 17 5/6

9. 8/21 – 5/21 =

(A) 1/21 (B) 1/7 (C) 3/21 (D) 2/7 (E) 3/7

10. 82 7/10 – 38 9/10 =

(A) 43 4/5 (B) 44 1/5 (C) 44 2/5 (D) 45 1/5 (E) 45 2/10

Finding the LCD

DIRECTIONS: Find the lowest common denominator of each group of fractions.

11. 2/3, 5/9, and 1/6.

(A) 9 (B) 18 (C) 27 (D) 54 (E) 162

12. 1/2, 5/6, and 3/4.

(A) 2 (B) 4 (C) 6 (D) 12 (E) 48

13. 7/16, 5/6, and 2/3.

(A) 3 (B) 6 (C) 12 (D) 24 (E) 48

14. 8/15, 2/5, and 12/25.

(A) 5 (B) 15 (C) 25 (D) 75 (E) 375

15. 2/3, 1/5, and 5/6.

(A) 15 (B) 30 (C) 48 (D) 90 (E) 120

16. 1/3, 9/42, and 4/21.

(A) 21 (B) 42 (C) 126 (D) 378 (E) 4,000

17. 4/9, 2/5, and 1/3.

(A) 15 (B) 17 (C) 27 (D) 45 (E) 135

18. 7/12, 11/36, and 1/9.

(A) 12 (B) 36 (C) 108 (D) 324 (E) 432

19. 3/7, 5/21, and 2/3.

(A) 21 (B) 42 (C) 31 (D) 63 (E) 441

20. 13/16, 5/8, and 1/4.

(A) 4 (B) 8 (C) 16 (D) 32 (E) 64

Adding Fractions with Different Denominators

DIRECTIONS: Add and write the answer in simplest form.

21. 1/3 + 5/12 =

(A) 2/5 (B) 1/2 (C) 9/12 (D) 3/4 (E) 1 1/3

22. 3 5/9 + 2 1/3 =

(A) 5 1/2 (B) 5 2/3 (C) 5 8/9 (D) 6 1/9 (E) 6 2/3

23. 12 9/16 + 17 3/4 + 8 1/8 =

(A) 37 7/16 (B) 38 7/16 (C) 38 1/2 (D) 38 2/3 (E) 39 3/16

24. 28 4/5 + 11 16/25 =

(A) 39 2/3 (B) 39 4/5 (C) 40 9/25 (D) 40 2/5 (E) 40 11/25

25. 2 1/8 + 1 3/16 + 5/12 =

(A) 3 35/48 (B) 3 3/4 (C) 3 19/24 (D) 3 13/16 (E) 4 1/12

Subtraction with Different Denominators

DIRECTIONS: Subtract and write the answer in simplest form.

26. 8 9/12 – 2 2/3 =

(A) 6 1/12 (B) 6 1/6 (C) 6 1/3 (D) 6 7/12 (E) 6 2/3

27. 185 11/15 – 107 2/5 =

(A) 77 2/15 (B) 78 1/5 (C) 78 3/10 (D) 78 1/3 (E) 78 9/15

28. 34 2/3 – 16 5/6 =

(A) 16 (B) 16 1/3 (C) 17 1/2 (D) 17 (E) 17 5/6

29. 3 11/48 – 2 3/16 =

(A) 47/48 (B) 1 1/48 (C) 1 1/24 (D) 1 8/48 (E) 1 7/24

30. 81 4/21 – 31 1/3 =

(A) 47 3/7 (B) 49 6/7 (C) 49 1/6 (D) 49 5/7 (E) 49 13/21

Multiplication

DIRECTIONS: Multiply and reduce the answer.

31. 2/3 × 4/5 =

(A) 6/8 (B) 3/4 (C) 8/15 (D) 10/12 (E) 6/5

32. 7/10 × 4/21 =

(A) 2/15 (B) 11/31 (C) 28/210 (D) 1/6 (E) 4/15

33. 5 1/3 × 3/8 =

(A) 4/11 (B) 2 (C) 8/5 (D) 5 1/8 (E) 5 17/24

34. 6 1/2 × 3 =

(A) 9 1/2 (B) 18 1/2 (C) 19 1/2 (D) 20 (E) 12 1/2

35. 3 1/4 × 2 1/3 =

(A) 5 7/12 (B) 6 2/7 (C) 6 5/7 (D) 7 7/12 (E) 7 11/12

Division

DIRECTIONS: Divide and reduce the answer.

36. 3/16 ÷ 3/4 =

(A) 9/64 (B) 1/4 (C) 6/16 (D) 9/16 (E) 3/4

37. 4/9 ÷ 2/3 =

(A) 1/3 (B) 1/2 (C) 2/3 (D) 7/11 (E) 8/9

38. 5 1/4 ÷ 7/10 =

(A) 2 4/7 (B) 3 27/40 (C) 5 19/20 (D) 7 1/2 (E) 8 1/4

39. 4 2/3 ÷ 7/9 =

(A) 2 24/27 (B) 3 2/9 (C) 4 14/27 (D) 5 12/27 (E) 6

40. 3 2/5 ÷ 1 7/10 =

(A) 2 (B) 3 4/7 (C) 4 7/25 (D) 5 1/10 (E) 5 2/7

Changing an Improper Fraction to a Mixed Number

DIRECTIONS: Write each improper fraction as a mixed number in simplest form.

41. 50/4

(A) 10 1/4 (B) 11 1/2 (C) 12 1/4 (D) 12 1/2 (E) 25

42. 17/5

(A) 3 2/5 (B) 3 3/5 (C) 3 4/5 (D) 4 1/5 (E) 4 2/5

43. 42/3

(A) 10 2/3 (B) 12 (C) 13 1/3 (D) 14 (E) 21 1/3

44. 85/6

(A) 9 1/6 (B) 10 5/6 (C) 11 1/2 (D) 12 (E) 14 1/6

45. 151/7

(A) 19 6/7 (B) 20 1/7 (C) 21 4/7 (D) 31 2/7 (E) 31 4/7

Changing a Mixed Number to an Improper Fraction

DIRECTIONS: Change each mixed number to an improper fraction in simplest form.

46. 2 3/5

(A) 4/5 (B) 6/5 (C) 11/5 (D) 13/5 (E) 17/5

47. 4 3/4

(A) 7/4 (B) 13/4 (C) 16/3 (D) 19/4 (E) 21/4

48. 6 7/6

(A) 13/6 (B) 43/6 (C) 19/36 (D) 42/36 (E) 48/6

49. 12 3/7

(A) 87/7 (B) 164/14 (C) 34/3 (D) 187/21 (E) 252/7

50. 21 1/2

(A) 11/2 (B) 22/2 (C) 24/2 (D) 42/2 (E) 43/2

---

3. DECIMALS

When we divide the denominator of a fraction into its numerator, the result is a **decimal**. The decimal is based upon a fraction with a denominator of 10, 100, 1,000, ... and is written with a **decimal point**. Whole numbers are placed to the left of the decimal point where the first place to the left is the units place; the second to the left is the tens; the third to the left is the hundreds, etc. The fractions are placed on the right where the first place to the right is the tenths; the second to the right is the hundredths, etc.

EXAMPLE

$$12\frac{3}{10} = 12.3 \qquad 4\frac{17}{100} = 4.17 \qquad \frac{3}{100} = .03$$

Since a **rational number** is of the form a/b, $b \neq 0$, then all rational numbers can be expressed as decimals by dividing b into a. The result is either a **terminating decimal**, meaning that b divides a with a remainder of 0 after a certain point; or **repeating decimal**, meaning that b continues to divide a so that the decimal has a repeating pattern of integers.

EXAMPLE

(A) $^1/_2 = .5$

(B) $^1/_3 = .333...$

(C) $^{11}/_{16} = .6875$

(D) $^2/_7 = .285714285714...$

(A) and (C) are terminating decimals; (B) and (D) are repeating decimals. This explanation allows us to define **irrational numbers** as numbers whose decimal form is non-terminating and non-repeating, e.g.,

$$\sqrt{2} = 1.414...$$

$$\sqrt{3} = 1.732...$$

PROBLEM

Express $-{}^{10}/_{20}$ as a decimal.

SOLUTION

$$-{}^{10}/_{20} = -{}^{50}/_{100} = -.5$$

PROBLEM

Convert the following fraction to a decimal: $3\frac{3}{4}$.

SOLUTION

Step 1 is to rewrite the fraction as a division problem, using only the fraction, not the integer.

$$\frac{3}{4} = 3 \div 4$$

Step 2 is to solve the division problem.

$$3 \div 4 = 4\overline{)3.00}$$

$$\begin{array}{r} .75 \\ 4\overline{)3.00} \\ \underline{2.8} \\ 20 \\ \underline{20} \\ 0 \end{array}$$

Step 3 is to add the decimal to the integer.

$$0.75 + 3 = 3.75$$

The correct answer is 3.75.

PROBLEM

Write ${}^{2}/_{7}$ as a repeating decimal.

SOLUTION

To write a fraction as a repeating decimal divide the numerator by the denominator until a pattern of repeated digits appears.

$$2 \div 7 = .285714285714\ldots$$

Identify the entire portion of the decimal which is repeated. The repeating decimal can then be written in the shortened form:

$${}^{2}/_{7} = .\overline{285714}$$

PROBLEM

Convert the following decimal to a fraction: 4.95.

SOLUTION

Step 1 is to write the decimal in terms of the fraction one-hundredth. Leave the integer "4" in place.

$$4 \text{ and } \frac{95}{100}$$

Step 2 is to simplify the fraction $\frac{95}{100}$.

$$\frac{95}{100} = \frac{19}{20}$$

Step 3 is to rewrite the fraction, using the integer. Simply join the integer and the fraction.

$$4 \text{ and } \frac{19}{20} = 4\frac{19}{20}$$

The correct answer is $4\frac{19}{20}$.

PROBLEM

Convert the following decimal to a fraction: $8.222\overline{2}$.

SOLUTION

$8.222\overline{2}$ is a repeating decimal. To convert repeating decimals into fractions, use the types of fractions that form repeating decimals.

Step 1 is to separate the integer and the fraction.

$$8.222\overline{2} = 8 \text{ and } 0.222\overline{2}$$

Step 2 is to determine which fraction forms the decimal $0.222\overline{2}$.

$$0.222\overline{2} = \frac{2}{9}$$

Step 3 is to rewrite the fraction, using the integer. Simply join the integer and the fraction.

$$8 \text{ and } \frac{2}{9} = 8\frac{2}{9}$$

The correct answer is $8\frac{2}{9}$.

OPERATIONS WITH DECIMALS

A) **To add numbers containing decimals,** write the numbers in a column making sure the decimal points are lined up, one beneath the other. Add the numbers as usual, placing the decimal point in the sum so that it is still in line with the others. It is important not to mix the digits in the tenths place with the digits in the hundredths place, and so on.

EXAMPLES

$2.558 + 6.391$ $\qquad\qquad$ $57.51 + 6.2$

$$\begin{array}{r} 2.558 \\ +\ 6.391 \\ \hline 8.949 \end{array} \qquad\qquad \begin{array}{r} 57.51 \\ +\ \ 6.20 \\ \hline 63.71 \end{array}$$

Similarly with subtraction,

$78.54 - 21.33$ $\qquad\qquad$ $7.11 - 4.2$

$$\begin{array}{r} 78.54 \\ -21.33 \\ \hline 57.21 \end{array} \qquad\qquad \begin{array}{r} 7.11 \\ -4.20 \\ \hline 2.91 \end{array}$$

Note that if two numbers differ according to the amount of digits to the right of the decimal point, zeros must be added.

$.63 - .214$ $\qquad\qquad$ $15.224 - 3.6891$

$$\begin{array}{r} .630 \\ -.214 \\ \hline .416 \end{array} \qquad\qquad \begin{array}{r} 15.2240 \\ -\ \ 3.6891 \\ \hline 11.5349 \end{array}$$

B) **To multiply numbers with decimals**, simply multiply as usual. Then, to figure out the number of decimal places that belong in the product, find the total number of decimal places in the numbers being multiplied.

EXAMPLES

$$\begin{array}{rl} 6.555 & \text{(3 decimal places)} \\ \times\ \ \ 4.5 & \text{(1 decimal place)} \\ \hline 32775 & \\ 26220\ \ & \\ \hline 294975 & \\ 29.4975 & \text{(4 decimal places)} \end{array} \qquad\qquad \begin{array}{rl} 5.32 & \text{(2 decimal places)} \\ \times\ \ .04 & \text{(2 decimal places)} \\ \hline 2128 & \\ 000\ & \\ \hline 2128 & \\ .2128 & \text{(4 decimal places)} \end{array}$$

C) **To divide numbers with decimals**, you must first make the divisor a whole number by moving the decimal point the appropriate number of places to the right. The decimal point of the dividend should also be moved the same number of places. Place a decimal point in the quotient, directly in line with the decimal point in the dividend.

EXAMPLES

$12.92 \div 3.4$ $\qquad\qquad$ $40.376 \div 7.21$

$$\begin{array}{r} 3.8 \\ 3.4.\overline{)12.9.2} \\ -102 \\ \hline 272 \\ -272 \\ \hline 0 \end{array} \qquad\qquad \begin{array}{r} 5.6 \\ 7.21.\overline{)40.37.6} \\ -3605 \\ \hline 4326 \\ -4326 \\ \hline 0 \end{array}$$

If the question asks to find the correct answer to two decimal places, simply divide until you have three decimal places and then round off. If the third decimal place is a 5 or larger, the number in the second decimal place is increased by 1. If the third decimal place is less than 5, that number is simply dropped.

PROBLEM

Find the answer to the following to 2 decimal places:

(1) 44.3 ÷ 3 (2) 56.99 ÷ 6

SOLUTION

(1)
```
     14.766
  3)44.300
   -3
    14
   -12
     23
    -21
      20
     -18
       20
      -18
        2
```

14.766 can be rounded off to 14.77

(2)
```
     9.498
  6)56.990
   -54
     29
    -24
     59
    -54
      50
     -48
       2
```

9.498 can be rounded off to 9.50

D) When comparing two numbers with decimals to see which is the larger, first look at the tenths place. The larger digit in this place represents the larger number. If the two digits are the same, however, take a look at the digits in the hundredths place, and so on.

EXAMPLES

.518 and .216

5 is larger than 2, therefore

.518 is larger than .216

.723 and .726

6 is larger than 3, therefore

.726 is larger than .723

PROBLEM

Round the following decimal to the nearest hundred: 134.22.

SOLUTION

Step 1 is to determine the digit that will be rounded.

134.22
 v v

Since the digit "1" is in the hundreds' place, it will be rounded.

Step 2 is to locate the digit to the right of "1."

134.22

The digit "3" is to the right of "1." Set all digits to the right of "3" equal to "0."

130.00

Step 3 is to determine if the decimal will be rounded up or down. Since "3" is less than 5, the decimal will be rounded down. To do this, set "3" equal to "0."

100.00

The correct answer is 100.

PROBLEM

Round the following decimal to the nearest thousandth: 0.9196.

SOLUTION

Step 1 is to determine the digit that will be rounded.

0.9196

Since the digit "9" is in the thousandths' place, it will be rounded.

Step 2 is to locate the digit to the right of "9."

0.9196

The digit "6" is to the right of "9." In this case, there are no digits to the right of "6" to set equal to "0."

0.9196

Step 3 is to determine if the decimal will be rounded up or down. Since "6" is greater than or equal to 5, the decimal will be rounded up. To do this, set "6" equal to "0," and increase "9" by one. Since "9" increased by one is ten, carry the addition over to the remaining decimal places.

0.9200

The correct answer is 0.920.

PROBLEM

Find the solution to the following operation.

6.11 + 3.251

SOLUTION

Step 1 is to align the decimal point of each number.

$$\begin{array}{r} 6.11 \\ \underline{+\,3.251} \end{array}$$

Step 2: since 3.251 has a digit in the thousandths' place, add a "0" in the thousandths' place in 6.11.

$$\begin{array}{r} 6.110 \\ \underline{+\ 3.251} \end{array}$$

Step 3 is to perform the addition.

$$\begin{array}{r} 6.110 \\ \underline{+\ 3.251} \\ 9.361 \end{array}$$

The solution is 9.361.

PROBLEM

Find the solution to the following operation.

0.0598 + 0.000031

SOLUTION

Step 1 is to align the decimal point of each number.

$$\begin{array}{r} 0.0598 \\ \underline{+\ 0.000031} \end{array}$$

Step 2: since 0.000031 has a digit in the hundred thousandths' and millionths' places, add a "0" in those places in 0.0598.

$$\begin{array}{r} 0.059800 \\ \underline{+\ 0.000031} \end{array}$$

Step 3 is to perform the addition.

$$\begin{array}{r} 0.059800 \\ \underline{+\ 0.000031} \\ 0.059831 \end{array}$$

The solution is 0.059831.

PROBLEM

Find the solution to the following operation. Round the solution to the nearest ten thousandth:

0.00031 – .0000043

SOLUTION

Step 1 is to align the decimal point of each number.

$$\begin{array}{r} 0.00031 \\ \underline{-\ 0.0000043} \end{array}$$

Step 2: since 0.0000043 has a digit in the millionths' and ten millionths' places, add a "0" in those places in 0.00031.

$$\begin{array}{r} 0.0003100 \\ -\ 0.0000043 \\ \hline \end{array}$$

Step 3 is to perform the subtraction.

$$\begin{array}{r} 0.0003100 \\ -\ 0.0000043 \\ \hline 0.0003057 \end{array}$$

The solution is 0.0003057. Rounded to the nearest ten thousandth, the solution becomes 0.0003.

PROBLEM

Find the solution to the following operation.

0.543×0.1012

SOLUTION

Step 1 is to align the decimals of the numbers together.

$$\begin{array}{r} 0.543 \\ \times\ 0.1012 \\ \hline \end{array}$$

Step 2; since 0.1012 has a digit in the ten thousandths' place, add a "0" in the ten thousandths' place in 0.543.

$$\begin{array}{r} 0.5430 \\ \times\ 0.1012 \\ \hline \end{array}$$

Step 3 is to perform the operation. Since there are four decimal places in 0.5430 and four decimal places in 0.1012, the result will have eight decimal places.

$$\begin{array}{r} 0.5430 \\ \times\ 0.1012 \\ \hline 0.05495160 \end{array}$$

The solution is 0.05495160.

PROBLEM

Find the solution to the following operation.

$0.0006 \div 0.0002$

SOLUTION

Step 1 is to set up the operation in the format that follows:

$$.0002\overline{).0006}$$

Step 2 is to convert the divisor into a whole number. To do this, move the decimal point four places to the right.

$0.0002 = 2$

Step 3 is to move the decimal point of the dividend the same number of places.

0.0006 = 6

Step 4 is to perform the operation.

$$\begin{array}{r} 3 \\ 2\overline{)6} \\ \underline{6} \\ 0 \end{array}$$

The solution is 3.

PROBLEM

Find the solution to the following problem.

$$(77.72 - 56.02) \div 3.1$$

SOLUTION

Step 1 is to perform the operation inside the parentheses.

$$\begin{array}{r} 77.72 \\ \underline{-\ 56.02} \\ 21.70 \end{array}$$

Step 2 is to divide 21.70 by 3.1.

$$\begin{array}{r} 7.0 \\ 3.1\overline{)21.7} \\ \underline{21.7} \\ 0 \end{array}$$

The answer is 7.0.

DRILL: DECIMALS

Addition

1. 1.032 + 0.987 + 3.07 =

(A) 4.089 (B) 5.089 (C) 5.189 (D) 6.189 (E) 13.972

2. 132.03 + 97.1483 =

(A) 98.4686 (B) 110.3513 (C) 209.1783
(D) 229.1486 (E) 229.1783

3. 7.1 + 0.62 + 4.03827 + 5.183 =

(A) 0.2315127 (B) 16.94127 (C) 17.57127
(D) 18.561 (E) 40.4543

4. $8 + 17.43 + 9.2 =$

(A) 34.63 (B) 34.86 (C) 35.63 (D) 176.63 (E) 189.43

5. $1036.173 + 289.04 =$

(A) 382.6573 (B) 392.6573 (C) 1065.077
(D) 1325.213 (E) 3926.573

Subtraction

6. $3.972 - 2.04 =$

(A) 1.932 (B) 1.942 (C) 1.976 (D) 2.013 (E) 2.113

7. $16.047 - 13.06 =$

(A) 2.887 (B) 2.987 (C) 3.041 (D) 3.141 (E) 4.741

8. $87.4 - 56.27 =$

(A) 30.27 (B) 30.67 (C) 31.1 (D) 31.13 (E) 31.27

9. $1046.8 - 639.14 =$

(A) 303.84 (B) 313.74 (C) 407.66 (D) 489.74 (E) 535.54

10. $10{,}000 - 842.91 =$

(A) 157.09 (B) 942.91 (C) 5236.09 (D) 9057.91 (E) 9157.09

Multiplication

11. $1.03 \times 2.6 =$

(A) 2.18 (B) 2.678 (C) 2.78 (D) 3.38 (E) 3.63

12. $93 \times 4.2 =$

(A) 39.06 (B) 97.2 (C) 223.2 (D) 390.6 (E) 3906

13. $0.04 \times 0.23 =$

(A) 0.0092 (B) 0.092 (C) 0.27 (D) 0.87 (E) 0.920

14. $0.0186 \times 0.03 =$

(A) 0.000348 (B) 0.000558 (C) 0.0548 (D) 0.0848 (E) 0.558

15. $51.2 \times 0.17 =$

(A) 5.29 (B) 8.534 (C) 8.704 (D) 36.352 (E) 36.991

Division

16. 123.39 ÷ 3 =

(A) 31.12 (B) 41.13 (C) 401.13 (D) 411.3 (E) 4,113

17. 1428.6 ÷ 6 =

(A) 0.2381 (B) 2.381 (C) 23.81 (D) 238.1 (E) 2,381

18. 25.2 ÷ 0.3 =

(A) 0.84 (B) 8.04 (C) 8.4 (D) 84 (E) 840

19. 14.95 ÷ 6.5 =

(A) 2.3 (B) 20.3 (C) 23 (D) 230 (E) 2,300

20. 46.33 ÷ 1.13 =

(A) 0.41 (B) 4.1 (C) 41 (D) 410 (E) 4,100

Comparing

21. Which is the **largest** number in this set — {0.8, 0.823, 0.089, 0.807, 0.852}?

(A) 0.8 (B) 0.823 (C) 0.089 (D) 0.807 (E) 0.852

22. Which is the **smallest** number in this set — {32.98, 32.099, 32.047, 32.5, 32.304}?

(A) 32.98 (B) 32.099 (C) 32.047 (D) 32.5 (E) 32.304

23. In which set below are the numbers arranged correctly from smallest to largest?

(A) {0.98, 0.9, 0.993}
(B) {0.113, 0.3, 0.31}
(C) {7.04, 7.26, 7.2}
(D) {0.006, 0.061, 0.06}
(E) {12.84, 12.801, 12.6}

24. In which set below are the numbers arranged correctly from largest to smallest?

(A) {1.018, 1.63, 1.368}
(B) {4.219, 4.29, 4.9}
(C) {0.62, 0.6043, 0.643}
(D) {16.34, 16.304, 16.3}
(E) {12.98, 12.601, 12.86}

25. Which is the **largest** number in this set — {0.87, 0.89, 0.889, 0.8, 0.987}?

(A) 0.87 (B) 0.89 (C) 0.889 (D) 0.8 (E) 0.987

Changing a Fraction to a Decimal

26. What is 1/4 written as a decimal?

(A) 1.4 (B) 0.14 (C) 0.2 (D) 0.25 (E) 0.3

27. What is 3/5 written as a decimal?

(A) 0.3 (B) 0.35 (C) 0.6 (D) 0.65 (E) 0.8

28. What is 7/20 written as a decimal?

(A) 0.35 (B) 0.4 (C) 0.72 (D) 0.75 (E) 0.9

29. What is 2/3 written as a decimal?

(A) 0.23 (B) 0.33 (C) 0.5 (D) 0.6 (E) $0.\overline{6}$

30. What is 11/25 written as a decimal?

(A) 0.1125 (B) 0.25 (C) 0.4 (D) 0.44 (E) 0.5

4. RATIO AND PROPORTION

Ratio

The results of observation or measurement often must be compared with some standard value in order to have any meaning. For example, to say that a man can read 400 words per minute has little meaning as it stands. However, when his rate is compared to the 250 words per minute of the average reader, one can see that he reads considerably faster than the average reader. How much faster? To find out, his rate is divided by the average rate, as follows:

$$\frac{400}{250} = \frac{8}{5}$$

Thus, for every 5 words read by the average reader, this man reads 8.

When the relationship between two numbers is shown in this way, they are compared as a **ratio**. A ratio is a comparison of two like quantities. It is the quotient obtained by dividing the first number of a comparison by the second.

Since a ratio is also a fraction, all the rules that govern fractions may be used in working with ratios. Thus, the terms may be reduced, increased, simplified, and so forth, according to the rules for fractions.

Proportion

Closely allied with the study of ratio is the subject of proportion. A **proportion** is nothing more than an equation in which the members are ratios. In other words, when two ratios are set equal to each other, a proportion is formed. The proportion may be written in three different ways, as in the following examples:

$$15:20 :: 3:4$$

$$15:20 = 3:4$$

$$\frac{15}{20} = \frac{3}{4}$$

The last two forms are the most common. All these forms are read, "15 is to 20 as 3 is to 4." In other words, 15 has the same ratio to 20 as 3 has to 4.

One reason for the extreme importance of proportions is that if any three of the terms are given, the fourth may be found by solving a simple equation. In science many chemical and physical relations are expressed as proportions. Consequently, a familiarity with proportions will provide one method for solving many applied problems. It is evident from the last form shown, $\frac{15}{20} = \frac{3}{4}$, that a proportion is really a fractional equation. Therefore, all the rules for fraction equations apply.

PROBLEM

Express the following phrases as a ratio of students in Mrs. Polly's class to students in Mr. Smith's:

Mrs. Polly teaches 30 students.

Mr. Smith teaches 10 students.

SOLUTION

Step 1 is to write the phrase in the following format:

30 students:10 students

Step 2 is to reduce the ratio to its lowest terms, if possible. To do this, write out each number as a product of its primes and find the greatest common divisor (GCD).

$$30 = 5 \times 6 \qquad 10 = 5 \times 2$$

$$= 5 \times 3 \times 2$$

The GCD is 10.

Next, divide each number by the GCD.

$$\frac{30}{10} = 3 \qquad \frac{10}{10} = 1$$

Step 3 is to rewrite the ratio.

3 students:1 student

The correct answer is 3 students:1 student.

PROBLEM

Express the following phrases as a ratio of Robin's apples to Tim's apples:

Robin has 5 apples.

Tim has 3 apples.

SOLUTION

Step 1 is to write the phrase in the following format:

5 apples:3 apples

Step 2 is to reduce the ratio to its lowest terms, if possible. To do this, write out each number as a product of its primes and find the greatest common divisor (GCD).

$5 = 5 \times 1 \qquad 3 = 3 \times 1$

Since there is no GCD, the ratio is already reduced to its lowest terms.

The correct answer is 5 apples:3 apples.

PROBLEM

Express the following phrases as a ratio of cats to dogs:

21 cats are in the house.

7 dogs are in the house.

SOLUTION

Step 1 is to write the phrase in the following format:

21 cats:7 dogs

Step 2 is to reduce the ratio to its lowest terms, if possible. To do this, write out each number as a product of its primes and find the greatest common divisor (GCD).

$21 = 7 \times 3 \quad 7 = 7 \times 1$

The GCD is 7.

Next, divide each number by the GCD.

$\frac{21}{7} = 3 \; \frac{7}{7} = 1$

Step 3 is to rewrite the ratio.

3 cats:1 dog

The correct answer is 3 cats:1 dog.

PROBLEM

Sandy has to walk 1 kilometer to get to school. John has to walk 400 meters to get to school. What is the ratio (in meters) of the distance walked by Sandy to that walked by John?

SOLUTION

Step 1 is to write the phrase in the following format:

1 kilometer:400 meters

Step 2 is to convert kilometers to meters.

1 kilometer = 1,000 meters

Step 3 is to rewrite the problem in meters.

1,000 meters:400 meters

Step 4 is to reduce the ratio to its lowest terms, if possible. To do this, write out each number as a product of its primes and find the greatest common divisor (GCD).

$1{,}000 = 10 \times 10 \times 10$
$= 5 \times 2 \times 5 \times 2 \times 5 \times 2$

$400 = 40 \times 10$
$= 4 \times 10 \times 10$
$= 2 \times 2 \times 5 \times 2 \times 5 \times 2$

The GCD is 200 or $2 \times 5 \times 2 \times 5 \times 2$.

Next, divide each number by the GCD.

$\frac{1{,}000}{200} = 5 \quad \frac{400}{200} = 2$

Step 5 is to rewrite the ratio.

5 meters:2 meters

The correct answer is 5 meters:2 meters.

PROBLEM

Express the following phrase as a ratio of dollars lost to dollars invested:

Stockbroker Brian lost $3 for every $6 invested.

SOLUTION

Step 1 is write the phrase in the following format:

–$3:$6

Step 2 is to reduce the ratio to its lowest terms, if possible. To do this, write out each number as a product of its primes and find the greatest common divisor (GCD).

$3 = 3 \times 1 \qquad 6 = 3 \times 2$

The GCD is 3.

Next, divide each number by the GCD.

$$\frac{3}{3} = 1 \qquad \frac{6}{3} = 2$$

Step 3 is to rewrite the ratio. Remember to carry the negative sign into the ratio.

–\$1:\$2

The correct answer is –\$1:\$2.

PROBLEM

Express the following fraction as a ratio: $\frac{54}{63}$.

SOLUTION

Step 1 is to rewrite the fraction in the following format:

$$\frac{54}{63} = 54:63$$

Step 2 is to simplify, if possible. To do this, write out each number as a product of its primes and find the greatest common divisor (GCD).

$$54 = 9 \times 6 \qquad 63 = 9 \times 7$$

$$= 3 \times 3 \times 3 \times 2 \qquad = 3 \times 3 \times 7$$

The GCD is 9.

Next, divide each number by the GCD.

$$\frac{54}{9} = 6 \qquad \frac{63}{9} = 7$$

Step 3 is to rewrite the ratio.

6:7

The correct answer is 6:7.

PROBLEM

Express the following fraction as a ratio: $\frac{8}{5}$.

SOLUTION

Step 1 is to rewrite the fraction in the following format:

$$\frac{8}{5} = 8:5$$

Step 2 is to simplify, if possible. To do this, write out each number as a product of its primes and find the greatest common divisor (GCD).

$$8 = 4 \times 2 \qquad 5 = 5 \times 1$$
$$= 2 \times 2 \times 2$$

Since there is no GCD, the ratio is already reduced to its lowest terms.

Step 3 is to rewrite the ratio.

8:5

The correct answer is 8:5.

PROBLEM

Express the following ratio as a fraction:

60:12.

SOLUTION

Step 1 is to determine the numerator and denominator.

The left side of the ratio becomes the numerator: 60.

The right side of the ratio becomes the denominator: 12.

Step 2 is to rewrite the ratio as a fraction.

$$60{:}12 = \frac{60}{12}$$

Step 3 is to simplify, if possible. To do this, write out each number as a product of its primes and find the greatest common divisor (GCD).

$$60 = 6 \times 10 \qquad 12 = 6 \times 2$$
$$= 6 \times 5 \times 2 \qquad = 3 \times 2 \times 2$$
$$= 3 \times 2 \times 5 \times 2$$

The GCD is 12.

Next, divide each number by the GCD.

$$\frac{60}{12} = 5 \qquad \frac{12}{12} = 1$$

Step 4 is to rewrite the fraction.

$$\frac{5}{1}$$

The correct answer is $\frac{5}{1}$.

PROBLEM

Express the following ratio as a fraction:

11:33.

SOLUTION

Step 1 is to determine the numerator and denominator.

The left side of the ratio becomes the numerator: 11.

The right side of the ratio becomes the denominator: 33.

Step 2 is to rewrite the ratio as a fraction.

$$11:33 = \frac{11}{33}$$

Step 3 is to simplify, if possible. To do this, write out each number as a product of its primes and find the greatest common divisor (GCD).

$$11 = 11 \times 1 \quad 33 = 11 \times 3$$

The GCD is 11.

Next, divide each number by the GCD.

$$\frac{11}{11} = 1 \qquad \frac{33}{11} = 3$$

Step 4 is to rewrite the fraction.

$$\frac{1}{3}$$

The correct answer is $\frac{1}{3}$.

PROBLEM

Find the solution to the proportion below.

$$\frac{3}{7} = \frac{?}{21}$$

SOLUTION

Step 1 is to put the proportion in the following format:

$$AD = BC \qquad 3(21) = ?(7)$$

Step 2 is to solve the left side of the proportion.

$$3(21) = 63$$

Step 3 is to rewrite the proportion.

$$63 = ?(7)$$

Step 4 is to find the missing integer that solves the proportion. To do this, divide both sides by the known mean, 7.

$$\frac{63}{7} = 9 \qquad \frac{?(7)}{7} = ?$$

Step 5 is to rewrite the proportion.

$$9 = ?$$

The solution is 9.

Step 6 is to check the answer.

$$\frac{3}{7} = \frac{?}{21}$$

Substituting 9 for ?,

$$\frac{3}{7} = \frac{9}{21}$$

Next, put the proportion in the following format:

$AD = BC \qquad 3(21) = 9(7)$

Solve both sides of the proportion.

$3(21) = 63 \quad 9(7) = 63$

Solution "9" checks out correctly.

PROBLEM

Find the solution to the proportion below.

$$\frac{?}{45} = -\frac{2}{5}$$

SOLUTION

Step 1 is to put the proportion in the following format:

$AD = BC \qquad ?(5) = -2(45)$

Step 2 is to solve the right side of the proportion.

$-2(45) = -90$

Step 3 is to rewrite the proportion.

$?(5) = -90$

Step 4 is to find the missing integer that solves the proportion. To do this, divide both sides by the known extreme, 5.

$$\frac{?(5)}{5} = ? \qquad -\frac{90}{5} = -18$$

Step 5 is to rewrite the proportion.

$? = -18$

The solution is –18.

Step 6 is to check the answer.

$$\frac{?}{45} = -\frac{2}{5}$$

Substituting –18 for ?,

$$-\frac{18}{45} = -\frac{2}{5}$$

Next, put the proportion in the following format.

$AD = BC$ $\quad\quad$ $-18(5) = -2(45)$

Solve both sides of the proportion.

$-18(5) = -90$ $\quad\quad$ $-2(45) = -90$

Solution "–18" checks out correctly.

PROBLEM

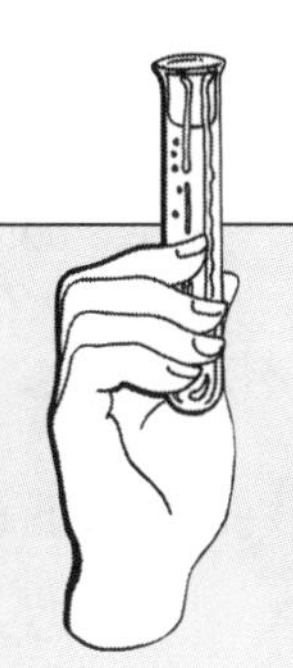

A chemist is preparing a chemical solution. She needs to add 3 parts sodium and 2 parts zinc to a flask of chlorine. If she has already placed 300 grams of sodium into the flask, how much zinc must she now add?

SOLUTION

Step 1 is to determine the ratio of sodium and zinc.

3 parts sodium, 2 parts zinc = 3:2

Step 2 is to write the problem as a proportion.

$$\frac{3}{2} = \frac{300}{?}$$

Step 3 is to put the proportion in the following format:

$AD = BC$ $\quad$ $3(?) = 2(300)$

Step 4 is to solve the right side of the proportion.

$2(300) = 600$

Step 5 is to rewrite the proportion.

$3(?) = 600$

Step 6 is to find the missing integer that solves the proportion. To do this, divide both sides by the known extreme, 3.

$$\frac{3(?)}{3} = ? \quad\quad \frac{600}{3} = 200$$

Step 7 is to rewrite the proportion.

$? = 200$

The solution is 200 grams of zinc.

PROBLEM

An automobile dealer has to sell 3.5 cars for every 1 truck to achieve the optimum profit. This year, it is estimated that 3,500 cars will be sold. How many trucks must he sell to achieve the optimum profit?

SOLUTION

Step 1 is to determine the ratio of cars to trucks.

3.5 cars, 1 truck = 3.5:1

Make both sides of the ratio an integer. To do this, multiply both sides of the ratio by 2.

2(3.5):2(1) = 7:2

Step 2 is to write the problem as a proportion.

$$\frac{7}{2} = \frac{3{,}500}{?}$$

Step 3 is to put the proportion in the following format:

$AD = BC$ $\quad$ 7(?) = 2(3,500)

Step 4 is to solve the right side of the proportion.

2(3,500) = 7,000

Step 5 is to rewrite the proportion.

7(?) = 7,000

Step 6 is to find the missing integer that solves the proportion. To do this, divide both sides by the known extreme, 7.

$$\frac{7(?)}{7} = ? \qquad \frac{7{,}000}{7} = 1{,}000$$

Step 7 is to rewrite the proportion.

? = 1,000

The solution is 1,000 trucks.

PROBLEM

Centralville's town council voted on a bill that would increase taxes. To pass the bill, 75% of the council has to vote for the bill. If all 56 council members voted, how many members would it take to pass the bill? If 30 members voted to pass the bill, will taxes be increased?

SOLUTION

Step 1 is to determine the ratio. Remember that fractions can also express ratios.

$$75\% = \frac{75}{100}$$

$$\frac{75}{100} = 3{:}4$$

Step 2 is to write the problem as a proportion.

$$\frac{3}{4} = \frac{?}{56}$$

Step 3 is to put the proportion in the following format:

$AD = BC$ $\quad 3(56) = 4(?)$

Step 4 is to solve the left side of the proportion.

$3(56) = 168$

Step 5 is to rewrite the proportion.

$168 = 4(?)$

Step 6 is to find the missing integer that solves the proportion. To do this, divide both sides by the known mean, 4.

$$\frac{4(?)}{4} = ? \qquad \frac{168}{4} = 42$$

Step 7 is to rewrite the proportion.

$? = 42$

The solution is that 42 votes are needed to pass the bill.

Step 8 is to determine if taxes will be increased.

$30 < 42$

Since only 30 council members voted for the bill, taxes will not be increased.

PROBLEM

A baker is making a new recipe for chocolate chip cookies. He decides that for every 6 cups of flour, he needs to add 1 cup of sugar. He puts 30 cups of flour and 2 cups of sugar into the batter. How much more sugar does he need?

SOLUTION

Step 1 is to determine the ratio of flour to sugar.

6 cups flour, 1 cup sugar = 6:1

Step 2 is to write the problem as a proportion.

$$\frac{6}{1} = \frac{30}{?}$$

Step 3 is to put the proportion in the following format:

$AD = BC \qquad 6(?) = 1(30)$

Step 4 is to solve the right side of the proportion.

$1(30) = 30$

Step 5 is to rewrite the proportion.

$6(?) = 30$

Step 6 is to find the missing integer that solves the proportion. To do this, divide both sides by the known extreme, 6.

$\frac{6(?)}{6} = ? \qquad \frac{30}{6} = 5$

Step 7 is to rewrite the proportion.

$? = 5$

The solution is that 5 cups of sugar must be added to the batter.

Step 8 is to determine how many more cups of sugar are needed.

$5 - 2 = 3$

Since only 2 cups have been added so far, the baker must still add 3 cups.

PROBLEM

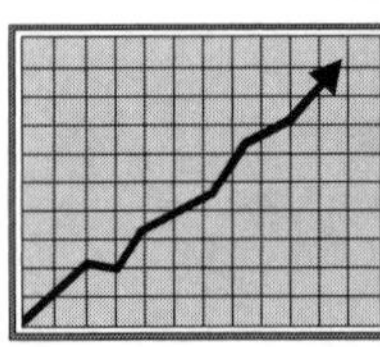

Interest Rates

Stacey, a stockbroker, is developing a portfolio for her client. She chooses a stock that will earn her client 8% interest. If her client invests $25,000, how much interest will be earned?

SOLUTION

Step 1 is to determine the ratio of interest earned per dollar invested. Remember that fractions can also express ratios.

$$8\% = \frac{8}{100}$$

$$\frac{8}{100} = 8:100$$

Step 2 is to divide both sides of the ratio by the GCD to simplify the expression.

$8 = 2 \times 4 \qquad\qquad 100 = 10 \times 10$

$\quad = 2 \times 2 \times 2 \qquad\qquad = 5 \times 2 \times 5 \times 2$

The GCD is 4; divide both sides of the proportion by 4.

$\frac{8}{4} = 2 \qquad \frac{100}{4} = 25$

The new proportion is 2:25.

Step 3 is to write the problem as a proportion.

$$\frac{2}{25} = \frac{?}{25,000}$$

Step 4 is to put the proportion in the following format:

$AD = BC \qquad 2(25,000) = ?(25)$

Step 5 is to solve the left side of the proportion.

$2(25,000) = 50,000$

Step 6 is to rewrite the proportion.

$50,000 = ?(25)$

Step 7 is to find the missing integer that solves the proportion. To do this, divide both sides by the known mean, 25.

$$\frac{50,000}{25} = 2,000 \qquad \frac{?(25)}{25} = ?$$

Step 8 is to rewrite the proportion.

$2,000 = ?$

The solution is that $2,000 of interest will be earned.

PROBLEM

Marina grew 4 inches during the last 10 months. If Michael grew at a proportional rate to Marina, how long did it take him to grow 1 inch?

SOLUTION

Step 1 is to determine the ratio of inches grown to months for Marina.

4 inches per 10 months = 4:10

Step 2 is to write the problem as a proportion.

$$\frac{4}{10} = \frac{1}{?}$$

Step 3 is to put the proportion in the following format:

$AD = BC \qquad 4(?) = 1(10)$

Step 4 is to solve the right side of the proportion.

$1(10) = 10$

Step 5 is to rewrite the proportion.

$4(?) = 10$

Step 6 is to find the missing integer that solves the proportion. To do this, divide both sides by the known extreme, 4.

$$\frac{4(?)}{4} = ? \qquad \frac{10}{4} = \frac{5}{2}$$

Step 7 is to rewrite the proportion.

$$? = \frac{5}{2}$$

Step 8 is to convert 5/2 to a mixed number.

$$\frac{5}{2} = 2\frac{1}{2}$$

The correct answer is $2\frac{1}{2}$ months.

PROBLEM

Mr. Miller owns a Christmas tree farm. Every year he plants 270 seedlings. During the summer, for every 9 seedlings he plants, 1 tree dies. How many seedlings survive after the summer?

SOLUTION

Step 1 is to determine the ratio of seedlings that live to those that die.

9 seedlings live, 1 seedling dies = 9:1

Step 2 is to write the problem as a proportion.

$$\frac{9}{1} = \frac{270}{?}$$

Step 3 is to put the proportion in the following format:

$AD = BC$ $\quad$ 9(?) = 1(270)

Step 4 is to solve the right side of the proportion.

1(270) = 270

Step 5 is to rewrite the proportion.

9(?) = 270

Step 6 is to find the missing integer that solves the proportion. To do this, divide both sides by the known extreme, 9.

$$\frac{9(?)}{9} = ? \qquad \frac{270}{9} = 30$$

Step 7 is to rewrite the proportion.

? = 30

Step 8 is to determine the number of trees that survive after the summer.

total seedlings – seedlings that die = surviving seedlings

270 – 30 = 240

The correct answer is that 240 seedlings survive after the summer.

PROBLEM

Elena works as a traveling salesperson for Knives Incorporated. Each year, she increases the number of customers proportionally (5:1). If in her first year she started with 100 customers, how many does she have by the end of the second year?

SOLUTION

Step 1 is to write the problem as a proportion.

$$\frac{5}{1} = \frac{?}{100}$$

Step 2 is to put the proportion in the following format:

$AD = BC \qquad 5(100) = 1(?)$

Step 3 is to solve the left side of the proportion.

$5(100) = 500$

Step 4 is to rewrite the proportion.

$500 = 1(?)$

Step 5 is to add the number of new customers to the original number of customers.

$100 + 500 = 600$

Therefore, after her first year, she has 600 customers. Repeat the above steps for the second year.

Step 6 is to rewrite the problem as a proportion.

$$\frac{5}{1} = \frac{?}{600}$$

Step 7 is to put the proportion in the following format:

$AD = BC \qquad 5(600) = 1(?)$

Step 8 is to solve the left side of the proportion.

$5(600) = 3{,}000$

Step 9 is to rewrite the proportion.

$3{,}000 = ?(1)$

Step 10 is to find the missing integer that solves the proportion. To do this, divide both sides by the known mean, 1.

$$\frac{3{,}000}{1} = 3{,}000 \qquad \frac{?(1)}{1} = ?$$

Step 11 is to rewrite the proportion.

$3{,}000 = ?$

Step 12 is to add the number of new customers to the number of customers acquired during the first year.

$600 + 3{,}000 = 3{,}600$

The correct answer is 3,600 customers.

5. PERCENTAGES

A **percent** is a way of expressing the relationship between part and whole, where whole is defined as 100%. A percent can be defined by a fraction with a denominator of 100. Decimals can also represent a percent. For instance,

$56\% = 0.56 = 56/100$

PROBLEM

Compute the value of

(1) 90% of 400

(2) 180% of 400

(3) 50% of 500

(4) 200% of 4

SOLUTION

The symbol % means per hundred, therefore $5\% = 5/100$

(1) 90% of $400 = 90/100 \times 400 = 90 \times 4 = 360$

(2) 180% of $400 = 180/100 \times 400 = 180 \times 4 = 720$

(3) 50% of $500 = 50/100 \times 500 = 50 \times 5 = 250$

(4) 200% of $4 = 200/100 \times 4 = 2 \times 4 = 8$

PROBLEM

What percent of

(1) 100 is 99.5

(2) 200 is 4

SOLUTION

(1) $99.5 = x \times 100$

$99.5 = 100x$

$.995 = x$; but this is the value of x per hundred. Therefore,

$x = 99.5\%$

(2) $4 = x \times 200$

$4 = 200x$

$.02 = x$. Again this must be changed to percent, so

$x = 2\%$

EQUIVALENT FORMS OF A NUMBER

Some problems may call for converting numbers into an equivalent or simplified form in order to make the solution more convenient.

1. Converting a fraction to a decimal:

   $^1/_2 = 0.50$

   Divide the numerator by the denominator:

$$\begin{array}{r} .50 \\ 2\overline{)1.00} \\ \underline{-10} \\ 00 \end{array}$$

2. Converting a number to a percent:

   $0.50 = 50\%$

   Multiply by 100:

   $0.50 = (0.50 \times 100)\% = 50\%$

3. Converting a percent to a decimal:

   $30\% = 0.30$

   Divide by 100:

   $30\% = 30/100 = 0.30$

4. Converting a decimal to a fraction:

   $0.500 = {}^1/_2$

   Convert .500 to 500/1000 and then simplify the fraction by dividing the numerator and denominator by common factors:

$$\frac{2 \times 2 \times 5 \times 5 \times 5}{2 \times 2 \times 2 \times 5 \times 5 \times 5}$$

   and then cancel out the common numbers to get $^1/_2$.

PROBLEM

Express

(1) 1.65 as a percentage of 100

(2) 0.7 as a fraction

(3) $-\,^{10}/_{20}$ as a decimal

(4) $^4/_2$ as an integer

SOLUTION

(1) $(1.65/100) \times 100 = 1.65\%$

(2) $0.7 = {}^{7}/_{10}$

(3) $-{}^{10}/_{20} = -0.5$

(4) ${}^{4}/_{2} = 2$

PROBLEM

Convert the following fraction to a percent: $\frac{17}{20}$.

SOLUTION

Step 1 is to write the problem as a proportion.

$$\frac{17}{20} = \frac{?}{100}$$

Step 2 is to rewrite the proportion in the following format:

$AD = BC \qquad 17(100) = ?(20)$

Step 3 is to solve the left side of the proportion.

$17(100) = 1{,}700$

Step 4 is to rewrite the proportion.

$1{,}700 = ?(20)$

Step 5 is to find the missing integer that solves the proportion. To do this, divide both sides by the known mean, 20.

$$\frac{1{,}700}{20} = 85 \qquad \frac{?(20)}{20} = ?$$

Step 6 is to rewrite the proportion.

$85 = ?$

The new fraction is $\frac{85}{100}$.

Step 7 is to change the numerator to a percent.

$85 = 85\%$

The answer is 85%.

PROBLEM

Convert the following fraction into a percent: $-\frac{8}{25}$.

SOLUTION

The negative sign will not affect the problem, but it must be carried through.

Step 1 is to write the problem as a proportion.

$$-\frac{8}{25} = \frac{?}{100}$$

Step 2 is to rewrite the proportion in the following format:

$AD = BC \qquad -8(100) = ?(25)$

Step 3 is to solve the left side of the proportion.

$-8(100) = -800$

Step 4 is to rewrite the proportion.

$-800 = ?(25)$

Step 5 is to find the missing integer that solves the proportion. To do this, divide both sides by the known mean, 25.

$$-\frac{800}{25} = -32 \qquad \frac{?(25)}{25} = ?$$

Step 6 is to rewrite the proportion.

$-32 = ?$

The new fraction is $\frac{-32}{100}$.

Step 7 is to change the numerator to a percent.

$-32 = -32\%$

The answer is -32%.

PROBLEM

Convert the following percent to a fraction: 68%.

SOLUTION

Step 1 is to convert 68% into a fraction with a denominator of 100.

$$68\% = \frac{68}{100}$$

Step 2 is to simplify the fraction.

$$\frac{68}{100} = \frac{17}{25}$$

The correct answer is $\frac{17}{25}$.

PROBLEM

Convert the following percent to a fraction: 119%.

SOLUTION

Step 1 is to convert 119% into a fraction with a denominator of 100.

$$119\% = \frac{119}{100}$$

Step 2 is to simplify the fraction.

$$\frac{119}{100} = \frac{119}{100}$$

Since the fraction cannot be simplified, the correct answer is $\frac{119}{100}$.

PROBLEM

Convert the following percent to a fraction: 225%.

SOLUTION

Step 1 is to convert 225% into a fraction with a denominator of 100.

$$225\% = \frac{225}{100}$$

Step 2 is to simplify the fraction.

$$\frac{225}{100} = \frac{9}{4}$$

The correct answer is $\frac{9}{4}$ or $2\frac{1}{4}$.

PROBLEM

Convert the following decimal to a percent: 0.99.

SOLUTION

Step 1 is to multiply the decimal by 100.

$$0.99 \times 100 = 99.00$$

Step 2 is to write the decimal as a percent.

$$99.00 = 99\%$$

The correct answer is 99%.

PROBLEM

Convert the following decimal to a percent: 12.69.

SOLUTION

Step 1 is to multiply the decimal by 100.

$$12.69 \times 100 = 1{,}269.00$$

Step 2 is to write the decimal as a percent.

$$1{,}269.00 = 1{,}269\%$$

The correct answer is 1,269%.

PROBLEM

Convert the following decimal to a percent: –0.26.

SOLUTION

The negative sign will have no effect on the problem. The steps for converting to a percent are the same, but the negative sign must be carried through the problem.

Step 1 is to multiply the decimal by 100.

$$-0.26 \times 100 = -26.00$$

Step 2 is to write the decimal as a percent.

$$-26.00 = -26\%$$

The correct answer is –26%.

PROBLEM

Convert the following percent to a decimal: 87%.

SOLUTION

Step 1 is to write the percent as a real number.

$$87\% = 87$$

Step 2 is to divide the real number by 100.

$$87 \div 100 = 0.87$$

The correct answer is 0.87.

PROBLEM

Convert the following percent to a decimal: 233%.

SOLUTION

Although 233% is greater than 100%, the steps for converting to a decimal remain the same.

Step 1 is to write the percent as a real number.

$$233\% = 233$$

Step 2 is to divide the real number by 100.

$$233 \div 100 = 2.33$$

The correct answer is 2.33.

PROBLEM

Convert the following percent to a decimal: 0.009%.

SOLUTION

Step 1 is to write the percent as a real number.

$0.009\% = 0.009$

Step 2 is to divide the real number by 100.

$0.009 \div 100 = 0.00009$

The correct answer is 0.00009.

PROBLEM

Perform the following operation. Convert the answer to a decimal.

$145\% + 2\%$

SOLUTION

Step 1 is to perform the operation.

$145\% + 2\% = 147\%$

Step 2 is to write the percent as a real number.

$147\% = 147$

Step 3 is to divide the real number by 100.

$147 \div 100 = 1.47$

The correct answer is 1.47.

PROBLEM

Which of the following statements is true?

a) $0.002\% = 0.200$

b) $1.967 = 196.7\%$

c) $-0.95 > -93\%$

d) $1.00 < 100\%$

SOLUTION

Statement a is incorrect because 0.200 is equivalent to 20%. Since 20% does not equal 0.002%, this statement cannot be true.

Statement b is correct. 1.967 is equivalent to 196.7%.

Statement c is incorrect because –0.95 is equivalent to –95%. Since –95% is not greater than –93%, this statement is not true.

Statement d is incorrect because 1.00 is equivalent to 100%, not less than 100%.

The only correct statement is statement b.

PROBLEM

What is 200% of 5?

SOLUTION

This problem has to be made into a multiplication operation.

Step 1 is to convert 200% into a decimal.

$200\% = 200 \div 100 = 2.00$

Step 2 is to set up the multiplication operation.

5×2.00

Step 3 is to perform the operation.

$5 \times 2.00 = 10$

The correct answer is 10.

PROBLEM

What is 99.9% of 56, rounded to the nearest tenth?

SOLUTION

This problem has to be made into a multiplication operation.

Step 1 is to convert 99.9% into a decimal.

$99.9\% = 99.9 \div 100 = 0.999$

Step 2 is to set up the multiplication operation.

56×0.999

Step 3 is to perform the operation.

$56 \times 0.999 = 55.9$

The correct answer is 55.9.

PROBLEM

70 is what percentage of 80?

SOLUTION

This problem has to be made into a division operation.

Step 1 is to set up the division operation.

$70 \div 80$

Step 2 is to perform the division.

$70 \div 80 = 0.875$

Step 3 is to convert the decimal into a percent.

$0.875 \times 100 = 87.5\%$

PROBLEM

15 is what percentage of 3,000?

SOLUTION

This problem has to be made into a division operation.

Step 1 is to set up the division operation.

$15 \div 3{,}000$

Step 2 is to perform the division.

$15 \div 3{,}000 = 0.005$

Step 3 is to convert the decimal into a percent.

$0.005 \times 100 = 0.5\%$

The correct answer is 0.5%.

PROBLEM

What is 25% of $\frac{1}{2}$?

SOLUTION

This problem has to be made into a multiplication operation.

Step 1 is to convert 25% into a decimal.

$25\% = 25 \div 100 = 0.25$

Step 2 is to set up the multiplication operation.

$\frac{1}{2} \times 0.25$ or $\frac{1}{2} \times \frac{1}{4}$

Step 3 is to perform the operation.

$\frac{1}{2} \times \frac{1}{4} = \frac{1}{8}$

The correct answer is $\frac{1}{8}$.

PROBLEM

Farmer Bob has planted 78% of his corn crop. If his total crop consists of 12,000 acres, how many acres of corn does he have left to plant?

SOLUTION

This problem has to be made into a multiplication operation.

Step 1 is to convert 78% into a decimal.

$78\% = 78 \div 100 = 0.78$

Step 2 is to set up the multiplication operation.

12,000 acres × 0.78

Step 3 is to perform the operation.

12,000 acres × 0.78 = 9,360 acres

Step 4 is to subtract 9,360 from 12,000 to find the number of acres left.

12,000 – 9,360 = 2,640 acres.

The correct answer is 2,640 acres.

PROBLEM

Ms. Harper is a used car salesperson. Every month she is required to sell 20 cars. This month, however, she has sold 110% of the monthly requirement. How many cars did she sell?

SOLUTION

This problem has to be made into a multiplication operation.

Step 1 is to convert 110% into a decimal.

110% = 110 ÷ 100 = 1.10

Step 2 is to set up the multiplication operation.

20 × 1.10

Step 3 is to perform the operation.

20 × 1.10 = 22

The correct answer is 22 cars.

PROBLEM

State University has a very competitive engineering program. After 1 week, 40% of the students drop the program. If 250 students start the program, how many are left after the first week?

SOLUTION

This problem has to be made into a multiplication operation.

Step 1 is to convert 40% into a decimal.

40% = 40 ÷ 100 = 0.40

Step 2 is to set up the multiplication operation.

250 × 0.40

Step 3 is to perform the operation.

250 × 0.40 = 100

Therefore, 100 students drop the program after 1 week.

Step 4 is to find how many students remain.

$$250 - 100 = 150$$

The correct answer is 150 students remain.

PROBLEM

A child is discovered to have a rare but non-fatal virus. Research has found that 0.0005% of people that come in contact with those who are infected with the virus will contract the virus as well. If 8 people were known to have contact with the child, how many people will get the virus?

SOLUTION

This problem has to be made into a multiplication operation.

Step 1 is to convert 0.0005% into a decimal.

$$0.0005\% = 0.0005 \div 100 = 0.000005$$

Step 2 is to set up the multiplication operation.

$$8 \times 0.000005$$

Step 3 is to perform the operation.

$$8 \times 0.000005 = 0.00004$$

The solution is 0.00004 people. This number is not realistic because 0.00004 people cannot exist. Therefore, we have to round the answer to the nearest whole number.

The correct answer is 0 people.

PROBLEM

Central City has a yearly marathon. 56 people signed up for the marathon last year but only 40 people actually competed. What percent of the people that signed up did not compete in the race? Round the answer to the nearest hundredth.

SOLUTION

This problem has to be made into a division operation.

Step 1 is to set up the division operation.

$$40 \div 56$$

Step 2 is to perform the division.

$$40 \div 56 = 0.71$$

Step 3 is to convert the decimal into a percent.

The correct answer is 87.5%.

$0.71 \times 100 = 71\%$

Therefore, 71% of those that signed up competed in the race.

Step 4: get the percent that did not compete, subtract 71% from 100%.

$100\% - 71\% = 29\%$

The correct answer is 29% of those that signed up did not compete in the race.

PROBLEM

Sam is a pitcher for her team in her town's softball league. Last year, her record as a pitcher was 9–6. What percent of the games did she win?

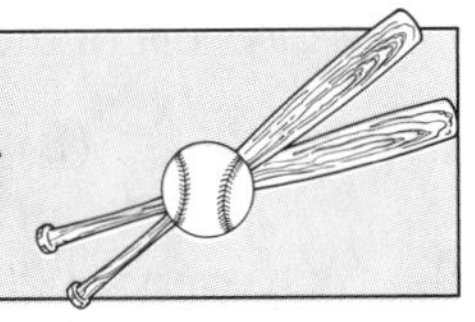

SOLUTION

This problem has to be made into a division operation.

Step 1 is to find the number of games that Sam pitched in.

$9 + 6 = 15$

Step 2 is to set up the division operation.

$9 \div 15$

Step 3 is to perform the division.

$9 \div 15 = 0.60$

Step 4 is to convert the decimal into a percent.

$0.60 \times 100 = 60\%$

The correct answer is that she won 60% of the games.

DRILL: PERCENTAGES

Finding Percents

1. Find 3% of 80.

(A) 0.24 (B) 2.4 (C) 24 (D) 240 (E) 2,400

2. Find 50% of 182.

(A) 9 (B) 90 (C) 91 (D) 910 (E) 9,100

3. Find 83% of 166.

(A) 0.137 (B) 1.377 (C) 13.778 (D) 137 (E) 137.78

4. Find 125% of 400.

(A) 425 (B) 500 (C) 525 (D) 600 (E) 825

5. Find 300% of 4.

(A) 12 (B) 120 (C) 1200 (D) 12,000 (E) 120,000

6. Forty-eight percent of the 1,200 students at Central High are males. How many male students are there at Central High?

(A) 57 (B) 576 (C) 580 (D) 600 (E) 648

7. For 35% of the last 40 days, there has been measurable rainfall. How many days out of the last 40 days have had measurable rainfall?

(A) 14 (B) 20 (C) 25 (D) 35 (E) 40

8. Of every 1,000 people who take a certain medicine, 0.2% develop severe side effects. How many people out of every 1,000 who take the medicine develop the side effects?

(A) 0.2 (B) 2 (C) 20 (D) 22 (E) 200

9. Of 220 applicants for a job, 75% were offered an initial interview. How many people were offered an initial interview?

(A) 75 (B) 110 (C) 120 (D) 155 (E) 165

10. Find 0.05% of 4,000.

(A) 0.05 (B) 0.5 (C) 2 (D) 20 (E) 400

Changing Percents to Fractions

11. What is 25% written as a fraction?

(A) 1/25 (B) 1/5 (C) 1/4 (D) 1/3 (E) 1/2

12. What is 33 1/3% written as a fraction?

(A) 1/4 (B) 1/3 (C) 1/2 (D) 2/3 (E) 5/9

13. What is 200% written as a fraction?

(A) 1/2 (B) 2/1 (C) 20/1 (D) 200/1 (E) 2000/1

14. What is 84% written as a fraction?

(A) 1/84 (B) 4/8 (C) 17/25 (D) 21/25 (E) 44/50

15. What is 2% written as a fraction?

(A) 1/50 (B) 1/25 (C) 1/10 (D) 1/4 (E) 1/2

Changing Fractions to Percents

16. What is 2/3 written as a percent?

(A) 23% (B) 32% (C) 33 1/3% (D) 57 1/3% (E) 66 2/3%

17. What is 3/5 written as a percent?

(A) 30% (B) 35% (C) 53% (D) 60% (E) 65%

18. What is 17/20 written as a percent?

(A) 17% (B) 70% (C) 75% (D) 80% (E) 85%

19. What is 45/50 written as a percent?

(A) 45% (B) 50% (C) 90% (D) 95% (E) 97%

20. What is 1 1/4 written as a percent?

(A) 114% (B) 120% (C) 125% (D) 127% (E) 133%

Changing Percents to Decimals

21. What is 42% written as a decimal?

(A) 0.42 (B) 4.2 (C) 42 (D) 420 (E) 422

22. What is 0.3% written as a decimal?

(A) 0.0003 (B) 0.003 (C) 0.03 (D) 0.3 (E) 3

23. What is 8% written as a decimal?

(A) 0.0008 (B) 0.008 (C) 0.08 (D) 0.80 (E) 8

24. What is 175% written as a decimal?

(A) 0.175 (B) 1.75 (C) 17.5 (D) 175 (E) 17,500

25. What is 34% written as a decimal?

(A) 0.00034 (B) 0.0034 (C) 0.034 (D) 0.34 (E) 3.4

Changing Decimals to Percents

26. What is 0.43 written as a percent?

(A) 0.0043% (B) 0.043% (C) 4.3% (D) 43% (E) 430%

27. What is 1 written as a percent?

(A) 1% (B) 10% (C) 100% (D) 111% (E) 150%

28. What is 0.08 written as a percent?

(A) 0.08% (B) 8% (C) 8.8% (D) 80% (E) 800%

29. What is 3.4 written as a percent?

(A) 0.0034% (B) 3.4% (C) 34% (D) 304% (E) 340%

30. What is 0.645 written as a percent?

(A) 64.5% (B) 65% (C) 69% (D) 70% (E) 645%

6. RADICALS

The **square root** of a number is a number that when multiplied by itself results in the original number. So, the square root of 81 is 9 since 9 × 9 = 81. However, –9 is also a root of 81 since (– 9) (– 9) = 81. Every positive number will have two roots. Yet, the principal root is the positive one. Zero has only one square root, while negative numbers do not have real numbers as their roots.

A **radical sign** indicates that the root of a number or expression will be taken. The **radicand** is the number of which the root will be taken. The **index** tells how many times the root needs to be multiplied by itself to equal the radicand. E.g.,

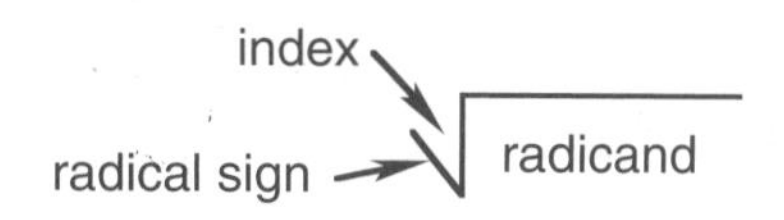

(1) $\sqrt[3]{64}$;

3 is the index and 64 is the radicand. Since $4 \times 4 \times 4 = 64, \sqrt[3]{64} = 4$

(2) $\sqrt[5]{32}$;

5 is the index and 32 is the radicand. Since $2 \times 2 \times 2 \times 2 \times 2 = 32, \sqrt[5]{32} = 2$

OPERATIONS WITH RADICALS

A) **To multiply two or more radicals**, we utilize the law that states,

$$\sqrt{a} \times \sqrt{b} = \sqrt{ab}.$$

Simply multiply the whole numbers as usual. Then, multiply the radicands and put the product under the radical sign and simplify. E.g.,

(1) $\sqrt{12} \times \sqrt{5} = \sqrt{60} = 2\sqrt{15}$

(2) $3\sqrt{2} \times 4\sqrt{8} = 12\sqrt{16} = \sqrt{48}$

(3) $2\sqrt{10} \times 6\sqrt{5} = 12\sqrt{50} = 60\sqrt{2}$

B) **To divide radicals**, simplify both the numerator and the denominator. By multiplying the radical in the denominator by itself, you can make the denominator a rational number. The numerator, however, must also be multiplied by this radical so that the value of the expression does not change. You must choose as many factors as necessary to rationalize the denominator. E.g.,

(1) $\frac{\sqrt{128}}{\sqrt{2}} = \frac{\sqrt{64} \times \sqrt{2}}{\sqrt{2}} = \frac{8\sqrt{2}}{\sqrt{2}} = 8$

(2) $\frac{\sqrt{10}}{\sqrt{3}} = \frac{\sqrt{10} \times \sqrt{3}}{\sqrt{3} \times \sqrt{3}} = \frac{\sqrt{30}}{3}$

(3) $\frac{\sqrt{8}}{2\sqrt{3}} = \frac{\sqrt{8} \times \sqrt{3}}{2\sqrt{3} \times \sqrt{3}} = \frac{\sqrt{24}}{2 \times 3} = \frac{2\sqrt{6}}{6} = \frac{\sqrt{6}}{3}$

C) **To add two or more radicals**, the radicals must have the same index and the same radicand. Only where the radicals are simplified can these similarities be determined.

EXAMPLE

(1) $6\sqrt{2} + 2\sqrt{2} = (6 + 2)\sqrt{2} = 8\sqrt{2}$

(2) $\sqrt{27} + 5\sqrt{3} = \sqrt{9}\sqrt{3} + 5\sqrt{3} = 3\sqrt{3} + 5\sqrt{3} = 8\sqrt{3}$

(3) $7\sqrt{3} + 8\sqrt{2} + 5\sqrt{3} = 12\sqrt{3} + 8\sqrt{2}$

Similarly to subtract,

(1) $12\sqrt{3} - 7\sqrt{3} = (12 - 7)\sqrt{3} = 5\sqrt{3}$

(2) $\sqrt{80} - \sqrt{20} = \sqrt{16}\sqrt{5} - \sqrt{4}\sqrt{5} = 4\sqrt{5} - 2\sqrt{5} = 2\sqrt{5}$

(3) $\sqrt{50} - \sqrt{3} = 5\sqrt{2} - \sqrt{3}$

PROBLEM

Find the solution to the following problem: $\sqrt[3]{8}$.

SOLUTION

The expression $\sqrt[3]{8}$ means to take the cube root of 8. A cube root is the reverse process of raising a number to the power of "3."

Step 1 is to determine what the base would be if you raise a number to the power of "3" to get 8.

Base = 2

Step 2 is to raise the base "2" to the exponent "3" to verify the solution.

$2 \times 2 \times 2 = 8$

The correct answer is 2.

PROBLEM

Find the solution to the following problem: $\sqrt[3]{-1}$.

SOLUTION

Step 1 is to determine what the base would be if you raise a number to the power of "3" to get –1.

Base = –1

Step 2 is to raise the base "–1" to the exponent "3" to verify the solution.

$-1 \times -1 \times -1 = -1$

The correct answer is –1.

PROBLEM

Find the solution to the following problem: $\sqrt[4]{625}$.

SOLUTION

Step 1 is to determine what the base would be if you raise a number to the power of "4" to get 625.

Base = 5

Step 2 is to raise the base "5" to the exponent "4" to verify the solution.

$5 \times 5 \times 5 \times 5 = 625$

The correct answer is 5.

PROBLEM

Find the solution to the following problem: $4^{1/2}$.

SOLUTION

Any time a base is raised to a fractional exponent, the problem should be rewritten as a root.

Step 1 is to rewrite the problem as a root. The denominator will determine what the *n* root will be. Since the denominator is 2, take the 2nd root (or square root).

$4^{1/2} = \sqrt{4}$

Step 2 is to determine what the base would be if you raise a number to the power of "2" to get 4.

Base = 2

Step 3 is to raise the base "2" to the exponent "2" to verify the solution.

$2 \times 2 = 4$

The correct answer is 2.

PROBLEM

Find the solution to the following problem: $\sqrt{12}$.

SOLUTION

Some square roots can be expressed as a product of two or more square roots.

Step 1 is to determine if $\sqrt{12}$ can be expressed as a product of two or more square roots.

$$\sqrt{12} = \sqrt{(4 \times 3)} = \sqrt{4}\sqrt{3}$$

Step 2 is to solve the square root that will not be an irrational number.

$$\sqrt{4} = 2$$

Step 3 is to rewrite the problem.

$$\sqrt{12} = 2\left(\sqrt{3}\right)$$

Step 4 is to determine if the problem is irrational. If so, then stop.

$2\left(\sqrt{3}\right)$ is irrational.

The correct answer is $2\left(\sqrt{3}\right)$.

PROBLEM

Find the solution to the following problem: $\sqrt{50}$.

SOLUTION

Some square roots can be expressed as a product of two or more square roots.

Step 1 is to determine if $\sqrt{50}$ can be expressed as a product of two or more square roots.

$$\sqrt{50} = \sqrt{(2 \times 25)} = \sqrt{2}\sqrt{25}$$

Step 2 is to solve the square root that will not be an irrational number.

$$\sqrt{25} = 5$$

Step 3 is to rewrite the problem.

$$\sqrt{50} = 5\left(\sqrt{2}\right)$$

Step 4 is to determine if the problem is irrational. If so, then stop.

$5\left(\sqrt{2}\right)$ is irrational.

The correct answer is $5\left(\sqrt{2}\right)$.

PROBLEM

Find the solution to the following problem: $\sqrt[3]{80}$.

SOLUTION

Some cube roots can be expressed as a product of two or more cube roots.

Step 1 is to determine if $\sqrt[3]{80}$ can be expressed as a product of two or more cube roots.

$$\sqrt[3]{80} = \sqrt[3]{(8 \times 5 \times 2)} = \left(\sqrt[3]{8}\right)\left(\sqrt[3]{5}\right)\left(\sqrt[3]{2}\right)$$

Step 2 is to solve the cube root that will not be an irrational number.

$$\sqrt[3]{8} = 2$$

Step 3 is to rewrite the problem.

$$\sqrt[3]{80} = (2)\left(\sqrt[3]{5}\right)\left(\sqrt[3]{2}\right)$$

Step 4 is to determine if the problem is irrational. If so, then stop.

$(2)\left(\sqrt[3]{5}\right)\left(\sqrt[3]{2}\right)$ is irrational.

The correct answer is $(2)\left(\sqrt[3]{5}\right)\left(\sqrt[3]{2}\right)$, or $(2)\left(\sqrt[3]{10}\right)$ if simplified.

DRILL 6: RADICALS

Multiplication

DIRECTIONS: Multiply and simplify each answer.

1. $\sqrt{6} \times \sqrt{5} =$

(A) $\sqrt{11}$ (B) $\sqrt{30}$ (C) $2\sqrt{5}$ (D) $3\sqrt{10}$ (E) $2\sqrt{3}$

2. $\sqrt{3} \times \sqrt{12} =$

(A) 3 (B) $\sqrt{15}$ (C) $\sqrt{36}$ (D) 6 (E) 8

3. $\sqrt{7} \times \sqrt{7} =$

(A) 7 (B) 49 (C) $\sqrt{14}$ (D) $2\sqrt{7}$ (E) $2\sqrt{14}$

4. $3\sqrt{5} \times 2\sqrt{5} =$

(A) $5\sqrt{5}$ (B) 25 (C) 30 (D) $5\sqrt{25}$ (E) $6\sqrt{5}$

5. $4\sqrt{6} \times \sqrt{2} =$

(A) $4\sqrt{8}$ (B) $8\sqrt{2}$ (C) $5\sqrt{8}$ (D) $4\sqrt{12}$ (E) $8\sqrt{3}$

Division

DIRECTIONS: Divide and simplify the answer.

6. $\sqrt{10} \div \sqrt{2} =$

(A) $\sqrt{8}$ (B) $2\sqrt{2}$ (C) $\sqrt{5}$ (D) $2\sqrt{5}$ (E) $2\sqrt{3}$

7. $\sqrt{30} \div \sqrt{15} =$

(A) $\sqrt{2}$ (B) $\sqrt{45}$ (C) $3\sqrt{5}$ (D) $\sqrt{15}$ (E) $5\sqrt{3}$

8. $\sqrt{100} \div \sqrt{25} =$

(A) $\sqrt{4}$ (B) $5\sqrt{5}$ (C) $5\sqrt{3}$ (D) 2 (E) 4

9. $\sqrt{48} \div \sqrt{8} =$

(A) $4\sqrt{3}$ (B) $3\sqrt{2}$ (C) $\sqrt{6}$ (D) 6 (E) 12

10. $3\sqrt{12} \div \sqrt{3} =$

(A) $3\sqrt{15}$ (B) 6 (C) 9 (D) 12 (E) $3\sqrt{36}$

Addition

DIRECTIONS: Simplify each radical and add.

11. $\sqrt{7} + 3\sqrt{7} =$

(A) $3\sqrt{7}$ (B) $4\sqrt{7}$ (C) $3\sqrt{14}$ (D) $4\sqrt{14}$ (E) $3\sqrt{21}$

12. $\sqrt{5} + 6\sqrt{5} + 3\sqrt{5} =$

(A) $9\sqrt{5}$ (B) $9\sqrt{15}$ (C) $5\sqrt{10}$ (D) $10\sqrt{5}$ (E) $18\sqrt{15}$

13. $3\sqrt{32} + 2\sqrt{2} =$

(A) $5\sqrt{2}$ (B) $\sqrt{34}$ (C) $14\sqrt{2}$ (D) $5\sqrt{34}$ (E) $6\sqrt{64}$

14. $6\sqrt{15} + 8\sqrt{15} + 16\sqrt{15} =$

(A) $15\sqrt{30}$ (B) $30\sqrt{45}$ (C) $30\sqrt{30}$ (D) $15\sqrt{45}$ (E) $30\sqrt{15}$

15. $6\sqrt{5} + 2\sqrt{45} =$

(A) $12\sqrt{5}$ (B) $8\sqrt{50}$ (C) $40\sqrt{2}$ (D) $12\sqrt{50}$ (E) $8\sqrt{5}$

Subtraction

DIRECTIONS: Simplify each radical and subtract.

16. $8\sqrt{5} - 6\sqrt{5} =$

(A) $2\sqrt{5}$ (B) $3\sqrt{5}$ (C) $4\sqrt{5}$ (D) $14\sqrt{5}$ (E) $48\sqrt{5}$

17. $16\sqrt{33} - 5\sqrt{33} =$

(A) $3\sqrt{33}$ (B) $33\sqrt{11}$ (C) $11\sqrt{33}$ (D) $11\sqrt{0}$ (E) $\sqrt{33}$

18. $14\sqrt{2} - 19\sqrt{2} =$

(A) $5\sqrt{2}$ (B) $-5\sqrt{2}$ (C) $-33\sqrt{2}$ (D) $33\sqrt{2}$ (E) $-4\sqrt{2}$

19. $10\sqrt{2} - 3\sqrt{8} =$

(A) $6\sqrt{6}$ (B) $-2\sqrt{2}$ (C) $7\sqrt{6}$ (D) $4\sqrt{2}$ (E) $-6\sqrt{6}$

20. $4\sqrt{3} - 2\sqrt{12} =$

(A) $-2\sqrt{9}$ (B) $-6\sqrt{15}$ (C) 0 (D) $6\sqrt{15}$ (E) $2\sqrt{12}$

---

7. EXPONENTS

When a number is multiplied by itself a specific number of times, it is said to be **raised to a power**. The way this is written is $a^n = b$ where a is the number or **base**, n is the **exponent** or **power** that indicates the number of times the base is to be multiplied by itself, and b is the product of this multiplication.

In the expression 3^2, 3 is the base and 2 is the exponent. This means that 3 is multiplied by itself 2 times and the product is 9.

An exponent can be either positive or negative. A negative exponent implies a fraction. Such that, if n is a positive integer

$$a^{-n} = \frac{1}{a^n}, a \neq 0. \text{ So, } 2^{-4} = \frac{1}{2^4} = \frac{1}{16}.$$

An exponent that is zero gives a result of 1, assuming that the base is not equal to zero.

$$a^0 = 1, a \neq 0.$$

An exponent can also be a fraction. If m and n are positive integers,

$$a^{\frac{m}{n}} = \sqrt[n]{a^m}.$$

The numerator remains the exponent of a, but the denominator tells what root to take. For example,

(1) $4^{\frac{3}{2}} = \sqrt[2]{4^3} = \sqrt{64} = 8$ (2) $3^{\frac{4}{2}} = \sqrt[2]{3^4} = \sqrt{81} = 9$

If a fractional exponent were negative, the same operation would take place, but the result would be a fraction. For example,

(1) $27^{-\frac{2}{3}} = \frac{1}{27^{2/3}} = \frac{1}{\sqrt[3]{27^2}} = \frac{1}{\sqrt[3]{729}} = \frac{1}{9}$

PROBLEM

Simplify the following expressions:

(1) -3^{-2}

(2) $(-3)^{-2}$

(3) $\frac{-3}{4^{-1}}$

SOLUTION

(1) Here the exponent applies only to 3. Since

$$x^{-y} = \frac{1}{x^y}, -3^{-2} = -(3)^{-2} = -\frac{1}{3^2} = -\frac{1}{9}$$

(2) In this case the exponent applies to the negative base. Thus,

$$(-3)^{-2} = \frac{1}{(-3)^2} = \frac{1}{(-3)(-3)} = \frac{1}{9}$$

(3) $$\frac{-3}{4^{-1}} = \frac{-3}{(\frac{1}{4})^1} = \frac{-3}{\frac{1^1}{4^1}} = \frac{-3}{\frac{1}{4}}$$

Division by a fraction is equivalent to multiplication by that fraction's reciprocal, thus

$$\frac{-3}{\frac{1}{4}} = -3 \times \frac{4}{1} = -12 \text{ and } \frac{-3}{4^{-1}} = -12$$

GENERAL LAWS OF EXPONENTS

A) $a^p a^q = a^{p+q}$

$4^2 4^3 = 4^{2+3} = 1{,}024$

B) $(a^p)^q = a^{pq}$

$(2^3)^2 = 2^6 = 64$

C) $\frac{a^p}{a^q} = a^{p-q}$

$\frac{3^6}{3^2} = 3^4 = 81$

D) $(ab)^p = a^p b^p$

$(3 \times 2)^2 = 3^2 \times 2^2 = (9)(4) = 36$

E) $\left(\frac{a}{b}\right)^p = \frac{a^p}{b^p},\ b \neq 0$

$\left(\frac{4}{5}\right)^2 = \frac{4^2}{5^2} = \frac{16}{25}$

PROBLEM

Find the solution to the following problem: $(-3^1)^2$.

SOLUTION

Step 1 is to identify the base and the exponents. In this problem, "–3" is the base. "1" and "2" are the exponents.

Step 2: since this problem raises an exponent to an exponent, multiply the exponents.

$1 \times 2 = 2$

Step 3 is to rewrite the problem.

-3^2

Step 4 is to set up the multiplication. Multiply the base, "–3," with itself.

-3×-3

Step 5 is to perform the operation.

$-3 \times -3 = 9$

The correct answer is 9.

PROBLEM

Find the solution to the following problem: $\frac{(6^6)}{(6^4)}$.

SOLUTION

Step 1 is to identify the base and the exponents. In this problem, "6" is the common base. "6" and "4" are the exponents.

Step 2: since the problem contains a common base, the exponents can be subtracted.

$6 - 4 = 2$

Step 3 is to rewrite the problem using the new exponent.

6^2

Step 4 is to perform the operation.

$6 \times 6 = 36$

The correct answer is 36.

PROBLEM

Find the solution to the following problem: $(7^{-3}) \div (7^{-5})$.

SOLUTION

Step 1 is to identify the bases and the exponents. In this problem, "7" is the common base. "–3" and "–5" are the exponents.

Step 2: since the problem contains a common base, the exponents can be subtracted.

$-3 - (-5) = 2$

Step 3 is to rewrite the problem using the new exponent.

7^2

Step 4 is to perform the operation.

$7 \times 7 = 49$

The correct answer is 49.

POWER TO A POWER

Consider the example $(3^2)^4$. Remembering that an exponent shows the number of times the base is to be taken as a factor and noting in this case that 3^2 is considered the base, we have

$$(3^2)^4 = 3^2 \times 3^2 \times 3^2 \times 3^2$$

Also in multiplication we add exponents. Thus,

$$3^2 \times 3^2 \times 3^2 \times 3^2 = 3^{(2+2+2+2)} = 3^8$$

Therefore,

$$(3^2)^4 = 3^{(4 \times 2)}$$
$$= 3^8$$

The laws of exponents for the power of a power may be stated as follows: To find the power of a power, multiply the exponents. It should be noted that this case is the only one in which multiplication of exponents is performed.

DRILL: EXPONENTS

Multiplication

Simplify

1. $4^6 \times 4^2 =$

(A) 4^4 (B) 4^8 (C) 4^{12} (D) 16^8 (E) 16^{12}

2. $2^2 \times 2^5 \times 2^3 =$

(A) 2^{10} (B) 4^{10} (C) 8^{10} (D) 2^{30} (E) 8^{30}

3. $6^6 \times 6^2 \times 6^4 =$

(A) 18^8 (B) 18^{12} (C) 6^{12} (D) 6^{48} (E) 18^{48}

4. $a^4b^2 \times a^3b =$

(A) ab (B) $2a^7b^2$ (C) $2a^{12}b$ (D) a^7b^3 (E) a^7b^2

5. $m^8n^3 \times m^2n \times m^4n^2 =$

(A) $3m^{16}n^6$ (B) $m^{14}n^6$ (C) $3m^{14}n^6$ (D) $3m^{14}n^5$ (E) m^2

Division

Simplify

6. $6^5 \div 6^3 =$

(A) 0 (B) 1 (C) 6 (D) 12 (E) 6^2

7. $11^8 \div 11^5 =$

(A) 1^3 (B) 11^3 (C) 11^{13} (D) 11^{40} (E) 88^5

8. $x^{10}y^8 \div x^7y^3 =$

(A) x^2y^5 (B) x^3y^4 (C) x^3y^5 (D) x^2y^4 (E) x^5y^3

9. $a^{14} \div a^9 =$

(A) 1^5 (B) a^5 (C) $2a^5$ (D) a^{23} (E) $2a^{23}$

10. $c^{17}d^{12}e^4 \div c^{12}d^8e =$

(A) $c^4d^5e^3$ (B) $c^4d^4e^3$ (C) $c^5d^8e^4$ (D) $c^5d^4e^3$ (E) $c^5d^4e^4$

Power to a Power

Simplify

11. $(3^6)^2 =$

(A) 3^4 (B) 3^8 (C) 3^{12} (D) 9^6 (E) 9^8

12. $(4^3)^5 =$

(A) 4^2 (B) 2^{15} (C) 4^8 (D) 20^3 (E) 4^{15}

13. $(a^4b^3)^2 =$

(A) $(ab)^9$ (B) a^8b^6 (C) $(ab)^{24}$ (D) a^6b^5 (E) $2a^4b^3$

14. $(r^3p^6)^3 =$

(A) r^9p^{18} (B) $(rp)^{12}$ (C) r^6p^9 (D) $3r^3p^6$ (E) $3r^9p^{18}$

15. $(m^6n^5q^3)^2 =$

(A) $2m^6n^5q^3$ (B) m^4n^3q (C) $m^8n^7q^5$

(D) $m^{12}n^{10}q^6$ (E) $2m^{12}n^{10}q^6$

8. SCIENTIFIC NOTATION

Technicians, engineers, and others engaged in scientific work are often required to solve problems involving very large and very small numbers. Problems such as

$$\frac{22{,}684 \times 0.00189}{0.0713 \times 83 \times 7}$$

are not uncommon. Solving such problems by the rules of ordinary arithmetic is laborious and time consuming. Moreover, the tedious arithmetic process lends itself to operational errors. Also, there is difficulty in locating the decimal point in the result. These difficulties can be greatly reduced by a knowledge of the powers of 10 and their use.

The laws of exponents form the basis for calculation using powers of 10. The following list includes several decimals and whole numbers expressed as powers of 10:

10,000	$= 10^4$	0.1	$= 10^{-1}$
1,000	$= 10^3$	0.01	$= 10^{-2}$
100	$= 10^2$	0.001	$= 10^{-3}$
10	$= 10^1$	0.0001	$= 10^{-4}$
1	$= 10^0$		

The concept of scientific notation may be demonstrated as follows:

$$60{,}000 = 6.0000 \times 10{,}000$$
$$= 6 \times 10^4$$
$$538 = 5.38 \times 100$$
$$= 5.38 \times 10^2$$

Notice that the final expression in each of the foregoing examples involves a number between 1 and 10, multiplied by a power of 10. Furthermore, in each case the exponent of the power of 10 is a number equal to the number of digits between the new position of the decimal point and the original position (understood) of the decimal point.

We apply this reasoning to write any number in scientific notation; that is, as a number between 1 and 10 multiplied by the appropriate power of 10. The appropriate power of 10 is found by the following mechanical steps:

1. Shift the decimal point to standard position, which is the position immediately to the right of the first nonzero digit.

2. Count the number of digits between the new position of the decimal point and its original position. This number indicates the value of the exponent for the power of 10.

3. If the decimal point is shifted to the left, the sign of the exponent of 10 is positive; if the decimal point is shifted to the right, the sign of the exponent is negative.

The validity of this rule, for those cases in which the exponent of 10 is negative, is demonstrated as follows:

$$\begin{aligned} 0.00657 &= 6.57 \times 0.001 \\ &= 6.57 \times 10^{-3} \\ 0.348 &= 3.48 \times 0.1 \\ &= 3.48 \times 10^{-1} \end{aligned}$$

Further examples of the use of scientific notation are given as follows:

$$\begin{aligned} 543{,}000{,}000 &= 5.43 \times 10^{8} \\ 186 &= 1.86 \times 10^{2} \\ 243.01 &= 2.4301 \times 10^{2} \\ 0.0000007 &= 7 \times 10^{-7} \\ 0.00023 &= 2.3 \times 10^{-4} \end{aligned}$$

9. AVERAGES

MEAN

The mean is the arithmetic average. It is the sum of the values divided by the total number of variables. For example:

$$\frac{4+3+8}{3} = 5$$

PROBLEM

Find the mean salary for four company employees who make \$5/hr., \$8/hr., \$12/hr., and \$15/hr.

SOLUTION

The mean salary is the average.

$$\frac{\$5+\$8+\$12+\$15}{4} = \frac{\$40}{4} = \$10/\text{hr}$$

PROBLEM

Find the mean length of five fish with lengths of 7.5 in, 7.75 in, 8.5 in, 8.5 in., 8.25 in.

SOLUTION

The mean length is the average length.

$$\frac{7.5+7.75+8.5+8.5+8.25}{5}=\frac{40.5}{5}=8.1\text{in}$$

MEDIAN

The median is the middle value in a set when there is an odd number of values. There is an equal number of values larger and smaller than the median. When the set is an even number of values, the average of the two middle values is the median. For example:

The median of (2, 3, 5, 8, 9) is 5.

The median of (2, 3, 5, 9, 10, 11) is $\frac{5+9}{2}=7$.

MODE

The mode is the most frequently occurring value in the set of values. For example the mode of 4, 5, 8, 3, 8, 2 would be 8, since it occurs twice while the other values occur only once.

PROBLEM

For this series of observations find the mean, median, and mode.

500, 600, 800, 800, 900, 900, 900, 900, 900, 1000, 1100

SOLUTION

The mean is the value obtained by adding all the measurements and dividing by the number of measurements.

$$\frac{500+600+800+800+900+900+900+900+900+1000+1100}{11}$$

$$=\frac{9300}{11}=845.45.$$

The median is the observation in the middle. We have 11 observations, so here the sixth, 900, is the median.

The mode is the observation that appears most frequently. That is also 900, since it has 5 appearances.

All three of these numbers are measures of central tendency. They describe the "middle" or "center" of the data.

PROBLEM

Nine rats run through a maze. The time each rat took to traverse the maze is recorded and these times are listed below.

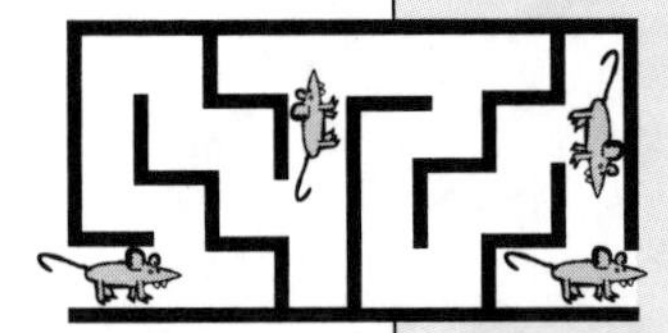

1 min, 2.5 min, 3 min, 1.5 min, 2 min, 1.25 min, 1 min, .9 min, 30 min

Which of the three measures of central tendency would be the most appropriate in this case?

SOLUTION

We will calculate the three measures of central tendency and then compare them to determine which would be the most appropriate in describing these data.

The mean is the sum of observations divided by the number of observations. In this case

$$\frac{1 + 2.5 + 3 + 1.5 + 2 + 1.25 + 1 + .9 + 30}{9} = \frac{43.15}{9} = 4.79.$$

The median is the "middle number" in an array of the observations from the lowest to the highest.

0.9, 1.0, 1.0, 1.25, 1.5, 2.0, 2.5, 3.0, 30.0

The median is the fifth observation in this array or 1.5. There are four observations larger than 1.5 and four observations smaller than 1.5.

The mode is the most frequently occurring observation in the sample. In this data set the mode is 1.0.

mean = 4.79

median = 1.5

mode = 1.0

The mean is not appropriate here. Only one rat took more than 4.79 minutes to run the maze and this rat took 30 minutes. We see that the mean has been distorted by this one large observation.

The median or mode seems to describe this data set better and would be more appropriate to use.

DRILL: AVERAGES

Mean

<u>DIRECTIONS</u>: Find the mean of each set of numbers:

1. 18, 25, and 32.

(A) 3 (B) 25 (C) 50 (D) 75 (E) 150

2. 4/9, 2/3, and 5/6.

(A) 11/18 (B) 35/54 (C) 41/54 (D) 35/18 (E) 54/18

3. 97, 102, 116, and 137.

(A) 40 (B) 102 (C) 109 (D) 113 (E) 116

4. 12, 15, 18, 24, and 31.

(A) 18 (B) 19.3 (C) 20 (D) 25 (E) 100

5. 7, 4, 6, 3, 11, and 14.

(A) 5 (B) 6.5 (C) 7 (D) 7.5 (E) 8

Median

DIRECTIONS: Find the median value of each set of numbers.

6. 3, 8, and 6.

(A) 3 (B) 6 (C) 8 (D) 17 (E) 20

7. 19, 15, 21, 27, and 12.

(A) 19 (B) 15 (C) 21 (D) 27 (E) 94

8. 1 2/3, 1 7/8, 1 3/4, and 1 5/6.

(A) 1 30/48 (B) 1 2/3 (C) 1 3/4 (D) 1 19/24 (E) 1 21/24

9. 29, 18, 21, and 35.

(A) 29 (B) 18 (C) 21 (D) 35 (E) 25

10. 8, 15, 7, 12, 31, 3, and 28.

(A) 7 (B) 11.6 (C) 12 (D) 14.9 (E) 104

Mode

DIRECTIONS: Find the mode(s) of each set of numbers.

11. 1, 3, 7, 4, 3, and 8.

(A) 1 (B) 3 (C) 7 (D) 4 (E) None

12. 12, 19, 25, and 42

(A) 12 (B) 19 (C) 25 (D) 42 (E) None

13. 16, 14, 12, 16, 30, and 28.

(A) 6 (B) 14 (C) 16 (D) 19.3 (E) None

14. 4, 3, 9, 2, 4, 5, and 2.

(A) 3 and 9 (B) 5 and 9 (C) 4 and 5 (D) 2 and 4 (E) None

15. 87, 42, 111, 116, 39, 111, 140, 116, 97, and 111.

(A) 111 (B) 116 (C) 39 (D) 140 (E) None

10. ESTIMATION

EXAMPLE

Jim is hosting a pizza party for 11 of his close friends. He wants to serve each guest a mini-pizza. Mini-pizzas cost $6.79 each. Estimate the total cost.

Round 11 to 10. Round $6.79 to $7.00. Now, multiply $7 times 10. The estimated cost is $70.00.

Note: The exact cost is $74.69.

EXAMPLE

The chart below shows the number of fishing lures a factory produced over a 6-year period.

Year	Fishing Lures
2000	6,257
2001	10,374
2002	5,890
2003	12,125
2004	9,642
2005	13,092

Estimate the total number of fishing lures produced from 2000 through 2005.

First, round each number to the thousands. Then, add the rounded numbers in order to find the estimate.

6,000 + 10,000 + 6,000 + 12,000 + 10,000 + 13,000 = 57,000 lures

Note: The actual number of lures is 57,380.

PROBLEM

A rectangular duck pond (43 feet by 47 feet) is on a lot that measures 108 feet by 96 feet. The rest of the lot is a flower garden. Estimate the size of the flower garden in square feet.

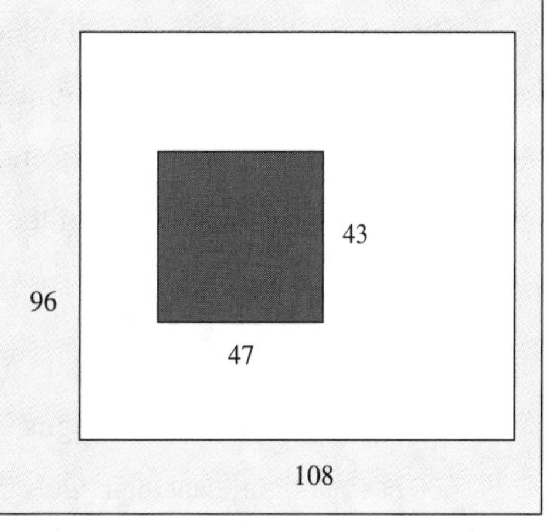

SOLUTION

First, estimate the area of the duck pond. Round 47 to 50, and round 43 to 40. Now, multiply: $50 \times 40 = 2{,}000$. The area of the duck pond is about 2,000 square feet. Now, calculate the area of the lot. Round 96 to 100, and 108 to 100. Multiply: $100 \times 100 = 10{,}000$. In order to find the approximate area of the flower garden, subtract the area of the duck pond from the area of the lot. $10{,}000 - 2{,}000 = 8{,}000$. The flower garden is about 8,000 square feet.

Note: The actual area of the flower garden is 8,347 square feet.

PROBLEM

Mr. Taylor drove his car to the gas station to fill it up. By the time he got there, the gas tank was empty. The total cost to fill up his gas tank was $31.45. Gasoline costs $1.46 per gallon. Estimate how many gallons Mr. Taylor's gas tank holds.

SOLUTION

Round $31.45 to $30.00 and $1.46 to $1.50. Divide: $30 \div 1.5 = 20$ gallons.

Note: The actual number of gallons that the tank holds is 21.54.

PROBLEM

A rancher owns 26,450 acres of land. According to the results from a recent geological survey and soil analysis, 38% of the land is farmable. Estimate the number of farmable acres.

SOLUTION

Round 26,450 to 25,000. Round 38% to 40%, which is 0.4 in decimal form. Then, multiply $25{,}000 \times 0.4 = 10{,}000$ acres.

Note: The actual number is 10,051 acres.

11. SIGNIFICANT DIGITS

The basic rules for significant digits are:

1) All nonzero digits are significant
2) All zeroes between significant digits are significant
3) Final zeroes to the right of the decimal point are significant

EXAMPLES

815.85 has five significant digits (All of the numbers give useful information.)

7.48629 has six significant digits

600 has one significant digit (Only the 6 is important. You don't know anything about the tens or the ones place, so the zeros are just seen as place holders.)

240 has two significant digits

542 has three significant digits

0.7500 has four significant digits (Final zeroes that lie to the right of the decimal point are significant.)

0.008 has one significant digit

5,006 has four significant digits (All zeroes between significant digits are significant.)

56,000 has two significant digits

PROBLEM

How many significant digits are in the number 720?

SOLUTION

There are two significant digits in the number 720. The zero is just a place-holder.

PROBLEM

How many significant digits are in the number 28.05

SOLUTION

There are four significant digits in the number 28.05. The zero counts because it is between two significant digits.

PROBLEM

How many significant digits are in the number 3,001?

SOLUTION

There are four significant digits in the number 3,001. The zeroes count because they are between two significant digits.

PROBLEM

How many significant digits are in the number 0.00017?

SOLUTION

There are two significant digits in the number 0.00071. The zeroes to the right of the decimal point are placeholders.

PROBLEM

How many significant digits are in the number 8,000?

SOLUTION

There is one significant digit in the number 8,000. Only the 8 is important. Without any other information about the measurement, the zeroes are seen as placeholders.

---

12. MONEY NOTATION

Certain types of notation are use to signify that a number represents money. The cent sign, symbolized by ¢, represents one-hundredth of a dollar. The dollar sign, symbolized by $, represents dollar amounts. The decimal place immediately to the right of the decimal point represents tenths of a dollar, while the decimal place to the right of this represents hundredths of a dollar.

EXAMPLES

1¢ = 1 cent	$0.01 = 1 cent
5¢ = 5 cents	$0.05 = 5 cents
32¢ = 32 cents	$0.32 = 32 cents
50¢ = 50 cents	$0.50 = 50 cents
89¢ = 89 cents	$0.89 = 89 cents
100¢ = 1 dollar	$1.00 = 1 dollar

$1.15 = one dollar and fifteen cents

$24.38 = twenty-four dollars and thirty-eight cents

$860.00 = eight hundred, sixty dollars

$5,743.66 = five thousand, seven hundred, forty-three dollars and sixty-six cents

$39,225.50 = thirty-nine thousand, two hundred, twenty-five dollars and fifty cents

$105,953.47 = one hundred and five thousand, nine hundred, fifty-three dollars and forty-seven cents

PROBLEM

Each student at Daisy Elementary School needs 10 pencils to start the year. There are 357 students and each pencil costs 4¢. How much will the pencils cost?

SOLUTION

$$\begin{array}{r} 357 \\ \times \quad 10 \\ \hline 3,570 \text{ pencils} \end{array} \qquad \begin{array}{r} 3,570 \\ \times \quad 0.04 \\ \hline \$142.80 \end{array}$$

The answer is $142.80.

PROBLEM

The will of a late real-estate tycoon is being settled in court. In the will, all assets are to be divided evenly among his remaining 20 relatives. The estate is worth $4,802,000. How much money will each heir receive?

SOLUTION

$$\begin{array}{r} 4,802,000 \\ \div \qquad 20 \\ \hline \$240,100 \end{array}$$

The answer is $240,100.

PROBLEM

A real-estate development company wants to build an office building and buys a piece of property in the downtown area for $2,175,000. A short time later, the company purchases the property next to it for $1,500,000 and the one to next that for $1,950,000. Then, the company pays construction costs of $6,000,000 to have the offices built. What is the total amount of money invested?

SOLUTION

$$\begin{array}{r} \$2,175,000 \\ \$1,500,000 \\ \$1,950,000 \\ + \$6,000,000 \\ \hline \$11,625,000 \end{array}$$

The total investment is $11,625,000.

PROBLEM

You go to the grocery store to buy fruits and vegetables. You buy green beans for $1.49, potatoes for $0.39, onions for $0.79, bananas for $0.49, tomatoes for $0.99, squash for $0.79, pears for $0.99, and asparagus for $2.59. If you pay with a $50 bill, how much change will you get back?

SOLUTION

$1.49 + 0.39 + 0.79 + 0.49 + 0.99 + 0.79 + 0.99 + 2.59 = \$8.52$ cost

Given $50

$$\begin{array}{r} \$50.00 \\ -\ 8.52 \\ \hline \$41.48 \end{array}$$

The answer is $41.48.

13. COUNTING AND SORTING

Counting and sorting involves both counting numbers and sorting them into categories. Creating a frequency distribution requires counting and sorting. A frequency distribution is two columns of numbers. The first column has scores in it and the second column has frequencies. Graphing a frequency distribution of scores allows you to see how frequently particular scores or groups of scores occur relative to the others.

The exact method for creating a frequency distribution depends on the range of the scores. If the scores have a narrow range of values, list this range from the minimum score to the maximum score. Put this list in a column labeled "Scores." Then, count the frequency of each score and enter these numbers into a column labeled "Frequency."

The other possibility occurs when the scores have a wide range of values. In this situation, the scores must be broken down into class intervals. Between 8 and 20 intervals are sufficient. These class intervals are listed in a column labeled "Scores." Then, count the frequency of scores in each interval and enter these numbers into a column labeled "Frequency."

EXAMPLE

Given a set of 20 scores, create a frequency distribution. Using this frequency distribution, draw a histogram. Since there is a small range of scores, the frequency of individual scores can be counted.

Scores: 7, 4, 8, 7, 9, 5, 4, 8, 6, 2, 2, 9, 5, 7, 4, 5, 7, 3, 6, 4

Score	Frequency
9	2
8	2
7	4
6	2
5	3
4	4
3	1
2	2
	20 scores

A histogram of this frequency distribution looks like this:

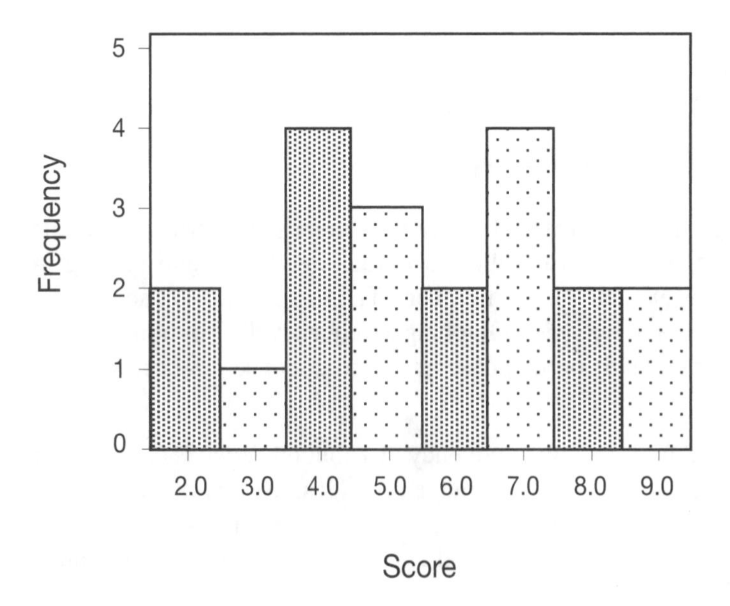

EXAMPLE

Given this set of 50 scores, create a frequency distribution.

92	42	61	15	7	31	80	21	72	50
97	81	33	43	9	73	52	65	25	82
98	53	27	76	44	33	82	67	28	77
84	35	68	44	53	85	35	77	46	56
85	56	78	36	39	56	85	89	57	88

Since the range of these scores is large, class intervals are used to divide the scores into groups.

Class Interval	Frequency
90 – 100	3
80 – 89	10
70 – 79	6
60 – 69	4
50 – 59	8
40 – 49	5
30 – 39	7
20 – 29	4
10 – 19	1
1 – 9	2
	50 scores

PROBLEM

A classroom of thirty eighth-grade students took a pop quiz in their math class. Their scores are listed below.

2	7	4	8	7	9
5	4	8	6	2	9
5	7	4	5	7	3
6	4	1	4	7	7
9	2	3	5	6	5

Create a frequency distribution.

SOLUTION

Since there is a small range of scores, the frequency of individual scores can be counted.

Score	Frequency
9	3
8	2
7	6
6	3
5	5
4	5
3	2
2	3
1	1
	30 scores

14. GAME STRATEGIES

A game strategy basically refers to discovering patterns. If we look at a graph, or a series of numbers or pictures, sometimes we can see a pattern.

EXAMPLE

Chelsea owns her own company. Chelsea notices a pattern in which her year in business is related to her profit that year. She made a table below based on her findings.

Year in Business	Company Profit
1	$30,000
2	$33,900
3	$38,300
4	$42,100

Describe the pattern that Chelsea found. How much money would Chelsea expect to make in her 5th year of business?

Chelsea earns about $4,000 more each year. Chelsea would expect to make about $46,000 during her 5th year of business.

EXAMPLE

The following figures form a pattern. What figure would continue this pattern?

 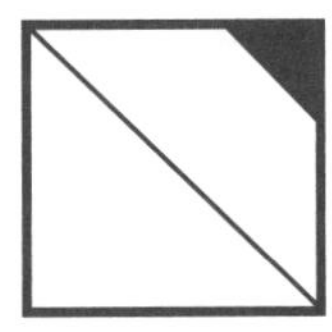 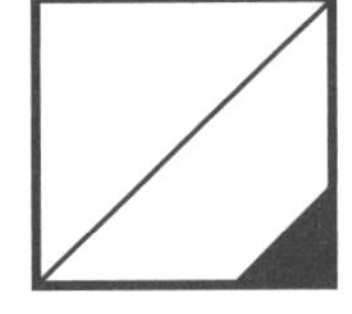  

The figures show a pattern of clockwise rotation. The figure that best continues this pattern is:

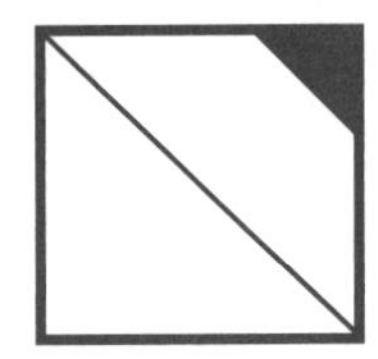

EXAMPLE

Carried out to 12 digits, $1 \div 7 = 0.142857142857$. The pattern here is that the numbers 142857 repeat themselves.

PROBLEM

The graph below shows the number of apples produced by a grower in California for the years 1997, 1999, and 2001. From this graph, which of the following was the most probable number of oranges produced by this grower in 1995?

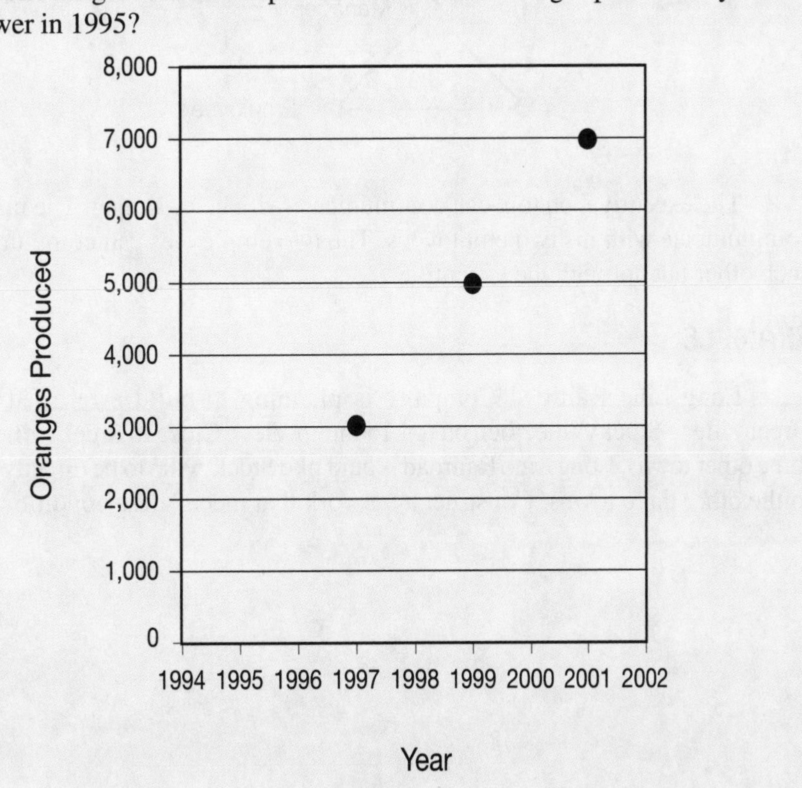

SOLUTION

Draw a line through the points given and extend this line downward. Next, find the year 1995 on the x-axis and go up until you hit the line that you have drawn. Finally, read the number of oranges produced on the y-axis. The answer is 1,000.

15. NETWORKS

A network is a system of interconnected people or things (for example, a system of interconnected computers). It can be system of intersecting lines or channels such as a railroad network or a network of canals. Networks can be graphed using vertices and lines. The vertices represent the person or thing, while the lines represent a connection.

EXAMPLE

An executive at big company wants to communicate with her manager. The manager wants to be able to communicate with his two employees. The two employees can communicate with each other but *not* with the executive. Draw a graph of a network that satisfies these requirements.

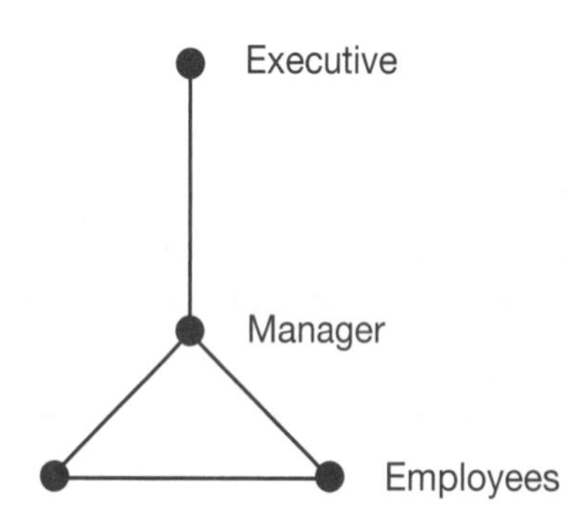

The executive, on top, can communicate with her manager. The manager can communicate with his two employees. The two employees can communicate with each other but not with the executive.

EXAMPLE

Long Line Railroad Company is planning to build a railroad line from Greenville to Stuckyville, then on to Mountain View. Close to Stuckyville, there are three other towns. Long Line Railroad would like Stuckyville to be directly connected to the other three towns. Construct a network that meets these conditions.

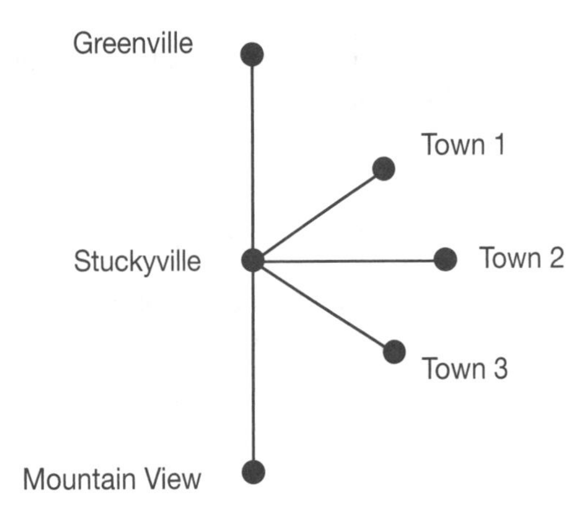

This network meets the conditions described. The railroad line can travel from Greenville to Stuckyville, then on to Mountain View. Also, Stuckyville is connected to the other three towns.

PROBLEM

An architect, a secretary, a foreman, and a carpenter all work at a construction company. Their computers are connected in the following way. The architect's computer is connected to both the secretary's computer and the foreman's computer. The secretary's computer is connected to everyone's computer. The carpenter's computer is connected to the foreman's. Draw a graph of a network that meets these conditions.

SOLUTION

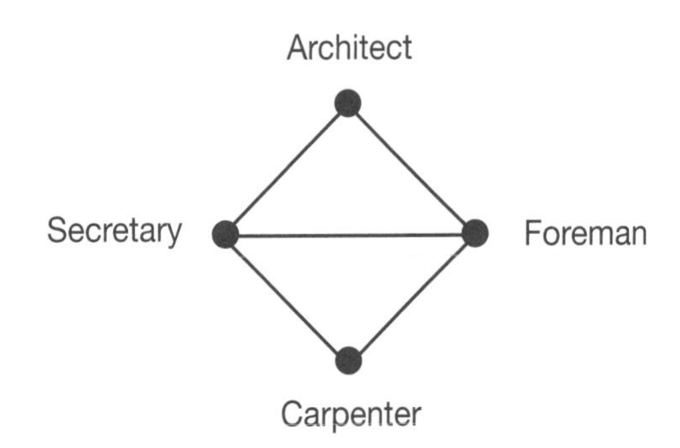

16. ALGORITHMS

An algorithm is a set of rules that always leads to the correct solution. There are different algorithms for different things.

EXAMPLE

The algorithm for calculating the total cost of an item at a store goes like this:

1. Multiply the price of the item times the sales tax expressed in decimal form to find the amount of the tax.
2. Add the sales tax to the price of the item to find the total cost.

EXAMPLE

The prime factorization of a number is when a number is expressed in terms of prime numbers. A prime number is a number that can only be divided evenly by one and itself. A factor is a number that can be divided evenly into another number. The process of prime factorization is breaking down a number into parts. The algorithm for finding the prime factored form of a number goes like this.

1) Divide the number by 2 as many times as possible.
2) Divide the remaining factor by 3 as many times as possible.
3) Divide the remaining factor by 5 as many times as possible.
4) Divide the remaining factor by 7 as many times as possible.
5) Continue factoring using prime numbers (for example, 11, 13, 17, 19, ...) until the number is expressed in terms of all its prime factors.

Using this algorithm, find the prime factored form of the number 2,520.

$2{,}520 \div 2 = 1{,}260$

$1{,}260 \div 2 = 630$

$630 \div 2 = 315$

$315 \div 3 = 105$

$105 \div 3 = 35$

$35 \div 5 = 7$

The prime factored form of 2,520 is $2 \times 2 \times 2 \times 3 \times 3 \times 5 \times 7$.

17. FLOW CHARTS

A flow chart is a graphical representation of the successive steps in solving a problem. A flow chart uses symbols connected by lines.

EXAMPLE

This flow chart shows the percent income tax paid depending upon how much income a person earned last year. Find the percent income tax for someone who earned $40,000 last year.

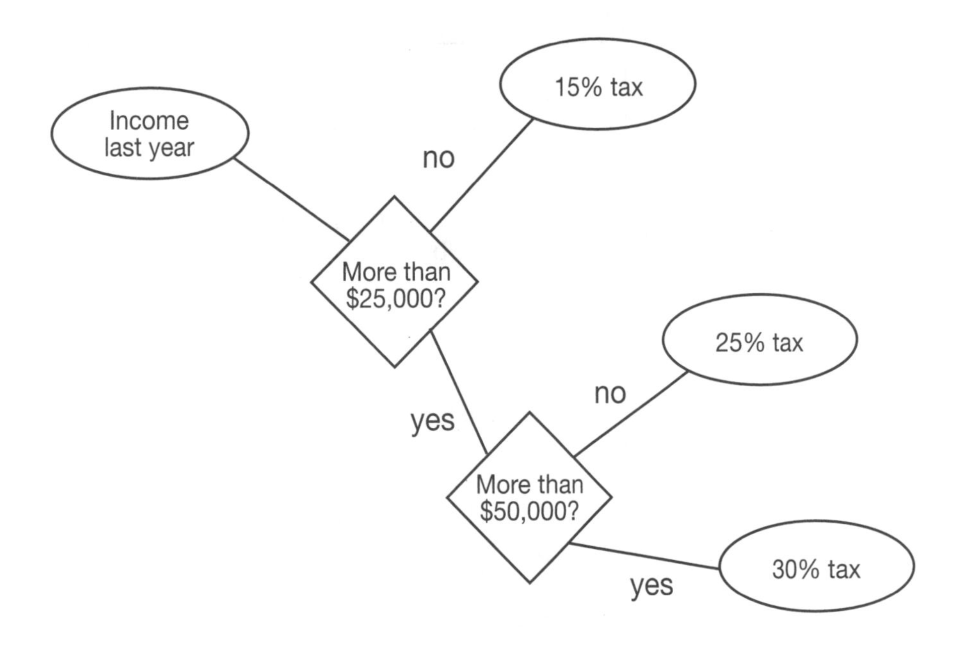

Consulting the flow chart, we see that a person with an income of $40,000 last year would have to pay 25% in income tax.

PROBLEM

This flow chart shows a series of mathematical operations. If the starting number is 100, what is the result?

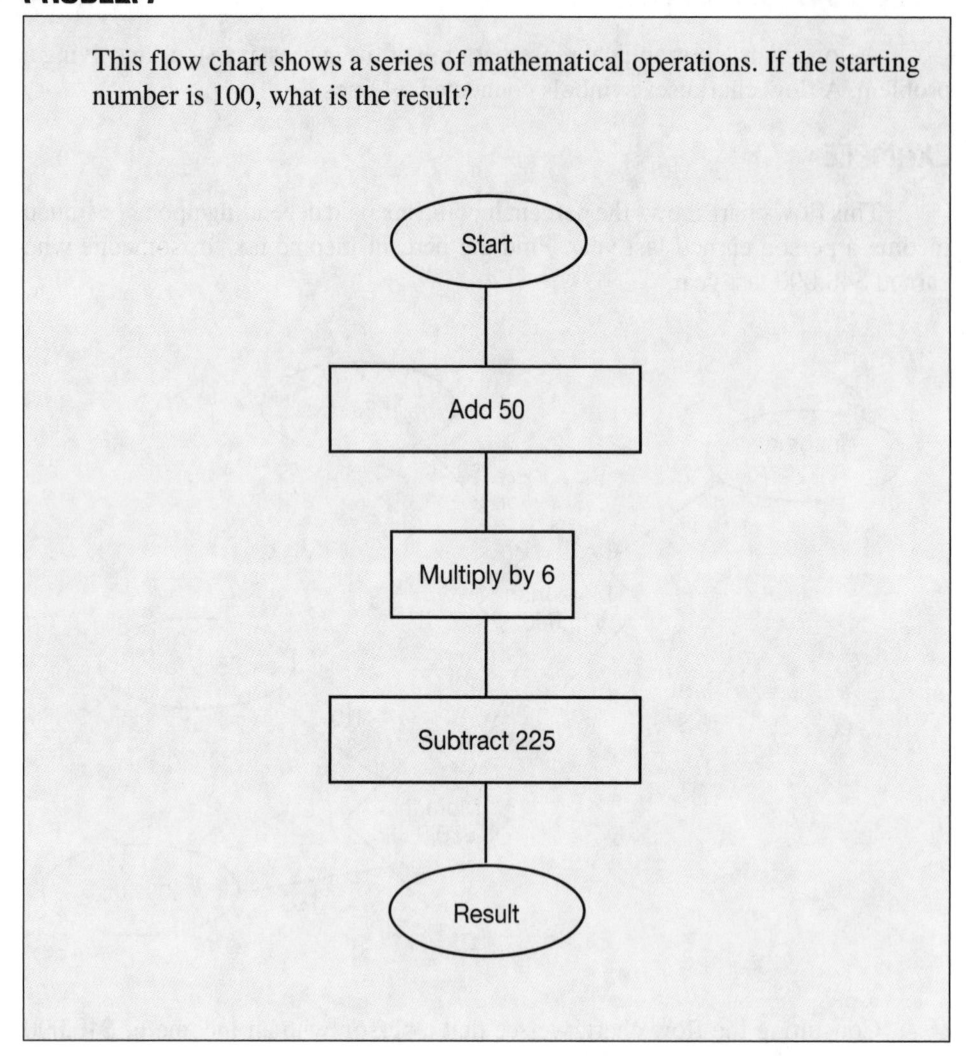

SOLUTION

Start with 100

$100 + 50 = 150$

$150 \times 6 = 900$

$900 - 225 = 675$

The result is 675.

18. MORE ESTIMATION

PROBLEM

Estimate $\frac{1}{9}$ of \$120.

SOLUTION

Round $\frac{1}{9}$ to $\frac{1}{10}$, then multiply: $\frac{1}{10} \times 120 = \$12$

Note: The actual value is \$13.33.

PROBLEM

In one state, the sales tax is 5.25%. If a man buys a lawnmower that costs \$195.50, how much tax will he pay?

SOLUTION

First, round 5.25% to 5% and express this percentage in decimal form, which is 0.05. Then, round \$195.50 to \$200.00. Finally, multiply: $0.05 \times 200 = \$10.00$.

Note: The actual tax is \$10.26.

PROBLEM

The water level of a lake is falling at the rate of 2.75 inches per day. At this rate, estimate how many inches the water level will fall in 21 days.

SOLUTION

Round 2.75 to 3 and round 21 to 20. Now, multiply: $3 \times 20 = 60$ inches.

Note: The actual change in water level is 57.75 inches.

ARITHMETIC DRILLS

DRILL—INTEGERS AND REAL NUMBERS

1.	(A)	9.	(E)	17.	(D)	25.	(D)
2.	(C)	10.	(B)	18.	(D)	26.	(D)
3.	(D)	11.	(B)	19.	(C)	27.	(B)
4.	(C)	12.	(E)	20.	(E)	28.	(C)
5.	(B)	13.	(C)	21.	(B)	29.	(A)
6.	(B)	14.	(A)	22.	(E)	30.	(C)
7.	(A)	15.	(B)	23.	(D)		
8.	(C)	16.	(B)	24.	(A)		

DRILL—FRACTIONS

1.	(D)	14.	(D)	27.	(D)	40.	(A)
2.	(E)	15.	(B)	28.	(E)	41.	(D)
3.	(C)	16.	(B)	29.	(C)	42.	(A)
4.	(A)	17.	(D)	30.	(B)	43.	(D)
5.	(C)	18.	(B)	31.	(C)	44.	(E)
6.	(B)	19.	(A)	32.	(A)	45.	(C)
7.	(C)	20.	(C)	33.	(B)	46.	(D)
8.	(A)	21.	(D)	34.	(C)	47.	(D)
9.	(B)	22.	(C)	35.	(D)	48.	(B)
10.	(A)	23.	(B)	36.	(B)	49.	(A)
11.	(B)	24.	(E)	37.	(C)	50.	(E)
12.	(D)	25.	(A)	38.	(D)		
13.	(E)	26.	(A)	39.	(E)		

DRILL—DECIMALS

1.	(B)	9.	(C)	17.	(D)	25.	(E)
2.	(E)	10.	(E)	18.	(D)	26.	(D)
3.	(B)	11.	(B)	19.	(A)	27.	(C)
4.	(A)	12.	(D)	20.	(C)	28.	(A)
5.	(D)	13.	(A)	21.	(E)	29.	(E)
6.	(A)	14.	(B)	22.	(C)	30.	(D)
7.	(B)	15.	(C)	23.	(B)		
8.	(D)	16.	(B)	24.	(D)		

DRILL—PERCENTAGES

1.	(B)	9.	(E)	17.	(D)	25.	(D)
2.	(C)	10.	(C)	18.	(E)	26.	(D)
3.	(E)	11.	(C)	19.	(C)	27.	(C)
4.	(B)	12.	(B)	20.	(C)	28.	(B)
5.	(A)	13.	(B)	21.	(A)	29.	(E)
6.	(B)	14.	(D)	22.	(B)	30.	(A)
7.	(A)	15.	(A)	23.	(C)		
8.	(B)	16.	(E)	24.	(B)		

DRILL—RADICALS

1.	(B)	6.	(C)	11.	(B)	16.	(A)
2.	(D)	7.	(A)	12.	(D)	17.	(C)
3.	(A)	8.	(D)	13.	(C)	18.	(B)
4.	(C)	9.	(C)	14.	(E)	19.	(D)
5.	(E)	10.	(B)	15.	(A)	20.	(C)

DRILL—EXPONENTS

1.	(B)	9.	(B)
2.	(A)	10.	(D)
3.	(C)	11.	(C)
4.	(D)	12.	(E)
5.	(B)	13.	(B)
6.	(E)	14.	(A)
7.	(B)	15.	(D)
8.	(C)		

DRILL—AVERAGES

1.	(B)	9.	(E)
2.	(B)	10.	(C)
3.	(D)	11.	(B)
4.	(C)	12.	(E)
5.	(D)	13.	(C)
6.	(B)	14.	(D)
7.	(A)	15.	(A)
8.	(D)		

ALGEBRA REVIEW

In algebra, letters or variables are used to represent numbers. A **variable** is defined as a placeholder, which can take on any of several values at a given time. A **constant**, on the other hand, is a symbol which takes on only one value at a given time. A **term** is a constant, a variable, or a combination of constants and variables. For example: 7.76, $3x$, xyz, $5z/x$, $(0.99)x^2$ are terms. If a term is a combination of constants and variables, the constant part of the term is referred to as the **coefficient** of the variable. If a variable is written without a coefficient, the coefficient is assumed to be 1.

EXAMPLE

$3x^2$	y^3
coefficient: 3	coefficient: 1
variable: x	variable: y

An **expression** is a collection of one or more terms. If the number of terms is greater than 1, the expression is said to be the sum of the terms.

EXAMPLE

$9,\ 9xy,\ 6x + x/3,\ 8yz - 2x$

An algebraic expression consisting of only one term is called a **monomial**, of two terms is called a **binomial**, of three terms is called a **trinomial**. In general, an algebraic expression consisting of two or more terms is called a **polynomial**.

PROBLEM

Verify that the algebraic expression below is true.

$$(50 - 45) \times \frac{3}{17} = x, \text{ if } x = -\frac{15}{17}$$

SOLUTION

The left side of the equation must be simplified.

Step 1 is to perform the operation inside the parentheses.

$50 - 45 = 5$

Step 2 is to rewrite the equation, substituting $-\frac{15}{17}$ for x.

$$5 \times \frac{3}{17} = -\frac{15}{17}$$

Step 3 is to perform the multiplication operation.

$$5 \times \frac{3}{17} = \frac{15}{17}$$

Step 4 is to rewrite the equation.

$$\frac{15}{17} = -\frac{15}{17}$$

Since $\frac{15}{17}$ does not equal $-\frac{15}{17}$, the equation is false.

PROBLEM

Verify that the algebraic expression below is true.

$$\frac{(3 \times 15)}{9} \geq x, \text{ if } x = 3$$

SOLUTION

The left side of the inequality must be simplified.

Step 1 is to perform the operation inside the parentheses.

$3 \times 15 = 45$

Step 2 is to rewrite the inequality, substituting 3 for x.

$$\frac{45}{9} \geq 3$$

Step 3 is to perform the division operation.

$$\frac{45}{9} = 5$$

Step 4 is to rewrite the inequality.

$5 \geq 3$.

Since $5 \geq 3$, the inequality is true.

1. OPERATIONS WITH POLYNOMIALS

A) **Addition of polynomials** is achieved by combining like terms, terms which differ only in their numerical coefficients. E.g.,

$$P(x) = (x^2 - 3x + 5) + (4x^2 + 6x - 3)$$

Note that the parentheses are used to distinguish the polynomials.

By using the commutative and associative laws, we can rewrite $P(x)$ as:

$$P(x) = (x^2 + 4x^2) + (6x - 3x) + (5 - 3)$$

Using the distributive law, $ab + ac = a(b + c)$, yields:

$$(1 + 4)x^2 + (6 - 3)x + (5 - 3) = 5x^2 + 3x + 2$$

B) **Subtraction of two polynomials** is achieved by first changing the sign of all terms in the expression which is being subtracted and then adding this result to the other expression. E.g.,

$(5x^2 + 4y^2 + 3z^2) - (4xy + 7y^2 - 3z^2 + 1)$

$= 5x^2 + 4y^2 + 3z^2 - 4xy - 7y^2 + 3z^2 - 1$

$= (5x^2) + (4y^2 - 7y^2) + (3z^2 + 3z^2) - 4xy - 1$

$= (5x^2) + (-3y^2) + (6z^2) - 4xy - 1$

C) **Multiplication of two or more polynomials** is achieved by using the laws of exponents, the rules of signs, and the commutative and associative laws of multiplication. Begin by multiplying the coefficients and then multiply the variables according to the laws of exponents. E.g.,

$(y^2)\,(5)\,(6y^2)\,(yz)\,(2z^2)$

$= (1)\,(5)\,(6)\,(1)\,(2)\,(y^2)\,(y^2)\,(yz)\,(z^2)$

$= 60[(y^2)\,(y^2)\,(y)]\,[(z)\,(z^2)]$

$= 60(y^5)\,(z^3)$

$= 60\,y^5z^3$

D) **Multiplication of a polynomial by a monomial** is achieved by multiplying each term of the polynomial by the monomial and combining the results. E.g.

$(4x^2 + 3y)\,(6xz^2)$

$= (4x^2)\,(6xz^2) + (3y)\,(6xz^2)$

$= 24x^3z^2 + 18xyz^2$

E) **Multiplication of a polynomial by a polynomial** is achieved by multiplying each of the terms of one polynomial by each of the terms of the other polynomial and combining the result. E.g.,

$(5y + z + 1)\,(y^2 + 2y)$

$[(5y)\,(y^2) + (5y)\,(2y)] + [(z)\,(y^2) + (z)\,(2y)] + [(1)\,(y^2) + (1)\,(2y)]$

$= (5y^3 + 10y^2) + (y^2z + 2yz) + (y^2 + 2y)$

$= (5y^3) + (10y^2 + y^2) + (y^2z) + (2yz) + (2y)$

$= 5y^3 + 11y^2 + y^2z + 2yz + 2y$

F) **Division of a monomial by a monomial** is achieved by first dividing the constant coefficients and the variable factors separately, and then multiplying these quotients. E.g.,

$6xyz^2 \div 2y^2z$

$= (6/2)\,(x/1)\,(y/y^2)\,(z^2/_z)$

$= 3xy^{-1}z$

$= 3xz/y$

G) **Division of a polynomial by a polynomial** is achieved by following the given procedure called Long Division.

Step 1: The terms of both the polynomials are arranged in order of ascending or descending powers of one variable.

Step 2: The first term of the dividend is divided by the first term of the divisor which gives the first term of the quotient.

Step 3: This first term of the quotient is multiplied by the entire divisor and the result is subtracted from the dividend.

Step 4: Using the remainder obtained from Step 3 as the new dividend, Steps 2 and 3 are repeated until the remainder is zero or the degree of the remainder is less than the degree of the divisor.

Step 5: The result is written as follows:

$$\frac{\text{dividend}}{\text{divisor}} = \text{quotient} + \frac{\text{remainder}}{\text{divisor}} \qquad \text{divisor} \neq 0$$

e.g. $(2x^2 + x + 6) \div (x + 1)$

$$\begin{array}{r} 2x - 1 \\ x + 1\overline{)2x^2 + x + 6} \\ \underline{-(2x^2 + 2x)} \\ -x + 6 \\ \underline{-(-x - 1)} \\ 7 \end{array}$$

The result is $(2x^2 + x + 6) \div (x + 1) = 2x - 1 + \frac{7}{x+1}$

DRILL: OPERATIONS WITH POLYNOMIALS

Addition

1. $9a^2b + 3c + 2a^2b + 5c =$

(A) $19a^2bc$ (B) $11a^2b + 8c$ (C) $11a^4b^2 + 8c^2$

(D) $19a^4b^2c^2$ (E) $12a^2b + 8c^2$

2. $14m^2n^3 + 6m^2n^3 + 3m^2n^3 =$

(A) $20m^2n^3$ (B) $23m^6n^9$ (C) $23m^2n^3$

(D) $32m^6n^9$ (E) $23m^8n^{27}$

3. $3x + 2y + 16x + 3z + 6y =$

(A) $19x + 8y$ (B) $19x + 11yz$ (C) $19x + 8y + 3z$

(D) $11xy + 19xz$ (E) $30xyz$

4. $(4d^2 + 7e^3 + 12f) + (3d^2 + 6e^3 + 2f) =$

(A) $23d^2e^3f$ (B) $33d^2e^2f$ (C) $33d^4e^6f^2$

(D) $7d^2 + 13e^3 + 14f$ (E) $23d^2 + 11e^3f$

5. $3ac^2 + 2b^2c + 7ac^2 + 2ac^2 + b^2c =$

(A) $12ac^2 + 3b^2c$ (B) $14ab^2c^2$ (C) $11ac^2 + 4ab^2c$

(D) $15ab^2c^2$ (E) $15a^2b^4c^4$

Subtraction

6. $14m^2n - 6m^2n =$

(A) $20m^2n$ (B) $8m^2n$ (C) $8m$ (D) 8 (E) $8m^4n^2$

7. $3x^3y^2 - 4xz - 6x^3y^2 =$

(A) $-7x^2y^2z$ (B) $3x^3y^2 - 10x^4y^2z$ (C) $-3x^3y^2 - 4xz$

(D) $-x^2y^2z - 6x^3y^2$ (E) $-7xyz$

8. $9g^2 + 6h - 2g^2 - 5h =$

(A) $15g^2h - 7g^2h$ (B) $7g^4h^2$ (C) $11g^2 + 7h$

(D) $11g^2 - 7h^2$ (E) $7g^2 + h$

9. $7b^3 - 4c^2 - 6b^3 + 3c^2 =$

(A) $b^3 - c^2$ (B) $-11b^2 - 3c^2$ (C) $13b^3 - c$

(D) $7b - c$ (E) 0

10. $11q^2r - 4q^2r - 8q^2r =$

(A) $22q^2r$ (B) q^2r (C) $-2\,q^2r$

(D) $-q^2r$ (E) $2\,q^2r$

Multiplication

11. $5p^2t \times 3p^2t =$

(A) $15p^2t$ (B) $15p^4t$ (C) $15p^4t^2$ (D) $8p^2t$ (E) $8p^4t^2$

12. $(2r + s)\,14r =$

(A) $28rs$ (B) $28r^2 + 14sr$ (C) $16r^2 + 14rs$

(D) $28r + 14sr$ (E) $17r^2s$

13. $(4m + p)(3m - 2p) =$

(A) $12m^2 + 5mp + 2p^2$ (B) $12m^2 - 2mp + 2p^2$ (C) $7m - p$

(D) $12m - 2p$ (E) $12m^2 - 5mp - 2p^2$

14. $(2a + b)(3a^2 + ab + b^2) =$

(A) $6a^3 + 5a^2b + 3ab^2 + b^3$ (B) $5a^3 + 3ab + b^3$

(C) $6a^3 + 2a^2b + 2ab^2$ (D) $3a^2 + 2a + ab + b + b^2$

(E) $6a^3 + 3a^2b + 5ab^2 + b^3$

15. $(6t^2 + 2t + 1)\,3t =$

(A) $9t^2 + 5t + 3$ (B) $18t^2 + 6t + 3$ (C) $9t^3 + 6t^2 + 3t$

(D) $18t^3 + 6t^2 + 3t$ (E) $12t^3 + 6t^2 + 3t$

Division

16. $(x^2 + x - 6) \div (x - 2) =$

(A) $x - 3$ (B) $x + 2$ (C) $x + 3$ (D) $x - 2$ (E) $2x + 2$

17. $24b^4c^3 \div 6b^2c =$

(A) $3b^2c^2$ (B) $4b^4c^3$ (C) $4b^3c^2$ (D) $4b^2c^2$ (E) $3b^4c^3$

18. $(3p^2 + pq - 2q^2) \div (p + q) =$

(A) $3p + 2q$ (B) $2q - 3p$ (C) $3p - q$

(D) $2q + 3p$ (E) $3p - 2q$

19. $(y^3 - 2y^2 - y + 2) \div (y - 2) =$

(A) $(y - 1)^2$ (B) $y^2 - 1$ (C) $(y + 2)(y - 1)$

(D) $(y + 1)^2$ (E) $(y + 1)(y - 2)$

20. $(m^2 + m - 14) \div (m + 4) =$

(A) $m - 2$ (B) $m - 3 + \frac{-2}{m+4}$ (C) $m - 3 + \frac{4}{m+4}$

(D) $m - 3$ (E) $m - 2 + \frac{-3}{m+4}$

2. SIMPLIFYING ALGEBRAIC EXPRESSIONS

To factor a polynomial completely is to find the prime factors of the polynomial with respect to a specified set of numbers.

The following concepts are important while factoring or simplifying expressions.

1. The factors of an algebraic expression consist of two or more algebraic expressions which when multiplied together produce the given algebraic expression.

2. A **prime factor** is a polynomial with no factors other than itself and 1. The **least common multiple (LCM)** for a set of numbers is the smallest quantity divisible by every number of the set. For algebraic expressions the least common numerical coefficients for each of the given expressions will be a factor.

3. The **greatest common factor (GCF)** for a set of numbers is the largest factor that is common to all members of the set. For algebraic expressions, the greatest common factor is the polynomial of highest degree and the largest numerical coefficient which is a factor of all the given expressions.

Some important formulae, useful for the factoring of polynomials, are listed below.

$$a(c + d) = ac + ad$$

$$(a + b)(a - b) = a^2 - b^2$$

$$(a + b)(a + b) = (a + b)^2 = a^2 + 2ab + b^2$$

$$(a - b)(a - b) = (a - b)^2 = a^2 - 2ab + b^2$$

$$(x + a)(x + b) = x^2 + (a + b)x + ab$$

$$(ax + b)(cx + d) = acx^2 + (ad + bc)x + bd$$

$$(a + b)(c + d) = ac + bc + ad + bd$$

$$(a + b)(a + b)(a + b) = (a + b)^3 = a^3 + 3a^2b + 3ab^2 + b^3$$

$$(a - b)(a - b)(a - b) = (a - b)^3 = a^3 - 3a^2b + 3ab^2 - b^3$$

$$(a - b)(a^2 + ab + b^2) = a^3 - b^3$$

$$(a + b)(a^2 - ab + b^2) = a^3 + b^3$$

$$(a + b + c)^2 = a^2 + b^2 + c^2 + 2ab + 2ac + 2bc$$

$$(a - b)(a^3 + a^2b + ab^2 + b^3) = a^4 - b^4$$

$$(a - b)(a^4 + a^3b + a^2b^2 + ab^3 + b^4) = a^5 - b^5$$

$$(a - b)(a^5 + a^4b + a^3b^2 + a^2b^3 + ab^4 + b^5) = a^6 - b^6$$

$$(a - b)(a^{n-1} + a^{n-2}b + a^{n-3}b^2 + \ldots + ab^{n-2} + b^{n-1}) = a^n - b^n$$

where n is any positive integer (1, 2, 3, 4, ...).

$$(a + b)(a^{n-1} - a^{n-2}b + a^{n-3}b^2 - \ldots - ab^{n-2} + b^{n-1}) = a^n + b^n$$

where n is any positive odd integer (1, 3, 5, 7, ...).

The procedure for factoring an algebraic expression completely is as follows:

Step 1: First find the greatest common factor if there is any. Then examine each factor remaining for greatest common factors.

Step 2: Continue factoring the factors obtained in Step 1 until all factors other than monomial factors are prime.

EXAMPLE

Factoring $4 - 16x^2$,

$$4 - 16x^2 = 4(1 - 4x^2) = 4(1 + 2x)(1 - 2x)$$

PROBLEM

Express each of the following as a single term.

(A) $3x^2 + 2x^2 - 4x^2$ (B) $5axy^2 - 7axy^2 - 3xy^2$

SOLUTION

(A) Factor x^2 in the expression.

$$3x^2 + 2x^2 - 4x^2 = (3 + 2 - 4)x^2 = 1x^2 = x^2.$$

(B) Factor xy^2 in the expression and then factor a.

$$\begin{aligned} 5axy^2 - 7axy^2 - 3xy^2 &= (5a - 7a - 3)xy^2 \\ &= [(5 - 7)a - 3]xy^2 \\ &= (-2a - 3)xy^2. \end{aligned}$$

PROBLEM

Simplify $\dfrac{\frac{1}{x-1} - \frac{1}{x-2}}{\frac{1}{x-2} - \frac{1}{x-3}}$.

SOLUTION

Simplify the expression in the numerator by using the addition rule:

$$\frac{a}{b} + \frac{c}{d} = \frac{ad + bc}{bd}$$

Notice bd is the Least Common Denominator, LCD. We obtain

$$\frac{(x-2)-(x-1)}{(x-1)(x-2)} = \frac{-1}{(x-1)(x-2)}$$

in the numerator.

Repeat this procedure for the expression in the denominator:

$$\frac{(x-3)-(x-2)}{(x-2)(x-3)} = \frac{-1}{(x-2)(x-3)}$$

We now have

$$\frac{\frac{-1}{(x-1)(x-2)}}{\frac{-1}{(x-2)(x-3)}},$$

which is simplified by inverting the fraction in the denominator and multiplying it by the numerator and cancelling like terms

$$\frac{-1}{(x-1)(x-2)} \times \frac{(x-2)(x-3)}{-1} = \frac{x-3}{x-1}.$$

COMMUTATIVE LAWS

The commutative laws refer to those situations in which the factors and terms of an expression are rearranged in a different order.

Addition

The algebraic form of the commutative law for addition is as follows:

$$a + b + c = a + c + b = c + b + a$$

Multiplication

The algebraic form of the commutative law for multiplication is as follows:

$$abc = acb = cba$$

In words, this law states that the product of two or more factors is the same regardless of the order in which the factors are arranged.

ASSOCIATIVE LAWS

The associative laws of addition and multiplication refer to the grouping (association) of terms and factors in a mathematical expression.

Addition

The algebraic form of the associative law for addition is as follows:

$$a + b + c = (a + b) + c = a + (b + c)$$

In words, this law states that the sum of three or more addends is the same regardless of the manner in which the addends are grouped.

Multiplication

The algebraic form of the associative law for multiplication is as follows:

$$a \times b \times c = (a \times b) \times c = a \times (b \times c)$$

In words, this law states that the product of three or more factors is the same regardless of the manner in which the factors are grouped.

DISTRIBUTIVE LAW

The distributive law refers to the distribution of factors among the terms of an additive expression. The algebraic form of this law is as follows:

$$a(b + c + d) = ab + ac + ad$$

In words, this law may be stated as follows: If the sum of two or more quantities is multiplied by a third quantity, the product is found by applying the multiplier to each of the original quantities separately and summing the resulting expressions.

PROBLEM

Which of the following statements illustrates the Commutative Property of Addition?

a) $12 + 24 = 12 + x$

c) $10x + 12y = 12y + 10x$

b) $69 + 3x = 3(x - 23)$

d) $2(2 + 3) = 10$

SOLUTION

The correct answer is statement c. Statement c correctly illustrates the Commutative Property of Addition. This property states that order is not relevant when performing addition. In statement c, $10x + 12y = 12y + 10x$, both sides of the equation will be equal, showing that order is not relevant in addition. Statements a, b, and d do not illustrate the Commutative Property of Addition.

PROBLEM

Using the Commutative Property of Addition, fill in the missing part of the equation below.

$78 + (95 - 2) = ?$

SOLUTION

The Commutative Property of Addition states that order is not relevant in addition. The format for this property is shown below.

$A + B = B + A$

Step 1 is to determine what terms represent A and B in the given equation.

A represents 78

B represents $(95 - 2)$

Step 2 is to rewrite the missing part of the given equation using the values for A and B.

$78 + (95 - 2) = (95 - 2) + 78$

The correct answer is $78 + (95 - 2) = (95 - 2) + 78$.

PROBLEM

Which of the following statements illustrates the Commutative Property of Multiplication?

a) $xy = x + y$

b) $2x(-45y) = -45y(2x)$

c) $35(2) = \frac{70}{1}$

d) $5^2 = 25$

SOLUTION

The correct answer is statement b. Statement b correctly illustrates the Commutative Property of Multiplication. This property states that order is not relevant when performing multiplication. In statement b, $2x(-45y) = -45y(2x)$, both sides of the equation will be equal, showing that order is not relevant in multiplication. Statements a, c, and d do not illustrate the Commutative Property of Multiplication.

PROBLEM

Using the Commutative Property of Multiplication, fill in the missing part of the equation below.

$$8\left(\frac{10}{12}\right) = ?$$

SOLUTION

The Commutative Property of Multiplication states that order is not relevant in multiplication. The format for this property is shown below.

$$A(B) = B(A)$$

Step 1 is to determine what terms represent A and B in the given equation.

A represents 8

B represents $\left(\frac{10}{12}\right)$

Step 2 is to rewrite the missing part of the given equation using the values for A and B.

$$8\left(\frac{10}{12}\right) = \left(\frac{10}{12}\right)8$$

The correct answer is $8\left(\frac{10}{12}\right) = \left(\frac{10}{12}\right)8$.

PROBLEM

Which of the following statements illustrates the Associative Property of Addition?

a) $(8x + 2) + 5y = 8x + (2 + 5y)$

b) $85 + 2 = 85y + 2y$

c) $\left(\frac{165}{5}\right) + 4x = \frac{(165 + 4x)}{5}$

d) $(5x - 2) + 20 = 20 + 5x - 2$

SOLUTION

The correct answer is statement a. Statement a correctly illustrates the Associative Property of Addition. This property states that grouping of terms is not relevant when performing addition. In statement a, $(8x + 2) + 5y = 8x + (2 + 5y)$, both sides of the equation will be equal, showing that grouping of terms is not relevant in addition. Statements b, c, and d do not illustrate the Associative Property of Addition.

PROBLEM

Using the Associative Property of Addition, fill in the missing part of the equation below.

$$(4x + 2y) + z = ?$$

SOLUTION

The Associative Property of Addition states that grouping is not relevant in addition. The format for this property is shown below. Notice that the order is the same, but the parentheses have changed position.

$$(A + B) + C = A + (B + C)$$

Step 1 is to determine what terms represent A, B, and C in the given equation.

A represents $4x$

B represents $2y$

C represents z

Step 2 is to rewrite the missing part of the given equation using the values for A, B, and C. Change the position of the parentheses but keep the order of the terms.

$$(4x + 2y) + z = 4x + (2y + z)$$

The correct answer is $(4x + 2y) + z = 4x + (2y + z)$.

PROBLEM

Which of the following statements illustrates the Associative Property of Multiplication?

a) $3(xy) = 3x + y$

b) $100(xy) = (100x)y$

c) $72y(3) - 25 = 8(9)y - 5(5)$

d) $10y = 5(2) + y$

SOLUTION

The correct answer is statement b. Statement b correctly illustrates the Associative Property of Multiplication. This property states that grouping of terms is not relevant when performing multiplication. In statement b, $100(xy) = (100x)y$, both sides of the equation will be equal, showing that grouping is not relevant in multiplication. Statements a, c, and d do not illustrate the Associative Property of Multiplication.

PROBLEM

Using the Associative Property of Multiplication, fill in the missing part of the equation below.

$8(x \times 3y) = ?$

SOLUTION

The Associative Property of Multiplication states that grouping is not relevant in multiplication. The format for this property is shown below. Notice that the order is the same, but the parentheses have changed position.

$A(BC) = AB(C)$

Step 1 is to determine what terms represent A, B, and C in the given equation.

A represents 8

B represents x

C represents $3y$

Step 2 is to rewrite the missing part of the given equation using the values for A, B, and C.

$8(x \times 3y) = (8x)3y$

The correct answer is $8(x \times 3y) = (8x)3y$.

PROBLEM

Which of the following statements illustrates the Distributive Property?

a) $10(x - y) = 10x - 10y$

b) $3x + 11y = 3(x + 11y)$

c) $\frac{5}{8x(8y)}(3) = \frac{5x}{y} + 3$

d) $x(y)(z) = x + y + z$

SOLUTION

The correct answer is statement a. Statement a correctly illustrates the Distributive Property. This property states that multiplying (or distributing) terms in parentheses will not change the solution. In statement a, $10(x - y) = 10x - 10y$, both sides of the equation will be equal, showing that multiplying terms will not change the solution. Statements b, c, and d do not illustrate the Distributive Property.

PROBLEM

Use the Distributive Property to solve the following problem.

$-(9y - 5)$

SOLUTION

The Distributive Property is shown in the format below.

$A(B + C) = AB + AC$

Step 1 is to determine what terms represent A, B, and C in the given equation.

A represents -1 (shown as "–")

B represents $9y$

C represents -5

Step 2 is to perform the multiplication operation.

$AB = -(9y) = -9y$

$AC = -(-5) = 5$

Step 3 is to rewrite the problem.

$-(9y - 5) = -9y + 5$

The answer is $-9y + 5$.

PROBLEM

Use the Distributive Property to solve the following problem.

$x(y + 2x)$

SOLUTION

The Distributive Property is shown in the format below.

$A(B + C) = AB + AC$

Step 1 is to determine what terms represent A, B, and C in the given equation.

A represents x

B represents y

C represents $2x$

Step 2 is to perform the multiplication operation.

$$AB = x(y) = xy$$

$$AC = x(2x) = 2x^2$$

Step 3 is to rewrite the problem.

$$x(y + 2x) = xy + 2x^2$$

The answer is $xy + 2x^2$.

PROBLEM

Simplify the following expression.

$$\frac{1}{2}x + \frac{3}{4}x + \frac{3}{2}x$$

SOLUTION

Step 1 is to group like terms. Since all the terms contain the variable x, all the terms in the expression are like terms. To make the problem easier, group the terms with the denominator "2" together.

$$\left(\frac{1}{2}x + \frac{3}{2}x\right) + \frac{3}{4}x$$

Step 2 is to perform the operation inside the parentheses.

$$\frac{1}{2}x + \frac{3}{2}x = \frac{4}{2}x = 2x$$

Step 3 is to rewrite the problem.

$$2x + \frac{3}{4}x$$

Step 4 is to perform the addition operation.

$$2x + \frac{3}{4}x = \frac{8}{4}x + \frac{3}{4}x = \frac{11}{4}x$$

The correct answer is $\frac{11}{4}x$.

PROBLEM

Simplify the following expression.

$$\frac{14y}{7} + 15 - 5y + 2$$

SOLUTION

Step 1 is to group like terms. Group the terms containing the variable y together. Group the constants together.

$$\left(\frac{14y}{7} - 5y\right) + (15 + 2)$$

Step 2 is to perform the operation inside the parentheses.

$$\left(\frac{14y}{7} - 5y\right) = 2y - 5y$$

$$2y - 5y = -3y$$

Step 3 is to rewrite the problem.

$$-3y + (15 + 2)$$

Step 4 is to perform the operation inside the parentheses.

$$15 + 2 = 17$$

Step 5 is to rewrite the problem.

$$-3y + 17$$

Since the remaining terms are not like terms, the problem cannot be simplified any further.

The correct answer is $-3y + 17$.

PROBLEM

Simplify the following expression.

$$3x^2 - 7 - 2x^2 + y$$

SOLUTION

Step 1 is to group like terms. Group the terms containing the variable x^2 together. Group the terms containing the variable x together. Group the terms containing the variable y together. Group the constants together.

$$(3x^2 - 2x^2) + y - 7$$

Step 2 is to perform the operation inside the parentheses.

$$3x^2 - 2x^2 = x^2$$

Step 3 is to rewrite the problem.

$$x^2 + y - 7$$

Since the remaining terms are not like terms, the problem cannot be simplified any further.

The correct answer is $x^2 + y - 7$.

PROBLEM

Simplify the following expression.

$$3(2x^2 + 6x - 1) + 2(5x + 1) + 4x^2$$

SOLUTION

Step 1 is to use the Distributive Property to rewrite the problem.

$$3(2x^2 + 6x - 1) = 3(2x^2) + 3(6x) - 3(1)$$

$$2(5x + 1) = 2(5x) + 2(1)$$

Step 2 is to perform the operations.

$$3(2x^2) + 3(6x) - 3(1) = 6x^2 + 18x - 3$$

$$2(5x) + 2(1) = 10x + 2$$

Step 3 is to rewrite the problem.

$$6x^2 + 18x - 3 + 10x + 2 + 4x^2$$

Step 4 is to group the like terms. Group the terms with the variable x^2 together. Group the terms with the variable x together. Group the constants together.

$$(6x^2 + 4x^2) + (18x + 10x) + (-3 + 2)$$

Step 5 is to perform the operation inside the parentheses.

$$6x^2 + 4x^2 = 10x^2$$

$$18x + 10x = 28x$$

$$-3 + 2 = -1$$

Step 6 is to rewrite the problem.

$$10x^2 + 28x - 1$$

Since the remaining terms are not like terms, the problem cannot be simplified any further.

The correct answer is $10x^2 + 28x - 5$.

PROBLEM

Factor the following expression.

$$x^2 - x^3 + x^4$$

SOLUTION

Step 1 is to find the greatest common factor (GCF) for x^2, x^3, and x^4. When looking for the GCF for the coefficients, the coefficient for each term is 1, so there is no need to divide.

Step 2 is to find the term with the lowest exponent of the same variable, x.

x^2 has the lowest exponent.

Step 3 is to divide each term by x^2.

$$\frac{x^2}{x^2} = 1$$

$$-\frac{x^3}{x^2} = -x$$

$$\frac{x^4}{x^2} = x^2$$

Step 4 is to rewrite the problem. Place the lowest exponent outside of the parentheses.

$x^2(x^2 - x + 1)$

The correct answer is $x^2(x^2 - x + 1)$.

PROBLEM

Factor the following expression.

$6x^3y + 8x^2y - 4xy$

SOLUTION

Step 1 is to find the greatest common factor (GCF) for $6x^3y$, $8x^2y$, and $4xy$. When looking for the GCF for a variable, only use the coefficient.

$6 = (1, 2, 3, 6)$

$8 = (1, 2, 4, 8)$

$4 = (1, 2, 4)$

The GCF is 2.

Step 2 is to divide each term by the GCF.

$$\frac{6x^3y}{2} = 3x^3y$$

$$\frac{8x^2y}{2} = 4x^2y$$

$$\frac{4xy}{2} = 2xy$$

Step 3 is to rewrite the problem. Place the GCF outside of the parentheses.

$2(3x^3y + 4x^2y - 2xy)$

Step 4 is to find the term with the lowest exponent of the same variable xy.

$2xy$ has the lowest exponent.

Step 5 is to divide each term by xy.

$$\frac{3x^3y}{xy} = 3x^2$$

$$\frac{4x^2y}{xy} = 4x$$

$$\frac{2xy}{xy} = 2$$

Step 6 is to rewrite the problem. Place the lowest exponent outside of the parentheses.

$$xy(2)(3x^2 + 4x - 2)$$

Step 7 is to multiply the GCF and the lowest exponent.

$$xy(2) = 2xy$$

The correct answer is $2xy(3x^2 + 4x - 2)$.

PROBLEM

Factor the following expression.

$$8x^3 - 2x^2 + 9x - 3y$$

SOLUTION

To solve this problem, the expression must be factored by grouping.

Step 1 is to group the first two terms and the last two terms together.

$$(8x^3 - 2x^2) + (9x - 3y)$$

Step 2 is to determine how each group will be factored.

Group 1 has a GCF of 2. Also, the variable x^2 can be factored since it appears in both terms.

Group 2 has a GCF of 3. Only the GCF can be factored.

Step 3 is to factor each group.

$$8x^3 - 2x^2 = 2x^2(4x - 1)$$

$$9x - 3y = 3(3x - y)$$

Step 4 is to rewrite the problem.

$$2x^2(4x - 1) + 3(3x - y)$$

The correct answer is $2x^2(4x - 1) + 3(3x - y)$.

PROBLEM

Factor the following trinomial.

$$2x^2 + 9x + 4$$

SOLUTION

Trinomials are in the format $ax^2 + bx + c$. Determine what terms represent a, b, and c in the given equation.

$a = 2$ $\quad b = 9$ $\quad c = 4$

Since the coefficient for x^2 does not equal 1, use the following steps.

Step 1 is to list pairs of factors for a and c.

$a = (2, 1)$

$c = (2, 2)\ (4, 1)$

Step 2 is to find factors of 4 that equal b when multiplied by factors of a and added. The factors can be made negative if necessary.

$4(2) + 1(1) = 9$

Step 3 is to put the factors in the following format. The numbers in parentheses become coefficients of x.

$(2x + 1)(x + 4)$

Step 4 is to verify the answer by using the First–Outer–Inner–Last (FOIL) method.

$$(2x + 1)(x + 4) = 2x(x) + 8(x) + x + 4$$
$$= 2x^2 + 9x + 4$$

The correct answer is $(2x + 1)(x + 4)$.

DRILL: SIMPLIFYING ALGEBRAIC EXPRESSIONS

1. $16b^2 - 25z^2 =$

(A) $(4b - 5z)^2$ (B) $(4b + 5z)^2$ (C) $(4b - 5z)\ (4b + 5z)$

(D) $(16b - 25z)^2$ (E) $(5z - 4b)\ (5z + 4b)$

2. $x^2 - 2x - 8 =$

(A) $(x - 4)^2$ (B) $(x - 6)\ (x - 2)$ (C) $(x + 4)\ (x - 2)$

(D) $(x - 4)\ (x + 2)$ (E) $(x - 4)\ (x - 2)$

3. $2c^2 + 5cd - 3d^2 =$

(A) $(c - 3d)\ (c + 2d)$ (B) $(2c - d)\ (c + 3d)$ (C) $(c - d)\ (2c + 3d)$

(D) $(2c + d)\ (c + 3d)$ (E) Not possible

4. $4t^3 - 20t =$

(A) $4t\ (t^2 - 5)$ (B) $4t^2(t - 20)$ (C) $4t(t + 4)\ (t - 5)$

(D) $2t(2t^2 - 10)$ (E) Not possible

5. $x^2 + xy - 2y^2 =$

(A) $(x - 2y)(x + y)$ (B) $(x - 2y)(x - y)$ (C) $(x + 2y)(x + y)$

(D) $(x + 2y)(x - y)$ (E) Not possible

6. $5b^2 + 17bd + 6d^2 =$

(A) $(5b + d)(b + 6d)$ (B) $(5b + 2d)(b + 3d)$ (C) $(5b - 2d)(b - 3d)$

(D) $(5b - 2d)(b + 3d)$ (E) Not possible

7. $x^2 + x + 1 =$

(A) $(x + 1)^2$ (B) $(x + 2)(x - 1)$ (C) $(x - 2)(x + 1)$

(D) $(x + 1)(x - 1)$ (E) Not possible

8. $3z^3 + 6z^2 =$

(A) $3(z^3 + 2z^2)$ (B) $3z^2(z + 2)$ (C) $3z(z^2 + 2z)$

(D) $z^2(3z + 6)$ (E) $3z^2(1 + 2z)$

9. $m^2p^2 + mpg - 6g^2 =$

(A) $(mp + 3g)(mp - 2g)$ (B) $mp(mp - 2q)(mp + 3q)$

(C) $mpq(1 - 6q)$ (D) $(mp + 2q)(mp + 3q)$

(E) Not possible

10. $2h^3 + 2h^2t - 4ht^2 =$

(A) $2(h^3 - t)(h + t)$ (B) $2h(h - 2t)^2$ (C) $4h(ht - t^2)$

(D) $2h(h + t) - 4ht^2$ (E) $2h(h + 2t)(h - t)$

3. EQUATIONS

An **equation** is defined as a statement that two separate expressions are equal.

A **solution** to the equation is a number that makes the equation true when it is substituted for the variable. For example, in the equation $3x = 18$, 6 is the solution since $3(6) = 18$. Depending on the equation, there can be more than one solution. Equations with the same solutions are said to be **equivalent equations**. An equation without a solution is said to have a solution set that is the **empty** or **null** set and is represented by ϕ.

Replacing an expression within an equation by an equivalent expression will result in a new equation with solutions equivalent to the original equation. Given the equation below

$$3x + y + x + 2y = 15$$

by combining like terms, we get,

$$3x + y + x + 2y = 4x + 3y$$

Since these two expressions are equivalent, we can substitute the simpler form into the equation to get

$$4x + 3y = 15$$

Performing the same operation to both sides of an equation by the same expression will result in a new equation that is equivalent to the original equation.

A) **Addition or subtraction**

$$y + 6 = 10$$

we can add (– 6) to both sides

$$y + 6 + (-6) = 10 + (-6)$$

to get $y + 0 = 10 - 6$; therefore $y = 4$.

B) **Multiplication or division**

$$3x = 6$$

$$3x/3 = 6/3$$

$$x = 2$$

$3x = 6$ is equivalent to $x = 2$.

C) **Raising to a power**

$$a = x^2y$$

$$a^2 = (x^2y)^2$$
$$a^2 = x^4y^2$$

This can be applied to negative and fractional powers as well. E.g.,

$$x^2 = 3y^4$$

If we raise both members to the – 2 power, we get

$$(x^2)^{-2} = (3y^4)^{-2}$$

$$\frac{1}{(x^2)^2} = \frac{1}{(3y^4)^2}$$
$$\frac{1}{x^4} = \frac{1}{9y^8}$$

If we raise both members to the $^1/_2$ power, which is the same as taking the square root, we get:

$$(x^2)^{1/2} = (3y^4)^{1/2}$$
$$x = \sqrt{3}y^2$$

D) The **reciprocal** of both members of an equation are equivalent to the original equation. Note: The reciprocal of zero is undefined.

$$\frac{2x+y}{z}=\frac{5}{2} \qquad \frac{z}{2x+y}=\frac{2}{5}$$

PROBLEM

Solve, justifying each step. $3x - 8 = 7x + 8$.

SOLUTION

	$3x - 8 = 7x + 8$
Adding 8 to both members,	$3x - 8 + 8 = 7x + 8 + 8$
Additive inverse property,	$3x + 0 = 7x + 16$
Additive identity property,	$3x = 7x + 16$
Adding $(-7x)$ to both members,	$3x - 7x = 7x + 16 - 7x$
Commuting,	$-4x = 7x - 7x + 16$
Additive inverse property,	$-4x = 0 + 16$
Additive identity property,	$-4x = 16$
Dividing both sides by -4,	$x = {}^{16}/_{-4}$
	$x = -4$

Check: Replacing x by -4 in the original equation:

$$3x - 8 = 7x + 8$$
$$3(-4) - 8 = 7(-4) + 8$$
$$-12 - 8 = -28 + 8$$
$$-20 = -20$$

THE EQUATION OF A LINE

The general equation for a line is $\boldsymbol{y = mx + b}$ where $\boldsymbol{m}$ is the slope and $\boldsymbol{b}$ is the y-intercept.

1. *Definition:* Slope measures the steepness of a line.
2. Slope $= \frac{rise}{run}$ = the amount of change y, for each one unit change in x.
3. The y-intercept is the point where the line crosses the y-axis.

EXAMPLE

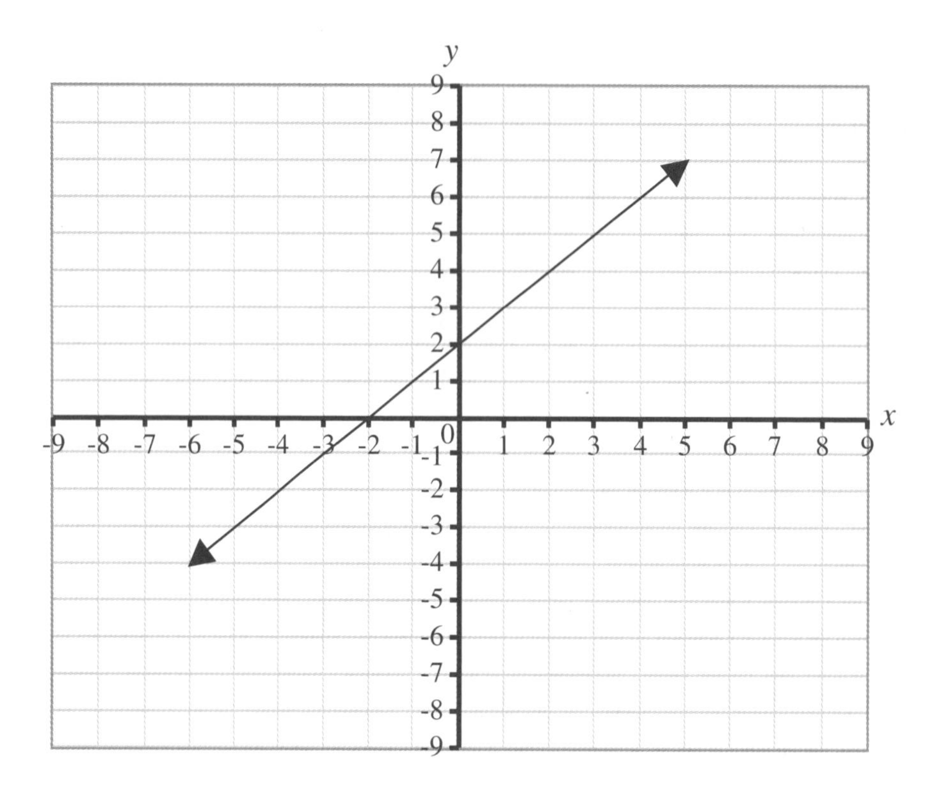

Determine the equation of the line shown above.

The general equation for a line is $y = mx + b$ where m is the slope and b is the y-intercept. The slope is equal to rise over run, so the slope of the line in the graph is equal to 1/1, or 1. The y-intercept is the point at which the line crosses the y-axis, so the y-intercept of the line is equal to 2. Therefore, the equation of the line is $y = x + 2$.

PROBLEM

The line graph shown demonstrates the relationship between the number of music CDs that someone buys and the total cost. How much would it cost to buy 4 music CDs? Also, how much does each CD cost?

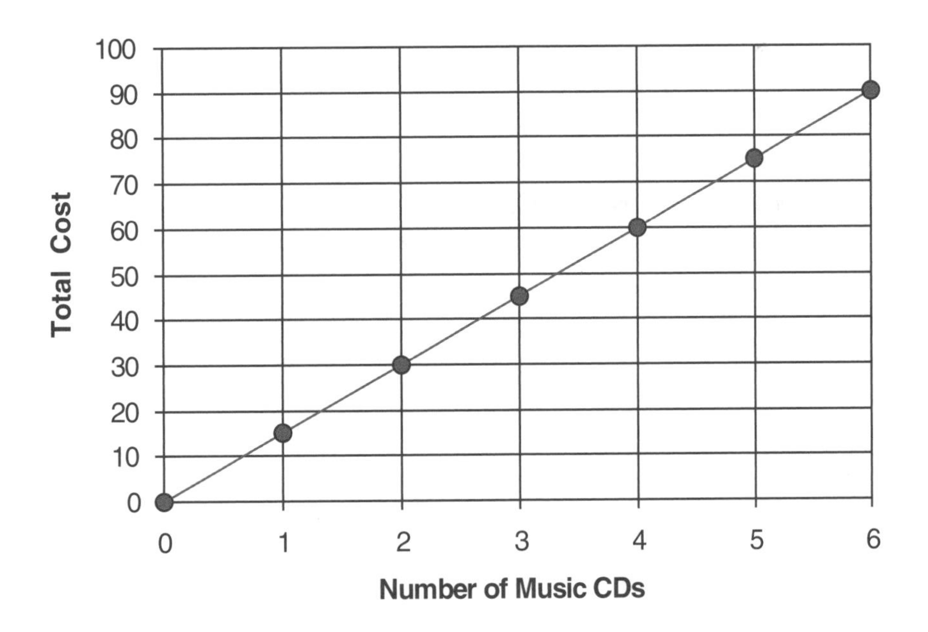

SOLUTION

Find 4 on the axis labeled "Number of Music CDs." Go straight up until you hit the line, then look left at the axis labeled "Total Cost." The cost for four CDs is \$60. The cost for one CD is actually the slope. By looking at the graph, we can see that for each one-unit change in number of CDs, the total cost goes up \$15.

The equation of this line is $y = 15x$, because the slope is 15 and the y-intercept is 0.

LINEAR EQUATIONS

A linear equation with one unknown is one that can be put into the form $ax + b = 0$, where a and b are constants, $a \neq 0$.

To solve a linear equation means to transform it in the form $x = {}^{-b}/_{a}$.

A) If the equation has unknowns on both sides of the equality, it is convenient to put similar terms on the same sides. E.g.,

$$4x + 3 = 2x + 9$$

$$4x + 3 - 2x = 2x + 9 - 2x$$

$$(4x - 2x) + 3 = (2x - 2x) + 9$$

$$2x + 3 = 0 + 9$$

$$\begin{aligned} 2x + 3 - 3 &= 0 + 9 - 3 \\ 2x &= 6 \\ {}^{2x}/_{2} &= {}^{6}/_{2} \\ x &= 3. \end{aligned}$$

B) If the equation appears in fractional form, it is necessary to transform it, using cross multiplication, and then repeating the same procedure as in A), we obtain:

$$\frac{3x+4}{3} \times \frac{7x+2}{5}$$

By using cross multiplication we would obtain:

$$3(7x + 2) = 5(3x + 4).$$

This is equivalent to:

$$21x + 6 = 15x + 20,$$

which can be solved as in A):

$$\begin{aligned} 21x + 6 &= 15x + 20 \\ 21x - 15x + 6 &= 15x - 15x + 20 \\ 6x + 6 - 6 &= 20 - 6 \\ 6x &= 14 \\ x &= {}^{14}/_{6} \\ x &= {}^{7}/_{3} \end{aligned}$$

C) If there are radicals in the equation, it is necessary to square both sides and then apply A)

$$\begin{aligned} \sqrt{3x+1} &= 5 \\ \left(\sqrt{3x+1}\right)^2 &= 5^2 \\ 3x + 1 &= 25 \\ 3x + 1 - 1 &= 25 - 1 \\ 3x &= 24 \\ x &= {}^{24}/_{3} \\ x &= 8 \end{aligned}$$

PROBLEM

Solve the equation $2(x + 3) = (3x + 5) - (x - 5)$.

SOLUTION

We transform the given equation to an equivalent equation where we can easily recognize the solution set.

$$2(x+3) = 3x+5-(x-5)$$

Distribute, $2x+6 = 3x+5-x+5$

Combine terms, $2x+6 = 2x+10$

Subtract $2x$ from both sides, $6 = 10$

Since $6 = 10$ is not a true statement, there is no real number which will make the original equation true. The equation is inconsistent and the solution set is ϕ, the empty set.

PROBLEM

Solve the equation $2(^2/_3\, y+5)+2(y+5)=130$.

SOLUTION

The procedure for solving this equation is as follows:

$^4/_3 y + 10 + 2y + 10 = 130,$		Distributive property
$^4/_3 y + 2y + 20 = 130,$		Combining like terms
$^4/_3 y + 2y = 110,$		Subtracting 20 from both sides
$^4/_3 y + {}^6/_3 y = 110,$		Converting $2y$ into a fraction with denominator 3
$^{10}/_3 y = 110,$		Combining like terms
$y = 110 \times {}^3/_{10} = 33,$		Dividing by $^{10}/_3$

Check: Replace y by 33 in the original equation,

$$2(^2/_3(33)+5)+2(33+5) = 130$$

$$2(22+5)+2(38) = 130$$

$$2(27)+76 = 130$$

$$54+76 = 130$$

$$130 = 130$$

Therefore the solution to the given equation is $y = 33$.

PROBLEM

Find the solution(s) to the following problem.

$$8z+24+(-7z)=0$$

SOLUTION

Step 1 is to simplify the left side of the equation.

$$8z + 24 + (-7z) = z + 24$$

Step 2 is to rewrite the equation.

$$z + 24 = 0$$

Step 3 is to subtract 24 from both sides of the equation to isolate z.

$$z + 24 - 24 = 0 - 24$$

Step 4 is to simplify the equation.

$$z = 0 - 24$$

Step 5 is to solve the right side of the equation.

$$0 - 24 = -24$$

Step 6 is to rewrite the equation.

$$z = -24$$

Step 7 is to verify the solution by substituting for z.

$$8(-24) + 24 + (-7[-24]) = 0$$

$$-192 + 24 + 168 = 0$$

The correct answer is $z = -24$.

PROBLEM

Find the solution(s) to the following problem.

$$x^4 + 1 = 82$$

SOLUTION

Step 1 is to subtract 1 from both sides of the equation to isolate x^4.

$$x^4 + 1 - 1 = 82 - 1$$

Step 2 is to simplify the equation.

$$x^4 = 82 - 1$$

Step 3 is to solve the right side of the equation.

$$82 - 1 = 81$$

Step 4 is to rewrite the equation.

$$x^4 = 81$$

Step 5 is to take the 4th root of both sides of the equation.

$$\sqrt[4]{x^4} = x \qquad \sqrt[4]{81} = 3 \text{ or } -3$$

Step 6 is to rewrite the equation.

$$x = 3 \text{ or } -3$$

Step 7 is to verify the solutions by substituting for x.

$(3)^4 + 1 = 82 \qquad (-3)^4 + 1 = 82$

The correct answer is $x = 3$ or $x = -3$.

PROBLEM

Find the solution(s) to the following problem.

$2y + x + 3 = 2y$

SOLUTION

Step 1 is to subtract 3 from both sides of the equation to isolate x.

$2y + x + 3 - 3 = 2y - 3$

Step 2 is to simplify the equation.

$2y + x = 2y - 3$

Step 3 is to subtract 2y from both sides of the equation to isolate x.

$2y + x - 2y = 2y - 3 - 2y$

Step 4 is to simplify both sides of the equation.

$x = -3$

Step 5 is to verify the solution by substituting for x.

$2y + (-3) + 3 = 2y$

$(-3) + 3 = 2y - 2y$

$(-3) + 3 = 0$

$0 = 0$

The correct answer is $x = -3$.

PROBLEM

Find the solution(s) to the following problem.

$x^2 + 7x + 6 = 0$

SOLUTION

Step 1 is to factor the equation.

$x^2 + 7x + 6 = (x + 6)(x + 1)$

Step 2 is to write each term separately.

$(x + 6) = 0$

$(x + 1) = 0$

Step 3 is to solve each term.

$x + 6 = 0 \qquad x + 1 = 0$

$x = -6 \qquad x = -1$

Step 4 is to verify each solution in the original equation.

$(-6)^2 + 7(-6) + 6 = 0 \qquad (-1)^2 + 7(-1) + 6 = 0$

$36 + (-42) + 6 = 0 \qquad 1 + (-7) + 6 = 0$

$0 = 0 \qquad 0 = 0$

The correct solutions are $x = -6$ and $x = -1$.

PROBLEM

Find the solution(s) to the following problem.

$x^2 - x - 2 = 0$

SOLUTION

Step 1 is to factor the equation.

$x^2 - x - 2 = (x - 2)(x + 1)$

Step 2 is to write each term separately.

$(x - 2) = 0$

$(x + 1) = 0$

Step 3 is to solve each term.

$x - 2 = 0 \qquad x + 1 = 0$

$x = 2 \qquad x = -1$

Step 4 is to verify each solution.

$(2)^2 - (2) - 2 = 0 \qquad (-1)^2 - (-1) - 2 = 0$

$4 - 2 - 2 = 0 \qquad 1 + 1 - 2 = 0$

$0 = 0 \qquad 0 = 0$

The correct solutions are $x = 2$ and $x = -1$.

DRILL: LINEAR EQUATIONS

Solve:

1. $4x - 2 = 10$

(A) -1 (B) 2 (C) 3 (D) 4 (E) 6

2. $7z + 1 - z = 2z - 7$

(A) -2 (B) 0 (C) 1 (D) 2 (E) 3

3. $^1/_3\, b + 3 = {}^1/_2\, b$

(A) 1/2 (B) 2 (C) 3 3/5 (D) 6 (E) 18

4. $0.4p + 1 = 0.7p - 2$

(A) 0.1 (B) 2 (C) 5 (D) 10 (E) 12

5. $4(3x + 2) - 11 = 3(3x - 2)$

(A) – 3 (B) – 1 (C) 2 (D) 3 (E) 7

4. TWO LINEAR EQUATIONS

Equations of the form $ax + by = c$, where a, b, c are constants and $a, b \neq 0$ are called **linear equations** with two unknown variables.

There are several ways to solve systems of linear equations in two variables:

Method 1: **Addition or subtraction** — if necessary, multiply the equations by numbers that will make the coefficients of one unknown in the resulting equations numerically equal. If the signs of equal coefficients are the same, subtract the equation, otherwise add.

The result is one equation with one unknown; we solve it and substitute the value into the other equations to find the unknown that we first eliminated.

Method 2: **Substitution** — find the value of one unknown in terms of the other, substitute this value in the other equation and solve.

Method 3: **Graph** — graph both equations. The point of intersection of the drawn lines is a simultaneous solution for the equations and its coordinates correspond to the answer that would be found analytically.

If the lines are parallel they have no simultaneous solution.

Dependent equations are equations that represent the same line, therefore every point on the line of a dependent equation represents a solution. Since there is an infinite number of points on a line there is an infinite number of simultaneous solutions, for example

$$\begin{cases} 2x + y = 8 \\ 4x + 2y = 16 \end{cases}$$

The equations above are dependent, they represent the same line, all points that satisfy either of the equations are solutions of the system.

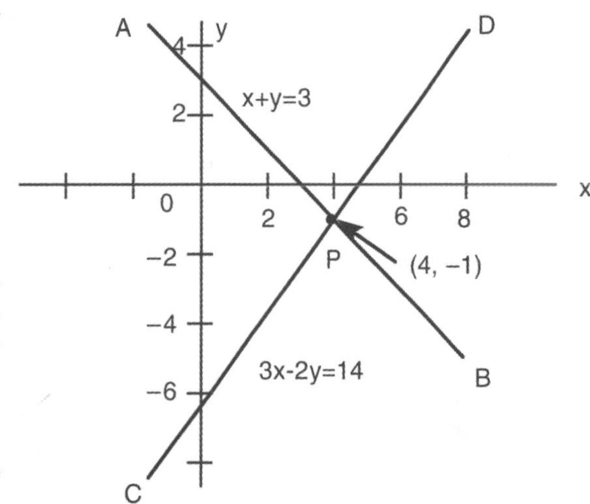

A system of linear equations is consistent if there

is only one solution for the system.

A system of linear equations is inconsistent if it does not have any solutions.

Example of a consistent system. Find the point of intersection of the graphs of the equations as shown in the previous figure.

$$x + y = 3,$$
$$3x - 2y = 14$$

To solve these linear equations, solve for y in terms of x. The equations will be in the form $y = mx + b$, where m is the slope and b is the intercept on the y-axis.

$x + y = 3$		
$y = 3 - x$	subtract x from both sides	
$3x - 2y = 14$	subtract $3x$ from both sides	
$-2y = 14 - 3x$	divide by -2.	
$y = -7 + {}^{3}/_{2}x$		

The graphs of the linear functions, $y = 3 - x$ and $y = -7 + {}^{3}/_{2}x$, can be determined by plotting only two points. For example, for $y = 3 - x$, let $x = 0$, then $y = 3$. Let $x = 1$, then $y = 2$. The two points on this first line are (0, 3) and (1, 2). For $y = -7 + {}^{3}/_{2}x$, let $x = 0$, then $y = -7$. Let $x = 1$, then $y = -5{}^{1}/_{2}$. The two points on this second line are $(0, -7)$ and $(1, -5{}^{1}/_{2})$.

To find the point of intersection P of

$$x + y = 3 \quad \text{and} \quad 3x - 2y = 14,$$

solve them algebraically. Multiply the first equation by 2. Add these two equations to eliminate the variable y.

$$\begin{array}{l} 2x + 2y = 6 \\ \underline{3x - 2y = 14} \\ 5x \qquad = 20 \end{array}$$

Solve for x to obtain $x = 4$. Substitute this into $y = 3 - x$ to get $y = 3 - 4 = -1$. P is $(4, -1)$. AB is the graph of the first equation, and CD is the graph of the second equation. The point of intersection P of the two graphs is the only point on both lines. The coordinates of P satisfy both equations and represent the desired solution of the problem. From the graph, P seems to be the point $(4, -1)$. These coordinates satisfy both equations, and hence are the exact coordinates of the point of intersection of the two lines.

To show that $(4, -1)$ satisfies both equations, substitute this point into both equations.

$$\begin{array}{rcl} x + y &=& 3 \\ 4 + (-1) &=& 3 \\ 4 - 1 &=& 3 \\ 3 &=& 3 \end{array} \qquad \begin{array}{rcl} 3x - 2y &=& 14 \\ 3(4) - 2(-1) &=& 14 \\ 12 + 2 &=& 14 \\ 14 &=& 14 \end{array}$$

Example of an inconsistent system. Solve the equations $2x + 3y = 6$ and $4x + 6y = 7$ simultaneously.

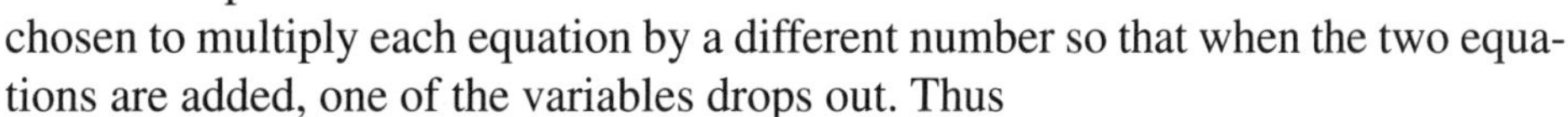

We have 2 equations in 2 unknowns,

$$2x + 3y = 6 \qquad (1)$$

and

$$4x + 6y = 7 \qquad (2)$$

There are several methods to solve this problem. We have chosen to multiply each equation by a different number so that when the two equations are added, one of the variables drops out. Thus

multiplying equation (1) by 2:	$4x + 6y = \ \ 12$	(3)
multiplying equation (2) by – 1:	$-4x - 6y = -7$	(4)
adding equations (3) and (4):	$0 = \ \ 5$	

We obtain a peculiar result!

Actually, what we have shown in this case is that if there were a simultaneous solution to the given equations, then 0 would equal 5. But the conclusion is impossible; therefore there can be no simultaneous solution to these two equations, hence no point satisfying both.

The straight lines which are the graphs of these equations must be parallel if they never intersect, but not identical, which can be seen from the graph of these equations (see the accompanying diagram).

Example of a dependent system. Solve the equations $2x + 3y = 6$ and $y = -(^{2x}/_{3}) + 2$ simultaneously.

We have 2 equations in 2 unknowns.

$$2x + 3y = 6 \qquad (1)$$

and

$$y = -(^{2x}/_{3}) + 2 \qquad (2)$$

There are several methods of solution for this problem. Since equation (2) already gives us an expression for y, we use the method of substitution. Substituting $-(^{2x}/_{3}) + 2$ for y in the first equation:

$$2x + 3(-^{2x}/_{3} + 2) = 6$$

Distributing,

$$2x - 2x + 6 = 6$$

$$6 = 6$$

Apparently we have gotten nowhere! The result $6 = 6$ is true, but indicates no solution. Actually, our work shows that no matter what real number x is, if y is determined by the second equation, then the first equation will always be satisfied.

The reason for this peculiarity may be seen if we take a closer look at the equation $y = -(^{2x}/_3) + 2$. It is equivalent to $3y = -2x + 6$, or $2x + 3y = 6$.

In other words, the two equations are equivalent. Any pair of values of x and y that satisfies one satisfies the other.

It is hardly necessary to verify that in this case the graphs of the given equations are identical lines, and that there are an infinite number of simultaneous solutions of these equations.

A system of three linear equations in three unknowns is solved by eliminating one unknown from any two of the three equations and solving them. After finding two unknowns substitute them in any of the equations to find the third unknown.

PROBLEM

Solve the system

$$2x + 3y - 4z = -8 \qquad (1)$$

$$x + y - 2z = -5 \qquad (2)$$

$$7x - 2y + 5z = 4 \qquad (3)$$

SOLUTION

We cannot eliminate any variable from two pairs of equations by a single multiplication. However, both x and z may be eliminated from equations (1) and (2) by multiplying equation 2 by -2. Then

$$2x + 3y - 4z = -8 \qquad (1)$$

$$-2x - 2y + 4z = 10 \qquad (4)$$

By addition, we have $y = 2$. Although we may now eliminate either x or z from another pair of equations, we can more conveniently substitute $y = 2$ in equations (2) and (3) to get two equations in two variables. Thus, making the substitution $y = 2$ in equations (2) and (3), we have

$$x - 2z = -7 \qquad (5)$$

$$7x + 5z = 8 \qquad (6)$$

Multiply (5) by 5 and multiply (6) by 2. Then add the two new equations. Then $x = -1$. Substitute x in either (5) or (6) to find z.

The solution of the system is $x = -1$, $y = 2$, and $z = 3$. Check by substitution.

A system of equations, as shown below, that has all constant terms $b_1, b_2, \ldots, b_n$ equal to zero is said to be a homogeneous system:

$$\begin{cases} a_{11}x_1 + a_{12}x_2 + \ldots + a_{1n}x_m = b_1 \\ a_{11}x_1 + a_{22}x_2 + \ldots + a_{2n}x_m = b_2 \\ \vdots \qquad \vdots \qquad \vdots \qquad \vdots \\ a_{n1}x_1 + a_{n2}x_2 + \ldots + a_{nm}x_m = b_n. \end{cases}$$

A homogeneous system always has at least one solution which is called the trivial solution that is $x_1 = 0, x_2 = 0, \ldots, x_m = 0$.

For any given homogeneous system of equations, in which the number of variables is greater than or equal to the number of equations, there are non-trivial solutions.

Two systems of linear equations are said to be equivalent if and only if they have the same solution set.

PROBLEM

Solve for x and y.

$$x + 2y = 8 \quad (1)$$

$$3x + 4y = 20 \quad (2)$$

SOLUTION

Solve equation (1) for x in terms of y:

$$x = 8 - 2y \quad (3)$$

Substitute $(8 - 2y)$ for x in (2):

$$3(8 - 2y) + 4y = 20 \quad (4)$$

Solve (4) for y as follows:

Distribute: $24 - 6y + 4y = 20$

Combine like terms and then subtract 24 from both sides:

$$24 - 2y = 20$$

$$24 - 24 - 2y = 20 - 24$$

$$-2y = -4$$

Divide both sides by -2:

$$y = 2$$

Substitute 2 for y in equation (1):

$$x + 2(2) = 8$$

$$x = 4$$

Thus, our solution is $x = 4, y = 2$.

Check: Substitute $x = 4, y = 2$ in equations (1) and (2):

$$4 + 2(2) = 8$$

$$8 = 8$$

$$3(4) + 4(2) = 20$$

$$20 = 20$$

PROBLEM

Solve algebraically:

$$4x + 2y = -1 \quad (1)$$

$$5x - 3y = 7 \quad (2)$$

SOLUTION

We arbitrarily choose to eliminate x first.

Multiply (1) by 5: $20x + 10y = -5$ (3)

Multiply (2) by 4: $20x - 12y = 28$ (4)

Subtract (3) – (4): $22y = -33$ (5)

Divide (5) by 22: $y = -\frac{33}{22} = -\frac{3}{2}$,

To find x, substitute $y = -\frac{3}{2}$ in either of the original equations. If we use Eq. (1), we obtain $4x + 2(-\frac{3}{2}) = -1$, $4x - 3 = -1$, $4x = 2$, $x = \frac{1}{2}$.

The solution $(\frac{1}{2}, -\frac{3}{2})$ should be checked in both equations of the given system.

Replacing $(\frac{1}{2}, -\frac{3}{2})$ in Eq. (1):

$$4x + 2y = -1$$

$$4(\tfrac{1}{2}) + 2(-\tfrac{3}{2}) = -1$$

$$\tfrac{4}{2} - 3 = -1$$

$$2 - 3 = -1$$

$$-1 = -1$$

Replacing $(\frac{1}{2}, -\frac{3}{2})$ in Eq. (2):

$$5x - 3y = 7$$

$$5(\tfrac{1}{2}) - 3(-\tfrac{3}{2}) = 7$$

$$\tfrac{5}{2} + \tfrac{9}{2} = 7$$

$$\tfrac{14}{2} = 7$$

$$7 = 7$$

(Instead of eliminating x from the two given equations, we could have eliminated y by multiplying Eq. (1) by 3, multiplying Eq. (2) by 2, and then adding the two derived equations.)

DRILL: TWO LINEAR EQUATIONS

DIRECTIONS: Find the solution set for each pair of equations.

1. $3x + 4y = -2$
   $x - 6y = -8$

(A) (2, – 1) (B) (1, – 2) (C) (– 2, – 1)
(D) (1, 2) (E) (– 2, 1)

2. $2x + y = -10$
$-2x - 4y = 4$

(A) (6, – 2) (B) (– 6, 2) (C) (– 2, 6)
(D) (2, 6) (E) (– 6, – 2)

3. $6x + 5y = -4$
$3x - 3y = 9$

(A) (1, – 2) (B) (1, 2) (C) (2, – 1)
(D) (– 2, 1) (E) (– 1, 2)

4. $4x + 3y = 9$
$2x - 2y = 8$

(A) (– 3, 1) (B) (1, – 3) (C) (3, 1)
(D) (3, – 1) (E) (– 1, 3)

5. $x + y = 7$
$x - y = -3$

(A) (5, 2) (B) (– 5, 2) (C) (2, 5)
(D) (– 2, 5) (E) (2, – 5)

6. $5x + 6y = 4$
$3x - 2y = 1$

(A) (3, 6) (B) (1/2, 1/4) (C) (– 3, 6)
(D) (2, 4) (E) (1/3, 3/2)

7. $x - 2y = 7$
$x + y = -2$

(A) (– 2, 7) (B) (3, – 1) (C) (– 7, 2)
(D) (1, – 3) (E) (1, – 2)

8. $4x + 3y = 3$
$-2x + 6y = 3$

(A) (1/2, 2/3) (B) (– 0.3, 0.6) (C) (2/3, – 1)
(D) (– 0.2, 0.5) (E) (0.3, 0.6)

9. $4x - 2y = -14$
$8x + y = 7$

(A) (0, 7) (B) (2, – 7) (C) (7, 0)

(D) (– 7, 2) (E) (0, 2)

10. $6x - 3y = 1$

$-9x + 5y = -1$

(A) (1, – 1) (B) (2/3, 1) (C) (1, 2/3)

(D) (– 1, 1) (E) (2/3, – 1)

5. QUADRATIC EQUATIONS

A second degree equation in x of the type $ax^2 + bx + c = 0$, where $a \neq 0$ and a, b and c are real numbers, is called a **quadratic equation.**

To solve a quadratic equation is to find values of x which satisfy $ax^2 + bx + c = 0$. These values of x are called **solutions**, or **roots**, of the equation.

A quadratic equation has a maximum of 2 roots. Methods of solving quadratic equations:

A) **Direct solution**: Given $x^2 - 9 = 0$.

We can solve directly by isolating the variable x:

$x^2 = 9$

$x = \pm 3.$

B) **Factoring**: Given a quadratic equation $ax^2 + bx + c = 0$ $(a, b, c \neq 0)$, to factor means to express it as the product $a(x - r_1)(x - r_2) = 0$, where r_1 and r_2 are the two roots.

Some helpful hints to remember are:

a) $r_1 + r_2 = -{}^b/_a$.

b) $r_1 r_2 = {}^c/_a$.

Given $x^2 - 5x + 4 = 0$.

Since $r_1 + r_2 = -{}^b/_a = -{}^{(-5)}/_1 = 5$, some possible solutions are (3, 2), (4, 1) and (5, 0). Also $r_1 r_2 = {}^c/_a = {}^4/_1 = 4$; this equation is satisfied only by the second pair, so $r_1 = 4$, $r_2 = 1$ and the factored form is $(x - 4)(x - 1) = 0$.

If the coefficient of x^2 is not 1, it is necessary to divide the equation by this coefficient and then factor.

Given $2x^2 - 12x + 16 = 0$

Dividing by 2, we obtain:

$x^2 - 6x + 8 = 0$

Since $r_1 + r_2 = -{}^b/_a = 6$, possible solutions are (6, 0), (5, 1), (4, 2), (3, 3). Also $r_1r_2 = 8$, so the only possible answer is (4, 2) and the expression $x^2 - 6x + 8 = 0$ can be factored as $(x - 4)(x - 2)$.

C) **Completing the Squares**:

If it is difficult to factor the quadratic equation using the previous method, we can complete the squares.

Given $x^2 - 12x + 8 = 0$.

We know that the two roots added up should be 12 because $r_1 + r_2 = -{}^b/_a = -{}^{(-12)}/_1 = 12$. Possible roots are (12, 0), (11, 1), (10, 2), (9, 3), (8,4), (7,5), (6, 6).

But none of these satisfy $r_1r_2 = 8$, so we cannot use (B).

To complete the square, it is necessary to isolate the constant term,

$$x^2 - 12x = -8.$$

Then take ${}^1/_2$ coefficient of x, square it and add to both sides.

$$x^2 - 12x + \left(\frac{-12}{2}\right)^2 = -8 + \left(\frac{-12}{2}\right)^2$$

$$x^2 - 12x + 36 = -8 + 36 = 28.$$

Now we can use the previous method to factor the left side: $r_1 + r_2 = 12$, $r_1r_2 = 36$ is satisfied by the pair (6, 6), so we have:

$$(x - 6)^2 = 28.$$

Now extract the root of both sides and solve for x.

$$(x - 6) = \pm\sqrt{28} = \pm 2\sqrt{7}$$

$$x = \pm 2\sqrt{7} + 6$$

So the roots are:

$$x = 2\sqrt{7} + 6, \quad x = -2\sqrt{7} + 6.$$

PROBLEM

Solve the equation $x^2 + 8x + 15 = 0$.

SOLUTION

Since $(x + a)(x + b) = x^2 + bx + ax + ab = x^2 + (a + b)x + ab$, we may factor the given equation, $0 = x^2 + 8x + 15$, replacing $a + b$ by 8 and ab by 15. Thus,

$$a + b = 8, \quad \text{and} \quad ab = 15.$$

We want the two numbers a and b whose sum is 8 and whose product is 15. We check all pairs of numbers whose product is 15:

(a) $1 \times 15 = 15$; thus $a = 1$, $b = 15$ and $ab = 15$.

$1 + 15 = 16$, therefore we reject these values because $a + b \neq 8$.

(b) $3 \times 5 = 15$, thus $a = 3$, $b = 5$, and $ab = 15$.

$3 + 5 = 8$. Therefore $a + b = 8$, and we accept these values.

Hence $x^2 + 8x + 15 = 0$ is equivalent to

$$0 = x^2 + (3 + 5)x + 3 \times 5 = (x + 3)(x + 5)$$

Hence, $x + 5 = 0$ or $x + 3 = 0$

since the product of these two numbers is zero, one of the numbers must be zero. Hence, $x = -5$, or $x = -3$, and the solution set is $x = \{-5, -3\}$.

The student should note that $x = -5$ or $x = -3$. We are certainly not making the statement, that $x = -5$, and $x = -3$. Also, the student should check that both these numbers do actually satisfy the given equations and hence are solutions.

Check: Replacing x by (-5) in the original equation:

$$\begin{aligned} x^2 + 8x + 15 &= 0 \\ (-5)^2 + 8(-5) + 15 &= 0 \\ 25 - 40 + 15 &= 0 \\ -15 + 15 &= 0 \\ 0 &= 0 \end{aligned}$$

Replacing x by (-3) in the original equation:

$$\begin{aligned} x^2 + 8x + 15 &= 0 \\ (-3)^2 + 8(-3) + 15 &= 0 \\ 9 - 24 + 15 &= 0 \\ -15 + 15 &= 0 \\ 0 &= 0. \end{aligned}$$

PROBLEM

Solve the following equations by factoring.

(a) $2x^2 + 3x = 0$

(b) $y^2 - 2y - 3 = y - 3$

(c) $z^2 - 2z - 3 = 0$

(d) $2m^2 - 11m - 6 = 0$

SOLUTION

(a) $2x^2 + 3x = 0$. Factoring out the common factor of x from the left side of the given equation,

$$x(2x + 3) = 0.$$

Whenever a product $ab = 0$, where a and b are any two numbers, either $a = 0$ or $b = 0$. Then, either

$$x = 0 \quad \text{or} \quad 2x + 3 = 0$$
$$2x = -3$$
$$x = {}^{-3}/_{2}$$

Hence, the solution set to the original equation $2x^2 + 3x = 0$ is: $\{^{-3}/_2, 0\}$.

(b) $y^2 - 2y - 3 = y - 3$. Subtract $(y - 3)$ from both sides of the given equation:

$$y^2 - 2y - 3 - (y - 3) = y - 3 - (y - 3)$$
$$y^2 - 2y - 3 - y + 3 = y - 3 - y + 3$$
$$y^2 - 2y - 3 - y + 3 = y - 3 - y + 3$$
$$y^2 - 3y = 0.$$

Factor out a common factor of y from the left side of this equation:

$$y(y - 3) = 0.$$

Thus, $y = 0$ or $y - 3 = 0$, $y = 3$.

Therefore, the solution set to the original equation $y^2 - 2y - 3 = y - 3$ is: $\{0, 3\}$.

(c) $z^2 - 2z - 3 = 0$. Factor the original equation into a product of two polynomials:

$$z^2 - 2z - 3 = (z - 3)(z + 1) = 0$$

Hence,

$$(z - 3)(z + 1) = 0; \text{ and } z - 3 = 0 \text{ or } z + 1 = 0$$
$$z = 3 \qquad z = -1$$

Therefore, the solution set to the original equation $z^2 - 2z - 3 = 0$ is: $\{-1, 3\}$.

(d) $2m^2 - 11m - 6 = 0$. Factor the original equation into a product of two polynomials:

$$2m^2 - 11m - 6 = (2m + 1)(m - 6) = 0$$

Thus,

$$2m + 1 = 0 \quad \text{or} \quad m - 6 = 0$$
$$2m = -1 \qquad m = 6$$
$$m = {}^{-1}/_{2}$$

Therefore, the solution set to the original equation $2m^2 - 11m - 6 = 0$ is $\{-^1/_2, 6\}$.

DRILL: QUADRATIC EQUATIONS

<u>DIRECTIONS</u>: Solve for all values of *x*.

1. $x^2 - 2x - 8 = 0$

(A) 4 and – 2 (B) 4 and 8 (C) 4

(D) – 2 and 8 (E) – 2

2. $x^2 + 2x - 3 = 0$

(A) – 3 and 2 (B) 2 and 1 (C) 3 and 1

(D) – 3 and 1 (E) – 3

3. $x^2 - 7x = -10$

(A) – 3 and 5 (B) 2 and 5 (C) 2

(D) – 2 and – 5 (E) 5

4. $x^2 - 8x + 16 = 0$

(A) 8 and 2 (B) 1 and 16 (C) 4

(D) – 2 and 4 (E) 4 and – 4

5. $3x^2 + 3x = 6$

(A) 3 and – 6 (B) 2 and 3 (C) – 3 and 2

(D) 1 and – 3 (E) 1 and – 2

6. $x^2 + 7x = 0$

(A) 7 (B) 0 and – 7 (C) – 7

(D) 0 and 7 (E) 0

7. $x^2 - 25 = 0$

(A) 5 (B) 5 and – 5 (C) 15 and 10

(D) – 5 and 10 (E) – 5

8. $2x^2 + 4x = 16$

(A) 2 and – 2 (B) 8 and – 2 (C) 4 and 8

(D) 2 and – 4 (E) 2 and 4

9. $6x^2 - x - 2 = 0$

(A) 2 and 3 (B) 1/2 and 1/3 (C) – 1/2 and 2/3

(D) 2/3 and 3 (E) 2 and – 1/3

10. $12x^2 + 5x = 3$

(A) 1/3 and – 1/4 (B) 4 and – 3 (C) 4 and 1/6

(D) 1/3 and – 4 (E) – 3/4 and 1/3

6. ABSOLUTE VALUE EQUATIONS

The absolute value of a, $| a |$, is defined as:

$$| a | = a \text{ when } a > 0, \; | a | = -a \text{ when } a < 0, \; | a | = 0 \text{ when } a = 0.$$

When the definition of absolute value is applied to an equation, the quantity within the absolute value symbol is considered to have two values. This value can be either positive or negative before the absolute value is taken. As a result, each absolute value equation actually contains two separate equations.

When evaluating equations containing absolute values, proceed as follows:

EXAMPLE

$| 5 - 3x | = 7$ is valid if either

$$\begin{aligned} 5 - 3x &= 7 & \text{or} \qquad 5 - 3x &= -7 \\ -3x &= 2 & -3x &= -12 \\ x &= -2/3 & x &= 4 \end{aligned}$$

The solution set is therefore $x = (-2/3, 4)$.

Remember, the absolute value of a number cannot be negative. So, for the equation $| 5x + 4 | = -3$, there would be no solution.

EXAMPLE

Solve for x in $| 2x - 6 | = | 4 - 5x |$

There are four possibilities here. $2x - 6$ and $4 - 5x$ can be either positive or negative. Therefore,

$$2x - 6 = 4 - 5x \qquad (1)$$

$$-(2x - 6) = 4 - 5x \qquad (2)$$

$$2x - 6 = -(4 - 5x) \qquad (3)$$

$$-(2x - 6) = -(4 - 5x) \qquad (4)$$

Equations (2) and (3) result in the same solution, as do equations (1) and (4). Therefore, it is necessary to solve only for equations (1) and (2). This gives:

$$\begin{aligned} 2x - 6 &= 4 - 5x & \text{or} \qquad -(2x - 6) &= 4 - 5x \\ 7x &= 10 & -2x + 6 &= 4 - 5x \\ & & 3x &= -2 \\ x &= 10/7 & x &= -2/3 \end{aligned}$$

The solution set is $(10/7, -2/3)$.

DRILL: ABSOLUTE VALUE EQUATIONS

1. $| 4x - 2 | = 6$

(A) -2 and -1 (B) -1 and 2 (C) 2

(D) $1/2$ and -2 (E) No solution

2. $|3 - 1/2y| = -7$

(A) -8 and 20 (B) 8 and -20 (C) 2 and -5

(D) 4 and -2 (E) No solution

3. $2|x + 7| = 12$

(A) -13 and -1 (B) -6 and 6 (C) -1 and 13

(D) 6 and -13 (E) No solution

4. $|5x| - 7 = 3$

(A) 2 and 4 (B) 4/5 and 3 (C) -2 and 2

(D) 2 (E) No solution

5. $\left|\frac{3}{4}m\right| = 9$

(A) 24 and -16 (B) 4/27 and $-4/3$ (C) 4/3 and 12

(D) -12 and 12 (E) No solution

7. INEQUALITIES

An inequality is a statement where the value of one quantity or expression is greater than (>), less than (<), greater than or equal to (≥), less than or equal to (≤), or not equal to (≠) that of another.

EXAMPLE

$5 > 4.$

The expression above means that the value of 5 is greater than the value of 4.

A **conditional inequality** is an inequality whose validity depends on the values of the variables in the sentence. That is, certain values of the variables will make the sentence true, and others will make it false. $3 - y > 3 + y$ is a conditional inequality for the set of real numbers, since it is true for any replacement less than zero and false for all others.

$x + 5 > x + 2$ is an **absolute inequality** for the set of real numbers, meaning that for any real value x, the expression on the left is greater than the expression on the right.

$5y < 2y + y$ is inconsistent for the set of non-negative real numbers. For any y greater than 0 the sentence is always false. A sentence is inconsistent if it is always false when its variables assume allowable values.

The solution of a given inequality in one variable x consists of all values of x for which the inequality is true.

The graph of an inequality in one variable is represented by either a ray or a line segment on the real number line.

The endpoint is not a solution if the variable is strictly less than or greater than a particular value.

EXAMPLE

$x > 2$

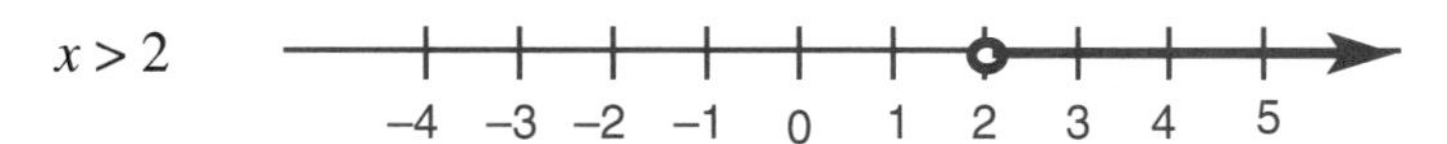

2 is not a solution and should be represented as shown.

The endpoint is a solution if the variable is either (1) less than or equal to or (2) greater than or equal to, a particular value.

EXAMPLE

$5 > x \geq 2$

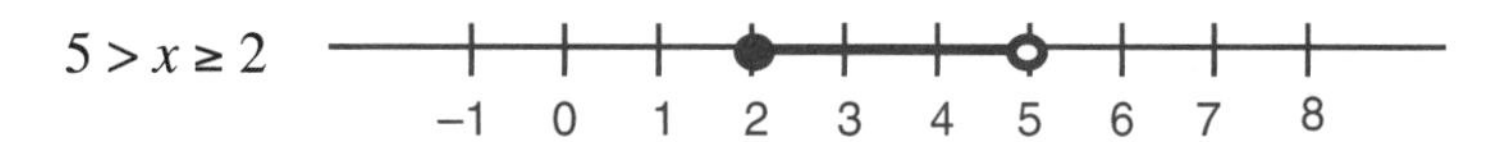

In this case 2 is the solution and should be represented as shown.

PROPERTIES OF INEQUALITIES

If x and y are real numbers then one and only one of the following statements is true.

$x > y, x = y,$ or $x < y$.

This is the **order property of real numbers**.

If a, b and c are real numbers:

A) If $a < b$ and $b < c$ then $a < c$.

B) If $a > b$ and $b > c$ then $a > c$.

This is the **transitive property of inequalities**.

If a, b and c are real numbers and $a > b$ then $a + c > b + c$ and $a - c > b - c$. This is the **addition property of inequality**.

Two inequalities are said to have the same **sense** if their signs of inequality point in the same direction.

The sense of an inequality remains the same if both sides are multiplied or divided by the same positive real number.

EXAMPLE

$4 > 3$

If we multiply both sides by 5 we will obtain:

$4 \times 5 > 3 \times 5$

$20 > 15$

The sense of the inequality does not change.

The sense of an inequality becomes opposite if each side is multiplied or divided by the same negative real number.

EXAMPLE

$4 > 3$

If we multiply both sides by – 5 we would obtain:

$4 \times -5 < 3 \times -5$

$-20 < -15$

The sense of the inequality becomes opposite.

If $a > b$ and a, b and n are positive real numbers, then:

$a^n > b^n$ and $a^{-n} < b^{-n}$

If $x > y$ and $q > p$ then $x + q > y + p$.

If $x > y > 0$ and $q > p > 0$ then $xq > yp$.

Inequalities that have the same solution set are called **equivalent inequalities**.

PROBLEM

Solve the inequality $2x + 5 > 9$.

SOLUTION

$2x + 5 + (-5) > 9 + (-5)$.	Adding – 5 to both sides.
$2x + 0 > 9 + (-5)$	Additive inverse property
$2x > 9 + (-5)$	Additive identity property
$2x > 4$	Combining terms
$^1/_2(2x) > {^1/_2} \times 4$	Multiplying both sides by $^1/_2$.
$x > 2$	

The solution set is

$X = \{x \mid 2x + 5 > 9\}$

$= \{x \mid x > 2\}$

(that is all x, such that x is greater than 2).

PROBLEM

Solve the inequality $4x + 3 < 6x + 8$.

SOLUTION

In order to solve the inequality $4x + 3 < 6x + 8$, we must find all values of x which make it true. Thus, we wish to obtain x alone on one side of the inequality.

Add – 3 to both sides:

$$\begin{array}{r} 4x + 3 < 6x + 8 \\ -3 \qquad -3 \\ \hline 4x < 6x + 5 \end{array}$$

Add – 6x to both sides:

$$\begin{array}{rr} 4x < & 6x + 5 \\ -6x & -6x \\ \hline -2x < & 5 \end{array}$$

In order to obtain x alone we must divide both sides by (– 2). Recall that dividing an inequality by a negative number reverses the inequality sign, hence

$$\frac{-2x}{-2} > \frac{5}{-2}$$

Cancelling $^{-2}/_{-2}$, we obtain $x > -\,^{5}/_{2}$.

Thus, our solution is $\{x : x > -\,^{5}/_{2}\}$ (the set of all x such that x is greater than $-^{5}/_{2}$).

PROBLEM

Find the solution to the following inequality.

$x + 35 > 100$

SOLUTION

Step 1 is to isolate the variable by subtracting 35 from both sides of the inequality.

$x + 35 - 35 > 100 - 35$

Step 2 is to simplify the left side of the inequality.

$x > 100 - 35$

Step 3 is to simplify the right side of the inequality.

$100 - 35 = 65$

Step 4 is to rewrite the inequality.

$x > 65$

Step 5 is to verify the solution. To do this, choose any number that is greater than 65 and substitute that for the variable x.

$70 + 35 > 100$

$105 > 100$

The solution to the inequality is $x > 65$.

PROBLEM

Find the solution to the following inequality.

$$\frac{1}{2} + z < 4$$

SOLUTION

Step 1 is to isolate the variable by subtracting $\frac{1}{2}$ from both sides of the inequality.

$$\frac{1}{2} + z - \frac{1}{2} < 4 - \frac{1}{2}$$

Step 2 is to simplify the left side of the inequality.

$$z < 4 - \frac{1}{2}$$

Step 3 is to simplify the right side of the inequality.

$$4 - \frac{1}{2} = \frac{7}{2}$$

Step 4 is to rewrite the inequality.

$$z < \frac{7}{2}$$

Step 5 is to verify the solution. To do this, choose any number that is less than $\frac{7}{2}$ and substitute that for the variable z.

$$\frac{1}{2} + \frac{2}{2} < 4$$

$$\frac{3}{2} < 4$$

The solution to the inequality is $z < \frac{7}{2}$.

PROBLEM

Find the solution to the following inequality.

$$y - 45 \geq 10$$

SOLUTION

Step 1 is to isolate the variable by adding 45 to both sides of the inequality.

$$y - 45 + 45 \geq 10 + 45$$

Step 2 is to simplify the left side of the inequality.

$$y \geq 10 + 45$$

Step 3 is to simplify the right side of the inequality.

$$10 + 45 = 55$$

Step 4 is to rewrite the inequality.

$$y \geq 55$$

Step 5 is to verify the solution. To do this, choose any number that is greater than or equal to 55 and substitute that for the variable y.

$$56 - 45 \geq 10$$

$$11 \geq 10$$

The solution to the inequality is $y \geq 55$.

PROBLEM

Find the solution to the following inequality.

$$5x + 7x + 3 \leq 25$$

SOLUTION

Step 1 is to isolate the variable by subtracting 3 from both sides of the inequality.

$$5x + 7x + 3 - 3 \leq 25 - 3$$

Step 2 is to simplify the left side of the inequality.

$$5x + 7x = 12x$$

Step 3 is to rewrite the inequality.

$$12x \leq 25 - 3$$

Step 4 is to simplify the right side of the inequality.

$$25 - 3 = 22$$

Step 5 is to rewrite the inequality.

$$12x \leq 22$$

Step 6 is to divide both sides of the inequality by 12.

$$\frac{12x}{12} = x$$

$$\frac{22}{12} = \frac{11}{6}$$

Step 7 is to rewrite the inequality.

$$x \leq \frac{11}{6}$$

Step 8 is to verify the solution. To do this, choose any number that is less than or equal to $\frac{11}{6}$ and substitute that for the variable x.

$$5(1) + 7(1) + 3 \leq 25$$

$$5 + 7 + 3 \leq 25$$

$15 \leq 25$

The solution to the inequality is $x \leq \frac{11}{6}$.

PROBLEM

Find the solution to the following inequality.

$-88 - 2y \geq -y$

SOLUTION

Step 1 is to isolate the variable $2y$ by adding 88 to both sides of the inequality.

$-88 - 2y + 88 \geq -y + 88$

Step 2 is to simplify the left side of the inequality.

$-2y \geq -y + 88$

Step 3 is to add y to both sides of the inequality to obtain the variable y.

$-2y + y \geq -y + 88 + y$

Step 4 is to simplify the left side of the inequality.

$-2y + y = -y$

Step 5 is to rewrite the inequality.

$-y \geq -y + 88 + y$

Step 6 is to simplify the right side of the inequality.

$-y + 88 + y = 88$

Step 7 is to rewrite the inequality.

$-y \geq 88$

Step 8 is to multiply both sides of the inequality by –1 to make the variable positive.

$-y(-1) \geq 88(-1)$

Step 9 is to perform the multiplication operation.

$y \geq -88$

In Step 10, since we multiplied by a negative number, the $\geq$ sign must be reversed.

$y \leq -88$

Step 11 is to verify the solution. To do this, choose any number that is less than or equal to –88 and substitute that for the variable y.

$-88 - 2(-88) \leq -(-88)$

$-88 - (-176) \leq 88$

$$-88 + 176 \le 88$$

$$88 \le 88$$

The solution to the inequality is $y \le -88$.

PROBLEM

Find the solution to the following problem.

$$\frac{y}{10} < 7$$

SOLUTION

Step 1 is to isolate the variable. To do this, multiply both sides of the inequality by 10.

$$\frac{y}{10}(10) < 7(10)$$

Step 2 is to simplify the left side of the inequality.

$$\frac{y}{10}(10) = y$$

Step 3 is to simplify the right side of the inequality.

$$7(10) = 70$$

Step 4 is to rewrite the inequality.

$$y < 70$$

Step 5 is to verify the solution. To do this, choose any number that is less than 70 and substitute that for the variable y.

$$\frac{1}{10}(50) < 7$$

$$5 < 7$$

The correct answer to the inequality is $y < 70$.

DRILL: INEQUALITIES

DIRECTIONS: Find the solution set for each inequality.

1. $3m + 2 < 7$

(A) $m \ge {}^5/_3$ (B) $m \le 2$ (C) $m < 2$

(D) $m > 2$ (E) $m < {}^5/_3$

2. ${}^1/_2\,x - 3 \le 1$

(A) $-4 \le x \le 8$ (B) $x \ge -8$ (C) $x \le 8$

(D) $2 \le x \le 8$ (E) $x \ge 8$

3. $-3p + 1 \geq 16$

(A) $p \geq -5$ (B) $p \geq \frac{-17}{3}$ (C) $p \leq \frac{-17}{3}$

(D) $p \leq -5$ (E) $p \geq 5$

4. $-6 < {}^2/_3\, r + 6 \leq 2$

(A) $-6 < r \leq -3$ (B) $-18 < r \leq -6$ (C) $r \geq -6$

(D) $-2 < r \leq {}^{-4}/_3$ (E) $r \leq -6$

5. $0 < 2 - y < 6$

(A) $-4 < y < 2$ (B) $-4 < y < 0$ (C) $-4 < y < -2$

(D) $-2 < y < 4$ (E) $0 < y < 4$

8. RATIOS AND PROPORTIONS

The ratio of two numbers x and y written $x : y$ is the fraction x / y where $y \neq 0$. A ratio compares x to y by dividing one by the other. Therefore, in order to compare ratios, simply compare the fractions.

A proportion is an equality of two ratios. The laws of proportion are listed below:

If $a/b = c/d$, then

(A) $ad = bc$

(B) $b/a = d/c$

(C) $a/c = b/d$

(D) $(a + b)/b = (c + d)/d$

(E) $(a - b)/b = (c - d)/d$

Given a proportion $a : b = c : d$, then a and d are called extremes, b and c are called the means, and d is called the fourth proportion to a, b, and c.

PROBLEM

Solve the proportion $\frac{x+1}{4} = \frac{15}{12}$.

SOLUTION

Cross multiply to determine x; that is, multiply the numerator of the first fraction by the denominator of the second, and equate this to the product of the numerator of the second and the denominator of the first.

$$(x + 1)\, 12 = 4 \times 15$$

$$12x + 12 = 60$$
$$x = 4.$$

PROBLEM

Find the ratios of $x : y : z$ from the equations

$$7x = 4y + 8z, \quad 3z = 12x + 11y.$$

SOLUTION

By transposition we have

$$7x - 4y - 8z = 0$$
$$12x + 11y - 3z = 0.$$

To obtain the ratio of $x : y$ we convert the given system into an equation in terms of just x and y. z may be eliminated as follows: Multiply each term of the first equation by 3, and each term of the second equation by 8, and then subtract the second equation from the first. We thus obtain:

$$\begin{aligned} 21x - 12y - 24z &= 0 \\ -(96x + 88y - 24z &= 0) \\ \hline -75x - 100y &= 0 \end{aligned}$$

Dividing each term of the last equation by 25 we obtain

$$-3x - 4y = 0$$

or,
$$-3x = 4y.$$

Dividing both sides of this equation by 4, and by -3, we have the proportion:

$$\frac{x}{4} = \frac{y}{-3}$$

We are now interested in obtaining the ratio of $y : z$. To do this, we convert the given system of equations into an equation in terms of just y and z, eliminating x as follows: Multiply each term of the first equation by 12, and each term of the second equation by 7, and then subtract the second equation from the first. We thus obtain:

$$\begin{aligned} 84x - 48y - 96z &= 0 \\ -(84x + 77y - 21z &= 0) \\ -125y - 75z &= 0. \end{aligned}$$

Dividing each term of the last equation by 25 we obtain

$$-5y - 3z = 0$$

or,
$$-3z = 5y.$$

Dividing both sides of this equation by 5, and by -3, we have the proportion:

$$\frac{z}{5} = \frac{y}{-3}.$$

From this result and our previous result we obtain:

$$\frac{x}{4} = \frac{y}{-3} = \frac{z}{5}$$

as the desired ratios.

DRILL: RATIOS AND PROPORTIONS

1. Solve for n : $\frac{4}{n} = \frac{8}{5}$

(A) 10 (B) 8 (C) 6 (D) 2.5 (E) 2

2. Solve for n: $\frac{2}{3} = \frac{n}{72}$

(A) 12 (B) 48 (C) 64 (D) 56 (E) 24

3. Solve for n: $n : 12 = 3 : 4$.

(A) 8 (B) 1 (C) 9 (D) 4 (E) 10

4. Four out of every five students at West High take a mathematics course. If the enrollment at West is 785, how many students take mathematics?

(A) 628 (B) 157 (C) 705 (D) 655 (E) 247

5. At a factory, three out of every 1,000 parts produced are defective. In a day, the factory can produce 25,000 parts. How many of these parts would be defective?

(A) 7 (B) 75 (C) 750 (D) 7,500 (E) 75,000

6. A summer league softball team won 28 out of the 32 games they played. What is the ratio of games won to games played?

(A) 4 : 5 (B) 3 : 4 (C) 7 : 8 (D) 2 : 3 (E) 1 : 8

7. A class of 24 students contains 16 males. What is the ratio of females to males?

(A) 1 : 2 (B) 2 : 1 (C) 2 : 3 (D) 3 : 1 (E) 3 : 2

8. A family has a monthly income of $1,250, but they spend $450 a month on rent. What is the ratio of the amount of income to the amount paid for rent?

(A) 16 : 25 (B) 25 : 9 (C) 25 : 16 (D) 9 : 25 (E) 36 : 100

9. A student attends classes 7.5 hours a day and works a part-time job for 3.5 hours a day. She knows she must get 7 hours of sleep a night. Write the ratio of the number of free hours in this student's day to the total number of hours in a day.

(A) 1 : 3 (B) 4 : 3 (C) 8 : 24 (D) 1 : 4 (E) 5 : 12

10. In a survey by mail, 30 out of 750 questionnaires were returned. Write the ratio of questionnaires returned to questionnaires mailed (write in simplest form).

(A) 30 : 750 (B) 24 : 25 (C) 3 : 75 (D) 1 : 4 (E) 1 : 25

9. PERMUTATION, COMBINATION, AND PROBABILITY

Permutation refers to the method of finding the number of ways things can be arranged.

PROBLEM

Evaluate each of the following symbols:

(a) $5!$ (b) $\frac{7!}{4!}$ (c) $P(6,2)$ (d) $P(9,2)$

SOLUTION

(a) Recalling $n! = n \times (n-1) \times (n-2) \times (n-3) \ldots 1$,

$$5! = 5 \times 4 \times 3 \times 2 \times 1 = 120$$

(b) Recalling $n! = n \times (n-1)! = n \times (n-1) \times (n-2)! = \ldots$

$$\frac{7!}{4!} = \frac{7 \times 6 \times 5 \times \cancel{4}!}{\cancel{4}!} = 210$$

(c) Recalling $P(n,r) = \frac{n!}{(n-r)!}$

$$P(6,2) = \frac{6!}{(6-2)!} = \frac{6!}{4!} = \frac{6 \times 5 \times \cancel{4}!}{\cancel{4}!} = 30$$

(d) Similarly $P(9,2) = \frac{9!}{(9-2)!} = \frac{9!}{7!} = \frac{9 \times 8 \times \cancel{7}!}{\cancel{7}!} = 72$

PROBLEM

Calculate the number of permutations of the letters a, b, c, d taken two at a time.

SOLUTION

The first of the two letters may be taken in 4 ways (a, b, c, or d). The second letter may therefore be selected from the remaining three letters in 3 ways. By the fundamental principle, the total number of ways of selecting two letters is equal to the product of the number of ways of selecting each letter. Hence

$$P(4,2) = 4 \times 3 = 12$$

The list of these permutations is:

ab	ba	ca	da
ac	bc	cb	db
ad	bd	cd	dc

PROBLEM

In how many ways may 3 books be placed next to each other on a shelf?

SOLUTION

We construct a pattern of 3 boxes to represent the places where the 3 books are to be placed next to each other on the shelf:

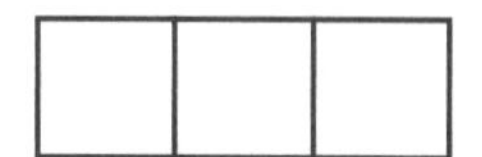

Since there are 3 books, the first place may be filled in 3 ways. There are then 2 books left, so that the second place may be filled in 2 ways. There is only 1 book left to fill the last place. Hence our boxes take the following form:

3	2	1

The Fundamental Principle of Counting states that if one thing can be done in a different ways and, when it is done in any one of these ways, a second thing can be done in b different ways, and a third thing can be done in c ways, ... then all the things in succession can be done in $a \times b \times c$ ... different ways. Thus the books can be arranged in $3 \times 2 \times 1 = 6$ ways.

This can also be seen using the following approach. Since the arrangement of books on the shelf is important, this is a permutations problem. Recalling the general formula for the number of pemutations of n things taken r at a time, ${}_nP_r = n!/(n-r)!$, we replace n by 3 and r by 3 to obtain

$${}_3P_3 = \frac{3!}{(3-3)!} = \frac{3!}{0!} = \frac{3 \times 2 \times 1}{1} = 6$$

PROBLEM

Candidates for 3 different political offices are to be chosen from a list of 10 people. In how many ways may this be done?

SOLUTION

There are 10 choices for the first office, and to go with each choice there are 9 choices for the second office, and to go with each of these there are 8 choices for the third office. The Fundamental Principle of Counting states that if one thing can be done in a different ways, a second thing can be done in b different ways, and a third thing can be done in c different ways ..., then all the things in succession can be done in $a \times b \times c$ ... different ways. Hence, there are $10 \times 9 \times 8 = 720$ ways of choosing the officers.

This can also be seen using the following approach. Since the arrangement of candidates on each slate is important (each arrangement represents people running for different political offices), this is a permutations problem. Recalling the general formula for the number of permutations of n things taken r at a time, ${}_nP_r = n!/(n-r)!$, we replace n by 10 and r by 3 to obtain

$${}_{10}P_3 = \frac{10!}{(10-3)!} = \frac{10!}{7!} = \frac{10 \times 9 \times 8 \times \not{7}!}{\not{7}!} = 720$$

PROBLEM

In how many ways may a party of four women and four men be seated at a round table if the women and men are to occupy alternate seats?

SOLUTION

If we consider the seats indistinguishable, then this is a problem in circular permutations, as opposed to linear permutations. In the standard linear permutations approach, each chair is distinguishable from the others. Thus, if a woman is seated first, she may be chosen 4 ways, then a man seated next to her may be chosen 4 ways, the next woman can be chosen 3 ways and the man next to her can be chosen 3 ways ... Our diagram to the linear approach shows the number of ways each seat can be occupied.

4	4	3	3	2	2	1	1

By the Fundamental Principle of Counting there are thus $4 \times 4 \times 3 \times 3 \times 2 \times 2 \times 1 \times 1 = 576$ ways to seat the people.

However, if the seats are indistinguishable, so long as each person has the same two people on each side, the seating arrangement is considered the same. Thus we may suppose one person, say a woman, is seated in a particular place, and then arrange the remaining three women and four men relative to her. Because of the alternate seating scheme, there are three possible places for the remaining three women, so that there are $3! = 6$ ways of seating them. There are four possible places for the four men, hence there are $4! = 24$ ways in which the men may be seated. Hence the total number of arrangements is $6 \times 24 = 144$. In general, the formula for circular permutations of n things and n other things which are alternating is $(n-1)!n!$. In our case, we have

$$(4-1)!4! = 3!4! = 3 \times 2 \times 4 \times 3 \times 2 = 144.$$

Combination refers to the method of finding the number of ways of arranging things when the *order* within the arrangements is not important.

PROBLEM

Evaluate each of the following symbols:

(a) C(6,3) (b) C(18,16)

SOLUTION

Recalling the general formula for the number of combinations of n different things taken r at a time, $C(n,r) = \dfrac{n!}{r!(n-r)!}$:

$$\text{(a)}\quad C(6,3) = \frac{6!}{3!(6-3)!} = \frac{6!}{3!3!} = \frac{\cancel{6}\times 5\times 4\times \cancel{3}!}{\cancel{3}\times\cancel{2}\times 1\times\cancel{3}!} = 20$$

$$\text{(b)}\quad C(18,16) = \frac{18!}{16!(18-16)!} = \frac{18!}{16!2!} = \frac{{}^{9}\cancel{18}\times 17\times\cancel{16}!}{\cancel{2}\times 1\times\cancel{16}!} = 153$$

PROBLEM

In how many different ways may a pair of dice fall?

SOLUTION

The Fundamental Principle of Counting states that if one thing can be done in a different ways, and when it is done in any one of these ways, a second thing can be done in b different ways, then both things in succession can be done in $a \times b$ different ways. A die has 6 sides, thus it may land in any of six ways. Since each die may land in 6 ways, by the Fundamental Principle both dice may fall in $6 \times 6 = 36$ ways. We can verify this result by enumerating all the possible ordered pairs of dice throws:

1,1	1,2	1,3	1,4	1,5	1,6
2,1	2,2	2,3	2,4	2,5	2,6
3,1	3,2	3,3	3,4	3,5	3,6
4,1	4,2	4,3	4,4	4,5	4,6
5,1	5,2	5,3	5,4	5,5	5,6
6,1	6,2	6,3	6,4	6,5	6,6

PROBLEM

In how many ways can we select a committee of 3 from a group of 10 people?

SOLUTION

The arrangement or order of people chosen is unimportant. Thus this is a combinations problem. Recalling the general formula for the number of combinations of n different things taken r at a time, $C(n,r) = \frac{n!}{r!(n-r)!}$, the number of committees of 3 from a group of 10 people is

$$C(10,3) = \frac{10!}{3!(10-3)!} = \frac{10!}{3!7!} = \frac{10 \times \overset{3}{\cancel{9}} \times \overset{4}{\cancel{8}} \times \cancel{7}!}{\cancel{3} \times \cancel{2} \times 1 \times \cancel{7}!} = 120$$

PROBLEM

How many baseball teams of nine members can be chosen from among twelve boys, without regard to the position played by each member?

SOLUTION

Since there is no regard to position, this is a combinations problem (if order or arrangement had been important it would have been a permutations problem). The general formula for the number of combinations of n things taken r at a time is $C(n,r) = \frac{n!}{r!(n-r)!}$.

We have to find the number of combinations of 12 things taken 9 at a time. Hence we have

$$C(12,9) = \frac{12!}{9!(12-9)!} = \frac{12!}{9!3!} = \frac{12 \times 11 \times 10 \times \cancel{9}!}{3 \times 2 \times 1 \times \cancel{9}!} = 220$$

Therefore, there are 220 possible teams.

PROBLEM

A man and his wife decide to entertain 24 friends by giving 4 dinners with 6 guests each. In how many ways can the first group be chosen?

SOLUTION

In the first group we are considering one dinner and there are 6 people out of 24 friends to be invited. We must find the number of ways to choose 6 out of 24. We are dealing with combinations. To select r things out of n objects, we use the definition of combinations:

$$\left(\frac{n}{r}\right) = \frac{n!}{r!(n-r!)} = c(n,r)$$

$$C(24,6) = \left(\frac{24}{6}\right) = \frac{24!}{6!18!} = \frac{24 \times 23 \times 22 \times 21 \times 20 \times 19 \times 18!}{6!18!}$$

$$= \frac{24 \times 23 \times 22 \times 21 \times 20 \times 19}{6 \times 5 \times 4 \times 3 \times 2 \times 1}$$

$$= 134{,}596$$

PROBLEM

A lady has 12 friends. She wishes to invite 3 of them to a bridge party. How many times can she entertain without having the same 3 people again?

SOLUTION

Since no reference has been made to order or arrangement (for example, the order in which the guests arrive or their seating at the bridge table), the problem is one of combinations rather than permutations. Recall the general formula for the number of combinations of n different things taken r at a time, $C(n,r) = \frac{n!}{r!(n-r)!}$. Thus the number of ways of selecting 3 friends from 12 is

$$C(12,3) = \frac{12!}{3!(12-3)!} = \frac{12!}{3!9!} = \frac{\overset{4}{\cancel{12}} \times 11 \times \overset{5}{\cancel{10}} \times \cancel{9}!}{\cancel{3} \times \cancel{2} \times 1 \times \cancel{9}!} = 220$$

In evaluating C(12,3), observe that the numerator and denominator of the fraction are first divided by the larger factorial in the denominator. (9! cancels in our fraction.) Thus, the lady can entertain 220 times without having the same 3 people.

Probability refers to the method of calculating the likelihood that an event will take place. Probability is expressed as a ratio between 0 and 1.

Probability is used when we cannot predict with certainty that an event will take place.

Probabilities can be expressed in the form of fractions, decimals, or percents.

PROBLEM

What is the probability of throwing a "six" with a single die?

SOLUTION

The die may land in any of 6 ways:

1, 2, 3, 4, 5, 6

The probability of throwing a six,

$$P(6) = \frac{\text{number of ways to get a six}}{\text{number of ways the die may land}}$$

Thus $P(6) = \frac{1}{6}$.

PROBLEM

A deck of playing cards is thoroughly shuffled and a card is drawn from the deck. What is the probability that the card drawn is the ace of diamonds?

SOLUTION

The probability of an event occurring is

$$\frac{\text{the number of ways the event can occur}}{\text{the number of possible outcomes}}$$

In our case there is one way the event can occur, for there is only one ace of diamonds and there are fifty-two possible outcomes (for there are 52 cards in the deck). Hence the probability that the card drawn is the ace of diamonds is 1/52.

PROBLEM

A box contains 7 red, 5 white, and 4 black balls. What is the probability of your drawing at random one red ball? One black ball?

SOLUTION

There are 7 + 5 + 4 = 16 balls in the box. The probability of drawing one red ball,

$$P(R) = \frac{\text{number of possible ways of drawing a red ball}}{\text{number of ways of drawing any ball}}$$

$$P(R) = \frac{7}{16}.$$

Similarly, the probability of drawing one black ball

$$P(B) = \frac{\text{number of possible ways of drawing a black ball}}{\text{number of ways of drawing any ball}}$$

Thus,

$$P(B) = \frac{4}{16} = \frac{1}{4}$$

PROBLEM

Calculate the probability of the Dodgers' winning the world series if the Yankees have won the first three games.

SOLUTION

We assume the probability of each team winning any specific game is 1/2. The only way the Dodgers can win is if they can take the next four games. Since the games can be considered independent events,

$$P(DDDD) = \frac{1}{2} \times \frac{1}{2} \times \frac{1}{2} \times \frac{1}{2} \text{ by the multiplication rule.}$$

$$P(DDDD) = \left(\frac{1}{2}\right)^4 = \frac{1}{16}$$

PROBLEM

A bag contains 4 white balls, 6 black balls, 3 red balls, and 8 green balls. If one ball is drawn from the bag, find the probability that it will be either white or green.

SOLUTION

The probability that it will be either white or green is: P(a white ball or a green ball) = P(a white ball) + P(a green ball). This is true because if we are given two mutually exclusive events A or B, then P(A or B) = P(A) + P(B). Note that two events, A and B, are mutually exclusive events if their intersection is null or empty set. In this case, the intersection of choosing a white ball and of choosing a green ball is the empty set. There are no elements in common.

$$P(\text{a white ball}) = \frac{\text{number of ways to choose a white ball}}{\text{number of ways to select a ball}}$$

$$= \frac{4}{21}$$

$$P(\text{a green ball}) = \frac{\text{number of ways to choose a green ball}}{\text{number of ways to select a ball}}$$

$$= \frac{8}{21}$$

Thus,

$$P(\text{a white ball or a green ball}) = \frac{4}{21} + \frac{8}{21} = \frac{12}{21} = \frac{4}{7}.$$

PROBLEM

A penny is to be flipped 3 times. What is the probability there will be 2 heads and 1 tail?

SOLUTION

We start this problem by constructing a set of all possible outcomes:

We can have heads on all 3 tosses:	(HHH)	
heads on first 2 tosses, tails on the third:	(HHT)	(1)
heads on first toss, tails on next two:	(HTT)	
etc.:	(HTH)	(2)
	(THH)	(3)
	(THT)	
	(TTH)	
	(TTT)	

Hence there are eight possible outcomes (2 possibilities on first toss $\times 2$ on second $\times$ 2 on third = $2 \times 2 \times 2 = 8$). We assume that these outcomes are all equally likely and assign the probability 1/8 to each. Now we look for the set of outcomes that produce 2 heads and 1 tail. We see there are 3 such outcomes out of the 8 possibilities (numbered (1), (2), (3) in our listing). Hence the probability of 2 heads and 1 tail is 3/8.

10. TREES

A list of all possible outcomes from an experiment is called a sample space. A tree diagram is helpful in determining these possible outcomes.

EXAMPLE

You flip a coin and you roll a six-sided die. Construct a tree diagram showing all possible outcomes from this experiment.

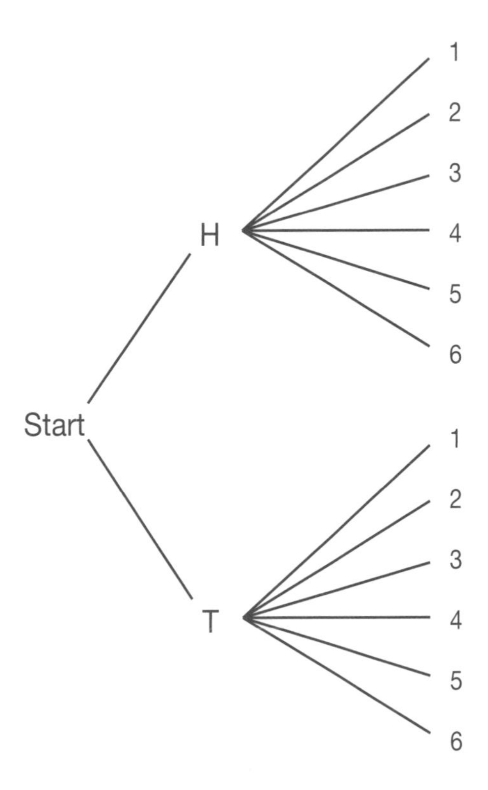

The sample space for this experiment looks like this:

{H1, H2, H3, H4, H5, H6, T1, T2, T3, T4, T5, T6}

There are twelve possible outcomes of this experiment.

EXAMPLE

You flip a coin three times. Construct a tree diagram showing all the possible outcomes of this experiment.

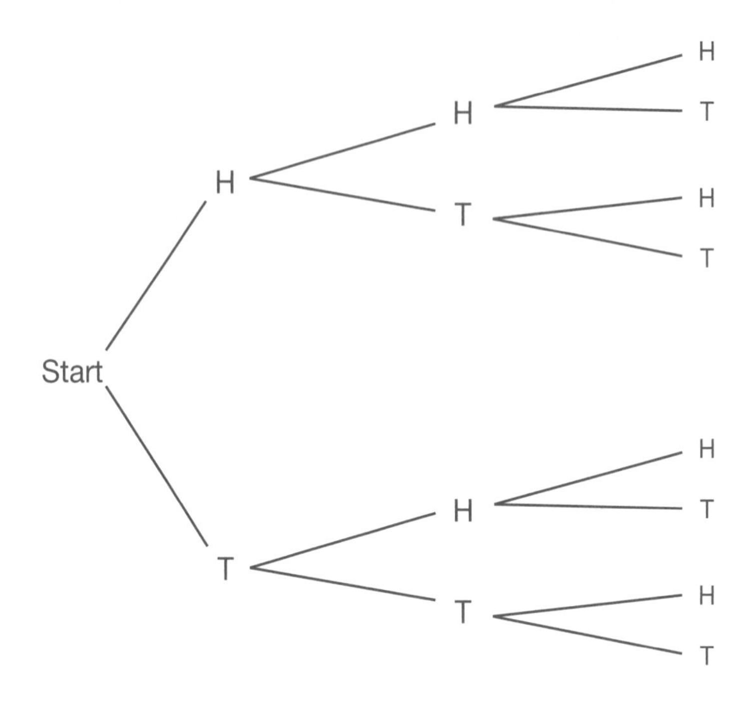

The sample space for this experiment looks like this:

{HHH, HHT, HTH, HTT, THH, THT, TTH, TTT}

There are eight possible outcomes of this experiment.

PROBLEM

There is a special at the school cafeteria: you get a hamburger or a veggie-burger along with your choice of one topping and one drink. The choices look like this:

Hamburger (H)	Lettuce (L)	Soda (S)
Veggie-burger (V)	Onion (O)	Juice (J)
	Tomato (T)	

What does a tree diagram of the cafeteria special look like?

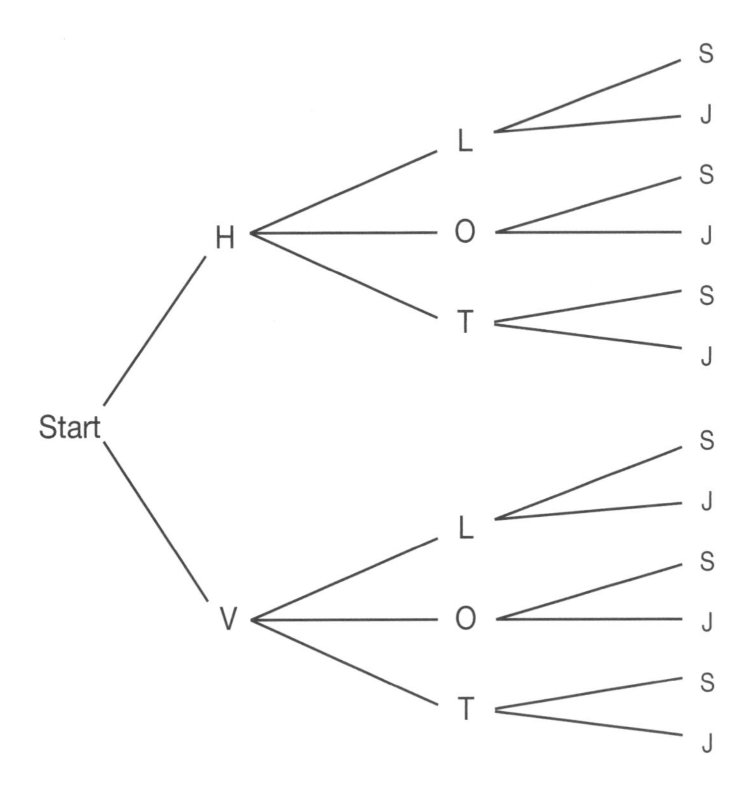

The sample space for this experiment looks like this:

{HLS, HLJ, HOS, HOJ, HTS, HTJ, VLS, VLJ, VOS, VOJ, VTS, VTJ}

There are twelve possible outcomes of this experiment.

11. ITERATION

Iteration means a repetition. In mathematics, iteration refers to the repetition of an operation. The output from this function can be used as the input for the next iteration.

EXAMPLE

Using the formula given below, calculate the value of the third iteration, if the starting number is 4.

$$\frac{x}{2}+4=$$

$$\frac{4}{2}+4=6$$

$$\frac{6}{2}+4=7$$ **one iteration**

$$\frac{7}{2}+4=7.5$$ **two iterations**

$\frac{7.5}{2} + 4 = 7.75$ **three iterations**

EXAMPLE

Using the formula given below, calculate the value of the fourth iteration, if the starting number is 0.

$3x - 1 =$

$3(0) - 1 = -1$

$3(-1) - 1 = -4$ **one iteration**

$3(-4) - 1 = -13$ **two iterations**

$3(-13) - 1 = -40$ **three iterations**

$3(-40) - 1 = -121$ **four iterations**

EXAMPLE

Using the formula given below, calculate the value of the second iteration, if the starting number is 1.

$x^2 + x =$

$1^2 + 1 = 2$

$2^2 + 2 = 6$ **one iteration**

$6^2 + 6 = 42$ **two iterations**

PROBLEM

Using the equation $2x + x + 1$, what is the value of the third iteration, if the starting number is 5?

SOLUTION

$2x + x + 1$

$2(5) + 5 + 1 = 16$

$2(16) + 16 + 1 = 49$ **one iteration**

$2(49) + 49 + 1 = 148$ **two iterations**

$2(148) + 148 + 1 = 445$ **three iterations**

PROBLEM

Using the formula $\frac{x+3}{2} =$, what is the value of the third iteration, if the starting number is 1?

SOLUTION

$$\frac{x+3}{2} =$$

$$\frac{1+3}{2} = 2$$

$$\frac{2+3}{2} = 2.5$$ **one iteration**

$$\frac{2.5+3}{2} = 2.75$$ **two iterations**

$$\frac{2.75+3}{2} = 2.875$$ **three iterations**

---

12. RECURSION

A recursive function is a procedure that calls a simpler version of itself to arrive at a solution.

EXAMPLE

The factorial function is an example of recursion. The factorial of a positive integer n, is written $n!$ This is the product of all positive integers from 1 up to and including n.

$$6! = 6 \times 5 \times 4 \times 3 \times 2 \times 1 = 720$$

$$5! = 5 \times 4 \times 3 \times 2 \times 1 = 120$$

$$4! = 4 \times 3 \times 2 \times 1 = 24$$

$$3! = 3 \times 2 \times 1 = 6$$

$$2! = 2 \times 1 = 2$$

$$1! = 1 \times 1 = 1$$

$$0! = 1$$

PROBLEM

What does 8! equal?

SOLUTION

$$8! = 8 \times 7 \times 6 \times 5 \times 4 \times 3 \times 2 \times 1 = 40{,}320$$

---

13. GROWTH

Growth is an increase in size or quantity over time. There are many different types of growth, depending on what you want to measure.

EXAMPLE

When a person puts money into a savings account, many banks pay compound interest. Compound interest is when a person earns interest on both the principle and on the interest previously accrued. The formula for compound interest is:

$$A = P(1 + r)^n$$

where

P is the principle (the money you start with)

r is the annual rate of interest in decimal form

n is the number of years you leave your money in the bank

A is the total amount of money you will have after n years, if interest is compounded once a year.

For example, if a person deposits $1,000 into a savings account that pays an interest rate of 5% compounded annually, how much money will be in this account after 2 years?

$$\begin{aligned} A &= P(1 + r)^n \\ &= 1{,}000(1 + 0.05)^2 \\ &= 1{,}000(1.05)^2 \\ &= 1{,}000(1.1025) \\ &= \$1{,}102.50 \end{aligned}$$

EXAMPLE

Exponential growth is when growth starts off slowly, but quickly increases. Bacteria are known to show exponential growth. Let's say that bacteria can divide in two about every 20 minutes. If you start off with 1 bacterium, calculate the number you will have at the end of 6 hours.

Since there are three 20-minute periods in an hour, you need $6 \times 3 = 18$ bacteria divisions starting from the single bacterium.

$1 \rightarrow 2 \rightarrow 4 \rightarrow 8 \rightarrow 16 \rightarrow 32 \rightarrow 64 \rightarrow 128 \rightarrow 256 \rightarrow 512 \rightarrow 1{,}024 \rightarrow 2{,}048 \rightarrow 4{,}096 \rightarrow 8{,}192 \rightarrow 16{,}384 \rightarrow 32{,}768 \rightarrow 65{,}536 \rightarrow 131{,}072 \rightarrow 262{,}144$

At the end of six hours, there will be 262,144 bacteria.

PROBLEM

Nevada is a fast-growing state. In 1990, the population was 770,280 people. In 2000, the population was 1,394,440. What is the percentage of population growth from 1990 to 2000? Round to the nearest percent.

SOLUTION

First, find out how many more people were living in Nevada in 2000 compared to 1990 by subtracting.

$$1{,}394{,}440 - 770{,}280 = 624{,}160$$

Divide the difference by the number of people living in Nevada in 1990. Lastly, change the decimal number into a percent.

$$624{,}160 \div 770{,}280 \approx 0.81 = 81\% \text{ growth}$$

14. GRAPHS OF FUNCTIONS

EXAMPLE

The equation $Y = 2X + 3$ is the equation of a line with a slope equal to 2. This means that for every one-unit change in X, the Y measurement goes up 2 units. The number 3 in this equation is called the y-intercept. The y-intercept is the point where the line crosses the y-axis. The y-intercept is the value of y when x equals 0.

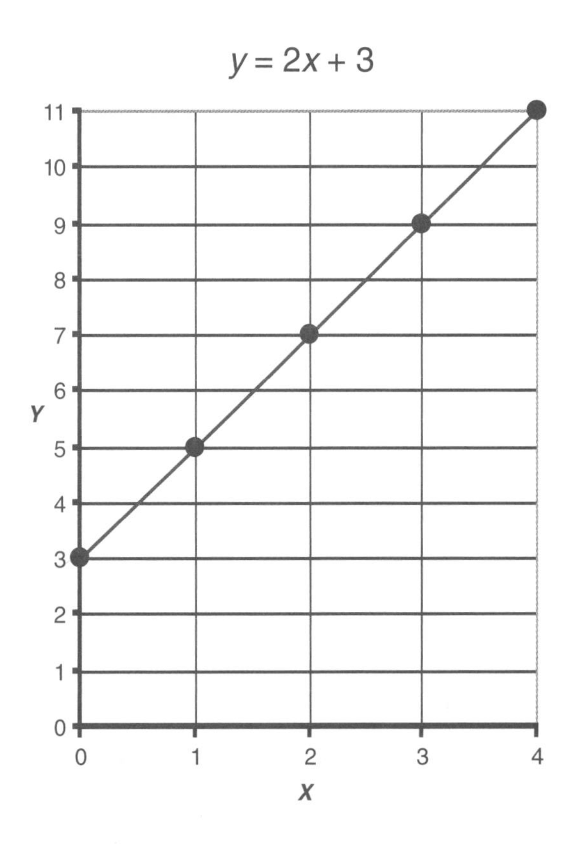

EXAMPLE

Graph the function $y = \frac{1}{2}x^2$.

When creating a graph, first make two columns and label the first column x and the second column y. Then, under the x column, put in the values 0, 1, –1, 3, and –3. Next, calculate the y values for each x value by plugging the x value into the equation. Finally, plot the (x, y) points on the graph and connect the points. *Note:* This curve is called a parabola.

x	y
0	0
1	0.5
–1	0.5
3	4.5
–3	4.5

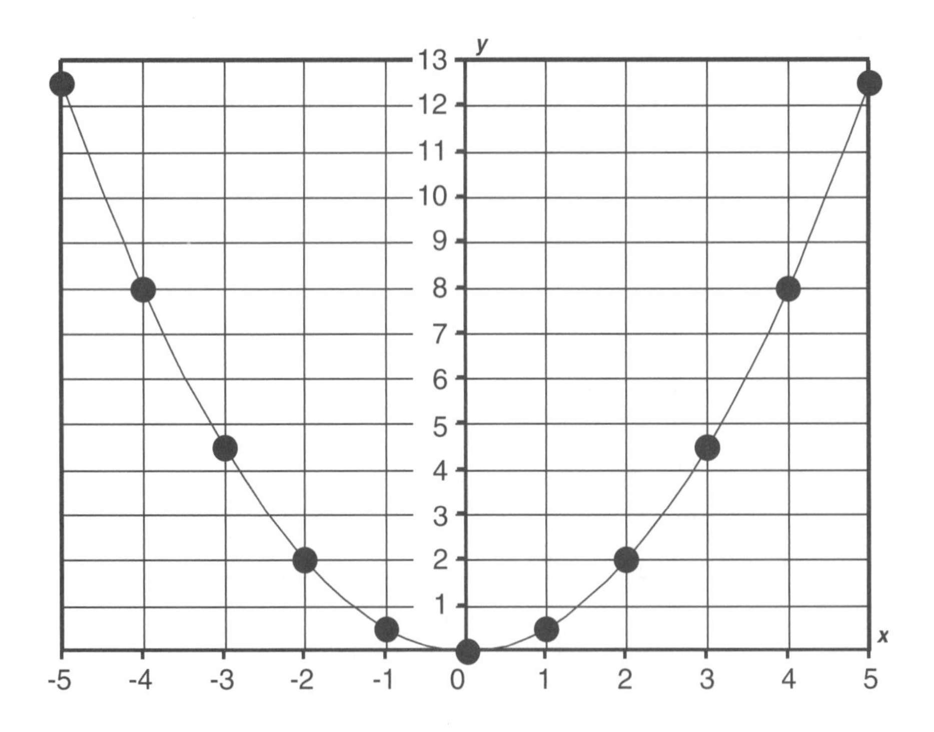

Note: In order to create a more detailed graph of this function, additional x values of 2, –2, 4, –4, 5, and –5 were used in the above graph.

ALGEBRA DRILLS

DRILL—OPERATIONS WITH POLYNOMIALS

1.	(B)	6.	(B)	11.	(C)	16.	(C)
2.	(C)	7.	(C)	12.	(B)	17.	(D)
3.	(C)	8.	(E)	13.	(E)	18.	(E)
4.	(D)	9.	(A)	14.	(A)	19.	(B)
5.	(A)	10.	(D)	15.	(D)	20.	(B)

DRILL—SIMPLIFYING ALGEBRAIC EXPRESSIONS

1.	(C)	6.	(B)
2.	(D)	7.	(E)
3.	(B)	8.	(B)
4.	(A)	9.	(A)
5.	(D)	10.	(E)

DRILL—LINEAR EQUATIONS

1.	(C)
2.	(A)
3.	(E)
4.	(D)
5.	(B)

DRILL—TWO LINEAR EQUATIONS

1.	(E)	6.	(B)
2.	(B)	7.	(D)
3.	(A)	8.	(E)
4.	(D)	9.	(A)
5.	(C)	10.	(B)

DRILL—QUADRATIC EQUATIONS

1.	(A)	6.	(B)
2.	(D)	7.	(B)
3.	(B)	8.	(D)
4.	(C)	9.	(C)
5.	(E)	10.	(E)

DRILL—ABSOLUTE VALUE EQUATIONS

1.	(B)	4.	(C)
2.	(E)	5.	(D)
3.	(A)		

DRILL—INEQUALITIES

1.	(E)	4.	(B)
2.	(C)	5.	(A)
3.	(D)		

DRILL—RATIOS AND PROPORTIONS

1.	(D)	4.	(A)	7.	(A)	10.	(E)
2.	(B)	5.	(B)	8.	(B)		
3.	(C)	6.	(C)	9.	(D)		

GEOMETRY REVIEW

1. POINTS, LINES, AND ANGLES

Geometry is built upon a series of undefined terms. These terms are those which we accept as known in order to define other undefined terms.

A) **Point**: Although we represent points on paper with small dots, a point has no size, thickness, or width.

B) **Line**: A line is a series of adjacent points which extends indefinitely. A line can be either curved or straight; however, unless otherwise stated, the term "line" refers to a straight line.

C) **Plane**: A plane is a collection of points lying on a flat surface, which extends indefinitely in all directions.

If *A* and *B* are two points on a line, then the **line segment** *AB* is the set of points on that line between *A* and *B* and including *A* and *B*, which are endpoints. The line segment is referred to as *AB*.

A **ray** is a series of points that lie to one side of a single endpoint.

PROBLEM

How many lines can be found that contain (a) one given point (b) two given points (c) three given points?

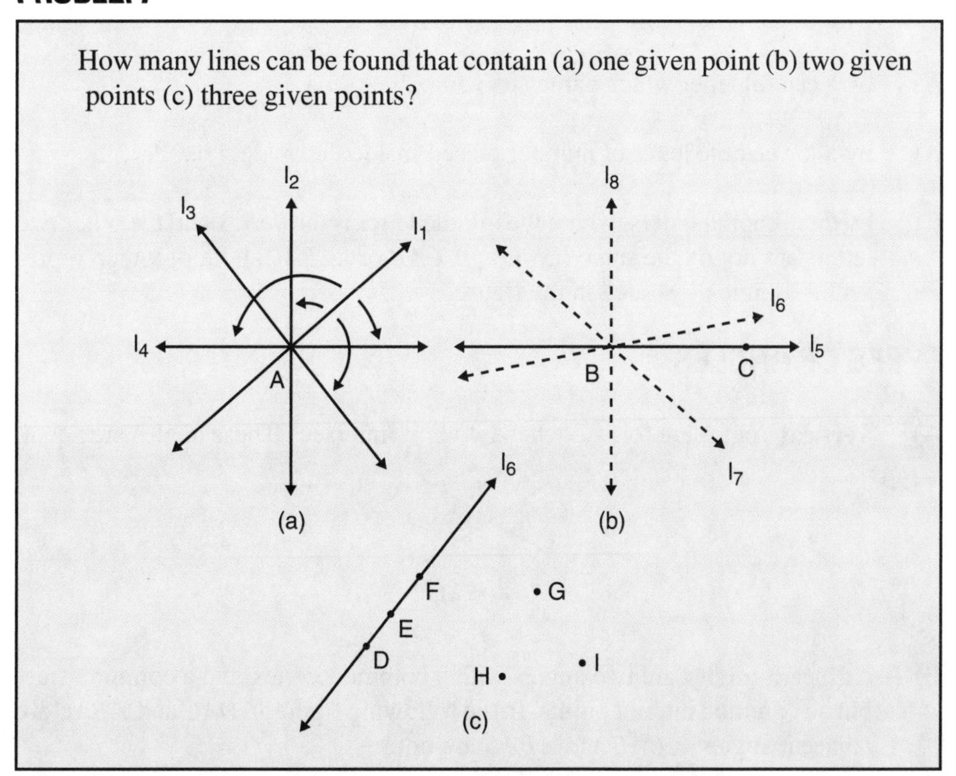

SOLUTION

(a) *Given one point A*, there are an infinite number of distinct lines that contain the given point. To see this, consider line l_1 passing through point A. By rotating l_1 around A like the hands of a clock, we obtain different lines l_2, l_3, etc. Since we can rotate l_1 in infinitely many ways, there are infinitely many lines containing A.

(b) *Given two distinct points B and C*, there is one and only one distinct line. To see this, consider all the lines containing point B; l_5, l_6, l_7 and l_8. Only l_5 contains both points B and C. Thus, there is only one line containing both points B and C. Since there is always at least one line containing two distinct points and never more than one, the line passing through the two points is said to be determined by the two points.

(c) *Given three distinct points*, there may be one line or none. If a line exists that contains the three points, such as D, E, and F, then the points are said to be **colinear**. If no such line exists — as in the case of points G, H, and I, then the points are said to be **noncolinear**.

INTERSECTION LINES AND ANGLES

An **angle** is a collection of points which is the union of two rays having the same endpoint. An angle such as the one illustrated below can be referred to in any of the following ways:

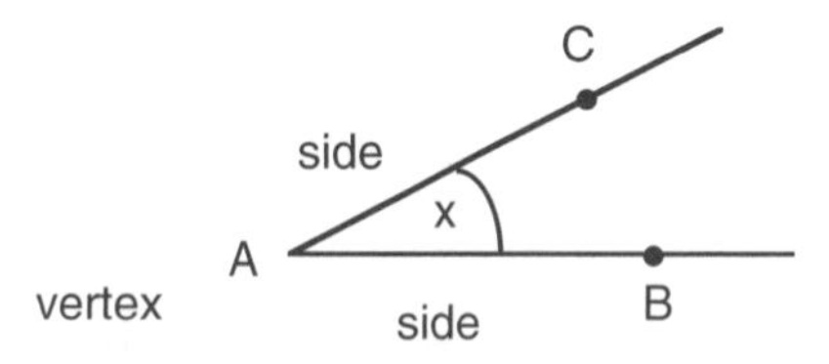

A) by a capital letter which names its vertex, i.e., $\angle A$;

B) by a lower-case letter or number placed inside the angle, i.e., $\angle x$;

C) by three capital letters, where the middle letter is the vertex and the other two letters are not on the same ray, i.e., $\angle CAB$ or $\angle BAC$, both of which represent the angle illustrated in the figure.

TYPES OF ANGLES

A) **Vertical angles** are formed when two lines intersect. These angles are equal.

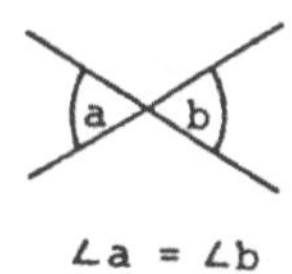

$\angle a = \angle b$

B) **Adjacent angles** are two angles with a common vertex and a common side, but no common interior points. In the following figure, $\angle DAC$ and $\angle BAC$ are adjacent angles. $\angle DAB$ and $\angle BAC$ are not.

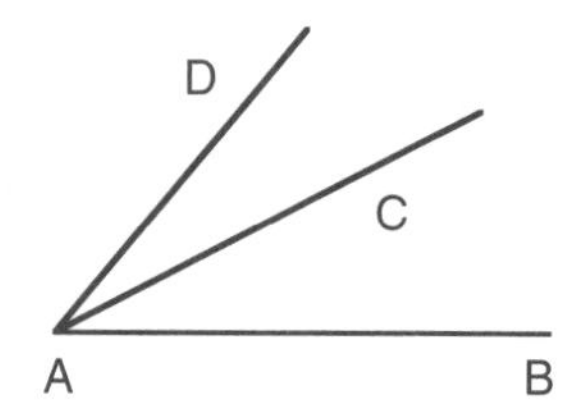

C) A **right angle** is an angle whose measure is 90°.

D) An **acute angle** is an angle whose measure is larger than 0° but less than 90°.

E) An **obtuse angle** is an angle whose measure is larger than 90° but less than 180°.

F) A **straight angle** is an angle whose measure is 180°. Such an angle is, in fact, a straight line.

G) A **reflex angle** is an angle whose measure is greater than 180° but less than 360°.

H) **Complementary angles** are two angles, the sum of the measures of which equals 90°.

I) **Supplementary angles** are two angles, the sum of the measures of which equals 180°.

J) **Congruent angles** are angles of equal measure.

PROBLEM

In the figure, we are given $\overline{AB}$ and triangle ABC. We are told that the measure of ∠ 1 is five times the measure of ∠ 2. Determine the measures of ∠ 1 and ∠ 2.

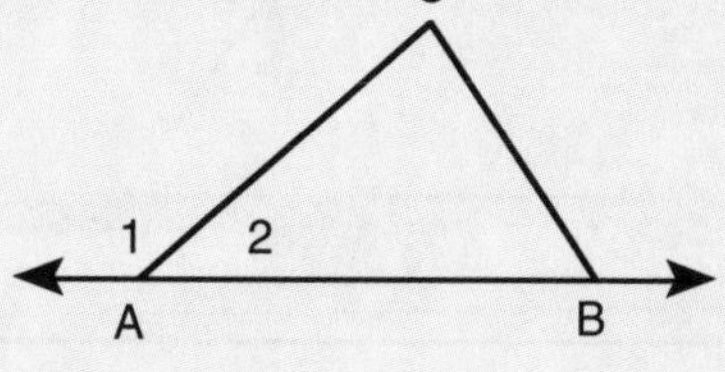

SOLUTION

Since ∠ 1 and ∠ 2 are adjacent angles whose non-common sides lie on a straight line, they are, by definition, supplementary. As supplements, their measures must sum to 180°.

If we let x = the measure of ∠2, then, $5x$ = the measure of ∠ 1.

To determine the respective angle measures, set $x + 5x = 180$ and solve for x. $6x = 180$. Therefore, $x = 30$ and $5x = 150$.

Therefore, the measure of ∠ 1 = 150 and the measure of ∠ 2 = 30.

PROBLEM

Measure the angle below. Determine if the angle is obtuse or acute.

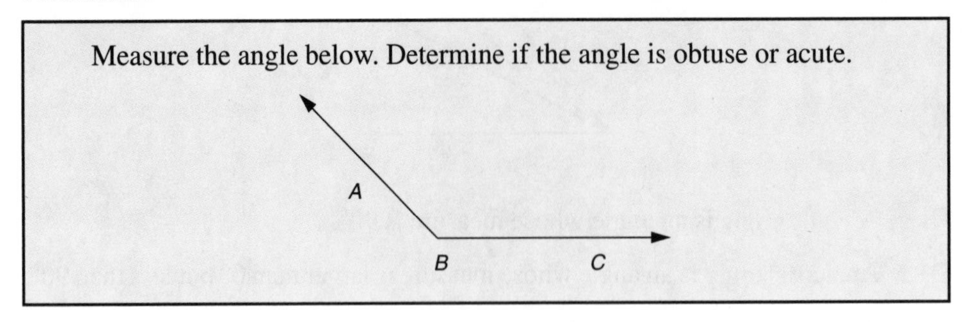

SOLUTION

Step 1 is to measure the angle by placing the center of the protractor on the vertex of the angle. Side BC of the angle should now be aligned with the straight edge of the protractor.

Step 2 is to follow side AB from the vertex toward the numbers on the curved edge of the protractor. Read the value where side AB crosses the numbers. This is the measure of $\angle ABC$.

The correct answer is 135 degrees. Since $\angle ABC$ is larger than 90 degrees, the angle is obtuse.

PROBLEM

Measure the angle below. Determine if the angle is obtuse or acute.

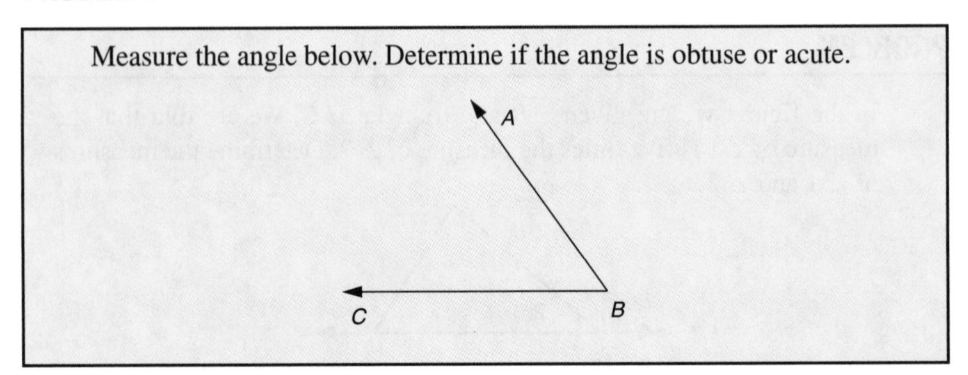

SOLUTION

Step 1 is to measure the angle by placing the center of the protractor on the vertex of the angle. Side BC of the angle should now be aligned with the straight edge of the protractor.

Step 2 is to follow side AB from the vertex toward the numbers on the curved edge of the protractor. Read the value where side AB crosses the numbers. This is the measure of $\angle ABC$.

The correct answer is 58 degrees. Since $\angle ABC$ is smaller than 90 degrees, the angle is acute.

PROBLEM

Measure the angles below. Determine if the angles are complementary.

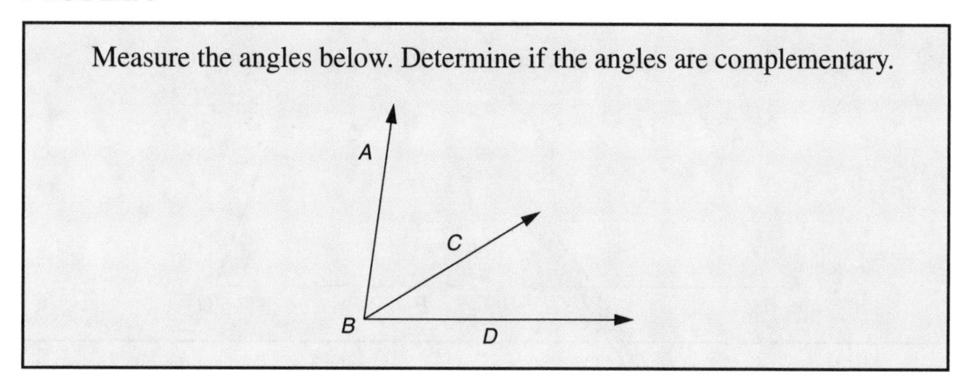

SOLUTION

Step 1 is to measure $\angle ABC$ by placing the center of the protractor on the vertex of the angle. Side *BC* of the angle should now be aligned with the straight edge of the protractor.

Step 2 is to follow side *AB* from the vertex toward the numbers on the curved edge of the protractor. Read the value where side *AB* crosses the numbers. The measure of $\angle ABC$ is 50 degrees.

Step 3 is to measure $\angle CBD$ by placing the center of the protractor on the vertex of the angle. Side *BD* of the angle should now be aligned with the straight edge of the protractor.

Step 4 is to follow side *CB* from the vertex toward the numbers on the curved edge of the protractor. Read the value where side *AB* crosses the numbers. The measure of $\angle CBD$ is 30 degrees.

Complementary angles are angles that sum to 90 degrees. Since $\angle ABC$ and $\angle CBD$ do not have a sum of 90, the angles are not complementary.

PROBLEM

The angles in the figure below are supplementary. If $\angle MNO$ is 117 degrees, what is the measurement of $\angle ONP$?

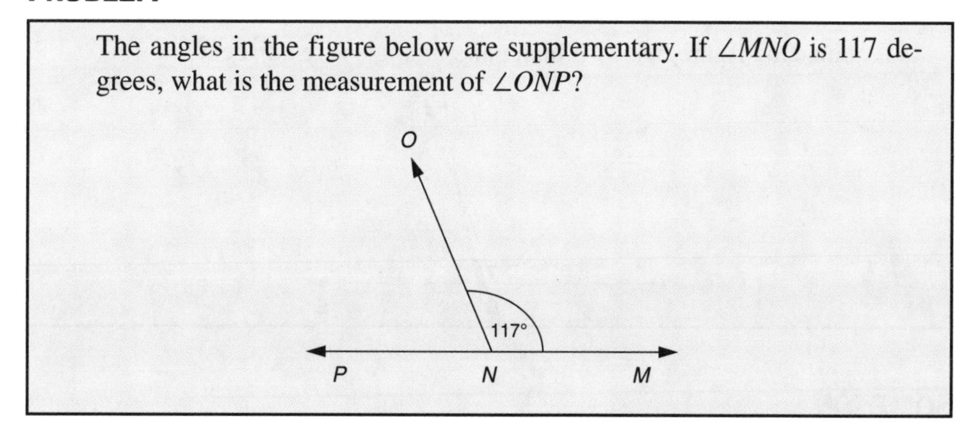

SOLUTION

It is known that the sum of supplementary angles equals 180 degrees. If $\angle MNO$ equals 117 degrees, then $\angle ONP$ equals 180 minus 117.

The correct answer is $\angle ONP$ is 63 degrees.

PROBLEM

Measure the angles below. Determine if the angles are congruent.

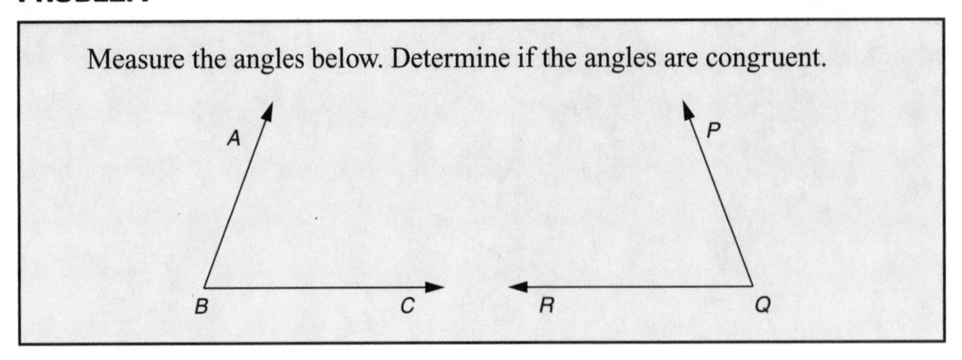

SOLUTION

Step 1 is to measure $\angle ABC$ by placing the center of the protractor on the vertex of the angle. Side BC of the angle should now be aligned with the straight edge of the protractor.

Step 2 is to follow side AB from the vertex toward the numbers on the curved edge of the protractor. Read the value where side AB crosses the numbers. The measure of $\angle ABC$ is 70 degrees.

Step 3 is to measure $\angle PQR$ by placing the center of the protractor on the vertex of the angle. Side QR of the angle should now be aligned with the straight edge of the protractor.

Step 4 is to follow side PQ from the vertex toward the numbers on the curved edge of the protractor. Read the value where side PQ crosses the numbers. The measure of $\angle PQR$ is 70 degrees.

Congruent angles are angles that are equal in measurement. Since $\angle ABC$ and $\angle PQR$ are equal, the angles are congruent.

PROBLEM

Examine the figure below. Which pairs of angles are adjacent?

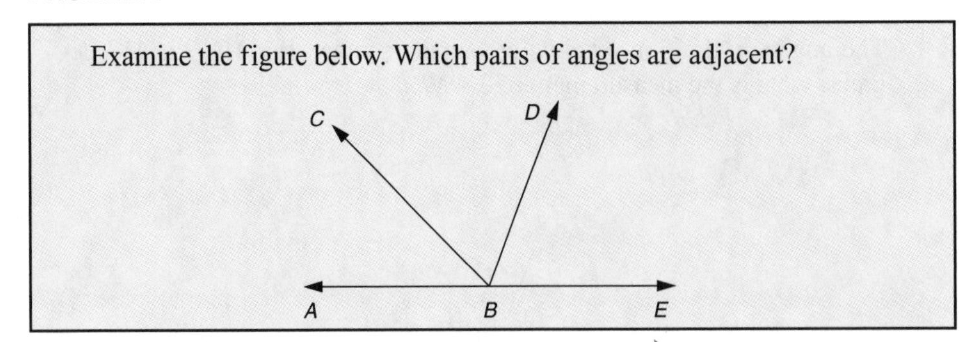

SOLUTION

Adjacent angles are angles that share a common side. The adjacent angles are $\angle ABC$ and $\angle CBD$, $\angle ABD$ and $\angle DBE$, $\angle CBD$ and $\angle DBE$, $\angle ABC$ and $\angle CBE$.

PERPENDICULAR LINES

Two lines are said to be **perpendicular** if they intersect and form right angles. The symbol for perpendicular (or, is therefore perpendicular to) is $\perp$; $\overline{AB}$ is perpendicular to $\overline{CD}$ is written $\overline{AB} \perp \overline{CD}$.

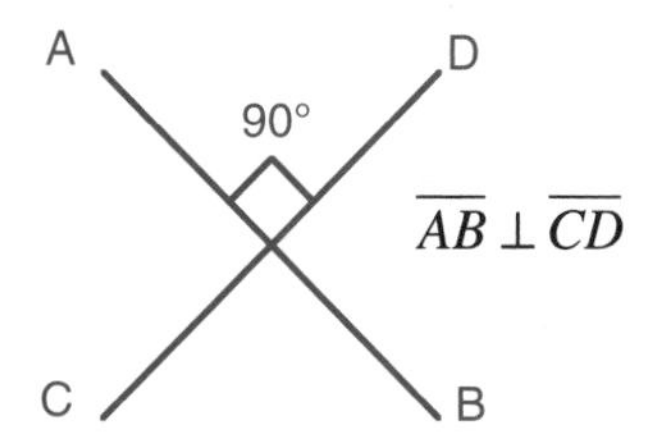

$\overline{AB} \perp \overline{CD}$

PROBLEM

> We are given straight lines $\overline{AB}$ and $\overline{CD}$ intersecting at point P. $\overline{PR} \perp \overline{AB}$ and the measure of $\angle APD$ is 170°. Find the measures of $\angle 1$, $\angle 2$, $\angle 3$, and $\angle 4$. (See figure below.)

SOLUTION

This problem will involve making use of several of the properties of supplementary and vertical angles, as well as perpendicular lines.

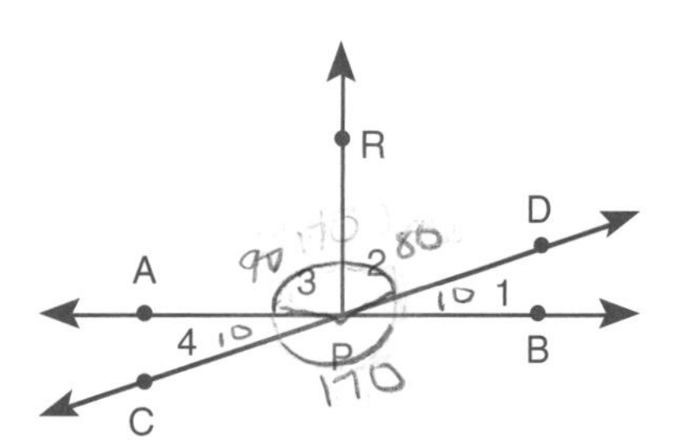

$\angle APD$ and $\angle 1$ are adjacent angles whose non-common sides lie on a straight line, $\overline{AB}$. Therefore, they are supplements and their measures sum to 180°.

$$m\angle APD + m\angle 1 = 180°.$$

We know $m\angle APD = 170°$. Therefore, by substitution, $170° + m\angle 1 = 180°$. This implies $m\angle 1 = 10°$.

$\angle 1$ and $\angle 4$ are vertical angles because they are formed by the intersection of two straight lines, $\overline{CD}$ and $\overline{AB}$, and their sides form two pairs of opposite rays. As vertical angles, they are, by theorem, of equal measure. Since $m\angle 1 = 10°$, then $m\angle 4 = 10°$.

Since $\overline{PR} \perp \overline{AB}$, at their intersection the angles formed must be right angles. Therefore, $\angle 3$ is a right angle and its measure is 90°. $m\angle 3 = 90°$.

The figure shows us that $\angle APD$ is composed of $\angle 3$ and $\angle 2$. Since the measure of the whole must be equal to the sum of the measures of its parts, $m\angle APD = m\angle 3 + m\angle 2$. We know the $m\angle APD = 170°$ and $m\angle 3 = 90°$, therefore, by substitution, we can solve for $m\angle 2$, our last unknown.

$$170° = 90° + m\angle 2$$

$$80° = m\angle 2$$

Therefore, $m\angle 1 = 10°$, $m\angle 2 = 80°$,

$m\angle 3 = 90°$, $m\angle 4 = 10°$.

PROBLEM

In the accompanying figure $\overline{SM}$ is the perpendicular bisector of $\overline{QR}$, and $\overline{SN}$ is the perpendicular bisector of $\overline{QP}$. Prove that $SR = SP$.

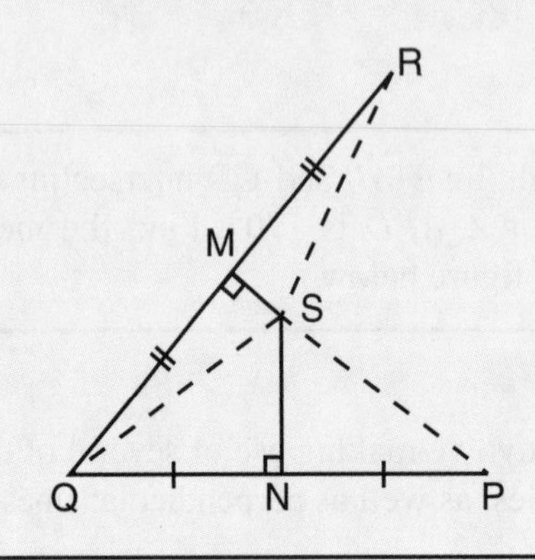

SOLUTION

Every point on the perpendicular bisector of a segment is equidistant from the endpoints of the segment.

Since point S is on the perpendicular bisector of $\overline{QR}$,

$$SR = SQ \qquad (1)$$

Also, since point S is on the perpendicular bisector of $\overline{QP}$,

$$SQ = SP \qquad (2)$$

By the transitive property (quantities equal to the same quantity are equal), we have:

$$SR = SP. \qquad (3)$$

PARALLEL LINES

Two lines are called **parallel lines** if, and only if, they are in the same plane (coplanar) and do not intersect. The symbol for parallel, or is parallel to, is $\parallel$; $\overline{AB}$ is parallel to $\overline{CD}$ is written $\overline{AB} \parallel \overline{CD}$.

The distance between two parallel lines is the length of the perpendicular segment from any point on one line to the other line.

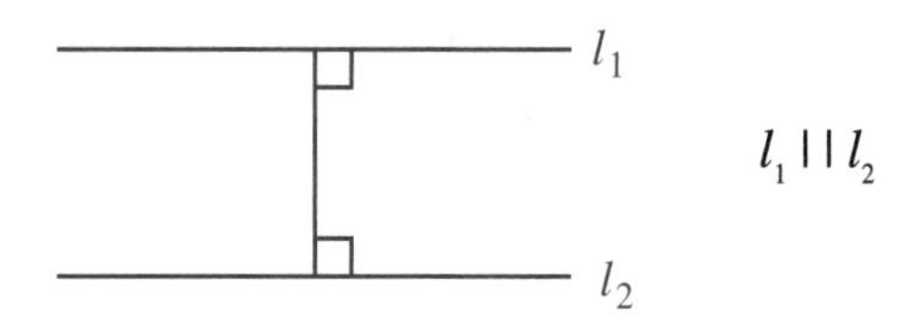

Given a line *l* and a point *P* not on line *l*, there is one and only one line through point *P* that is parallel to line *l*.

Two coplanar lines are either intersecting lines or parallel lines.

If two (or more) lines are perpendicular to the same line, then they are parallel to each other.

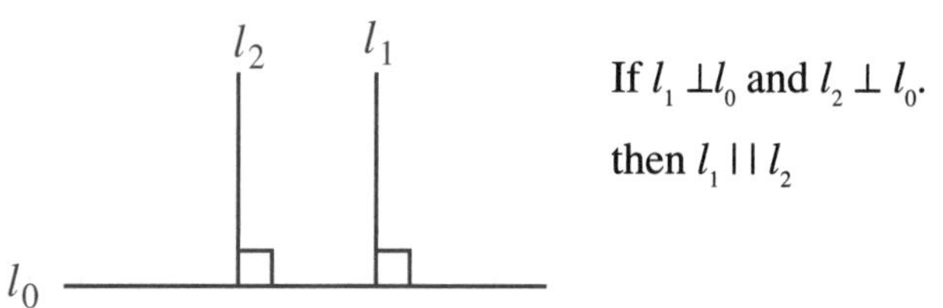

If two lines are cut by a transversal so that alternate interior angles are equal, the lines are parallel.

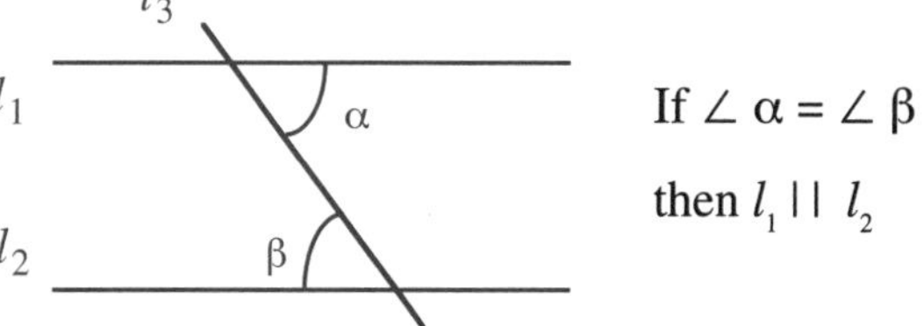

If two lines are parallel to the same line, then they are parallel to each other.

l_2

l_0

If $l_1 \parallel l_0$ and $l_2 \parallel l_0$

then $l_1 \parallel l_2$

If a line is perpendicular to one of two parallel lines, then it is perpendicular to the other line, too.

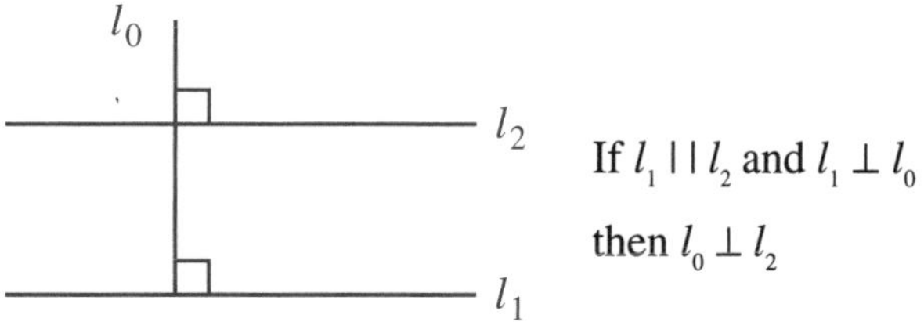

If two lines being cut by a transversal form congruent corresponding angles, then the two lines are parallel.

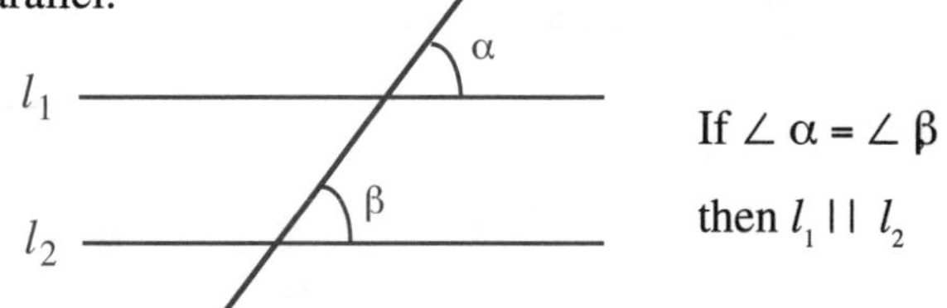

If two lines being cut by a transversal form interior angles on the same side of the transversal that are supplementary, then the two lines are parallel.

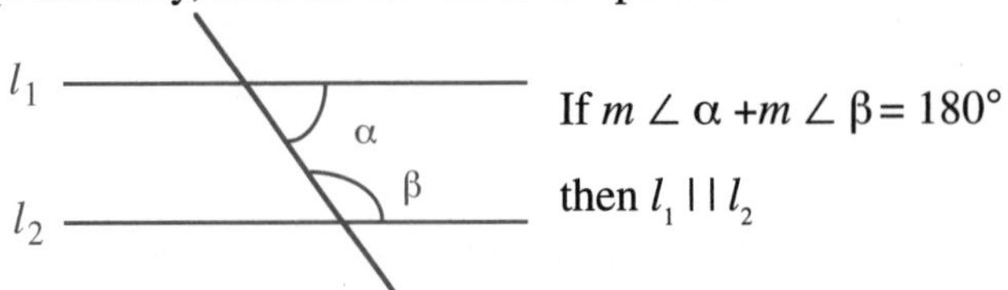

If a line is parallel to one of two parallel lines, it is also parallel to the other line.

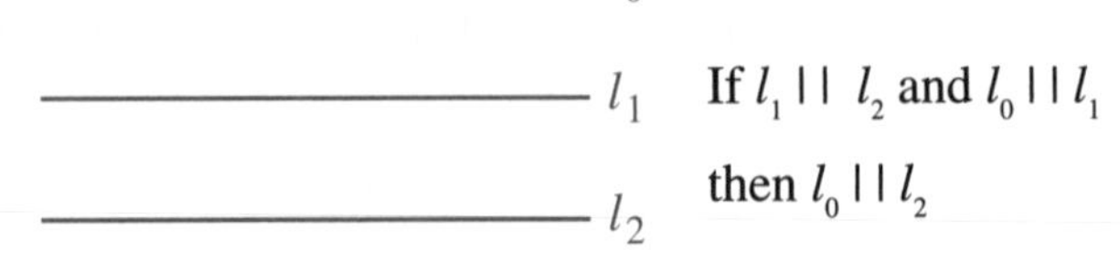

If two parallel lines are cut by a transversal, then:

A) The alternate interior angles are congruent.

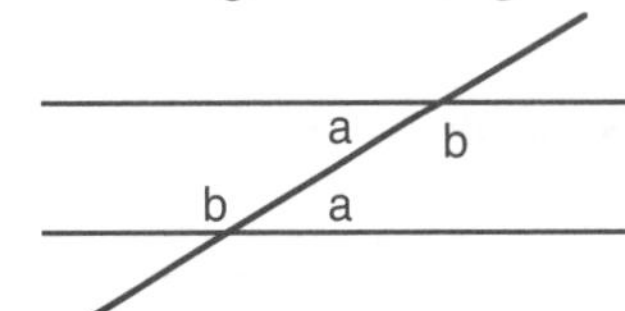

B) The corresponding angles are congruent.

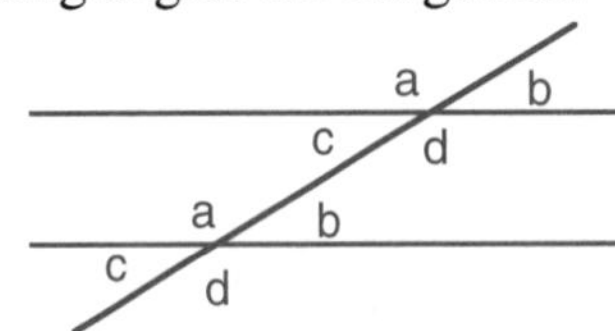

C) The consecutive interior angles are supplementary.

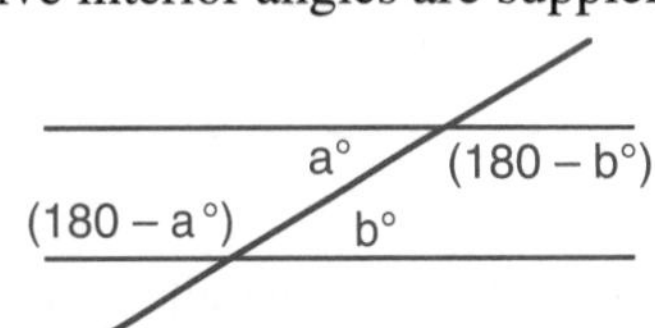

D) The alternate exterior angles are congruent.

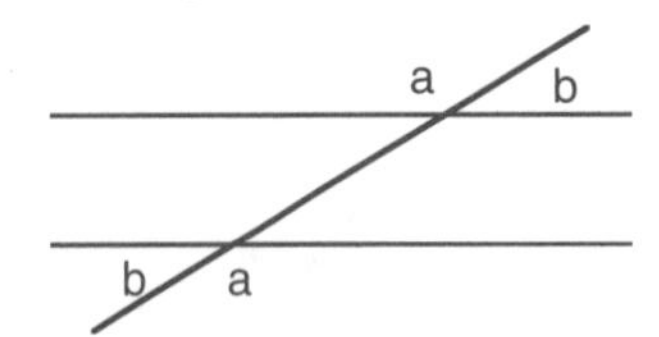

PROBLEM

Given: $\angle 2$ is supplementary to $\angle 3$.

Prove: $l_1 \parallel l_2$.

t

3

l_1

1

l_2

2

SOLUTION

Given two lines intercepted by a transversal, if a pair of corresponding angles are congruent, then the two lines are parallel. In this problem, we will show that since $\angle 1$ and $\angle 2$ are supplementary and $\angle 2$ and $\angle 3$ are supplementary, $\angle 1$ and $\angle 3$ are congruent. Since corresponding angles $\angle 1$ and $\angle 3$ are congruent, it follows $l_1 \parallel l_2$.

	Statement		Reason
1.	$\angle 2$ is supplementary to $\angle 3$.	1.	Given.
2.	$\angle 1$ is supplementary to $\angle 2$.	2.	Two angles that form a linear pair are supplementary.
3.	$\angle 1 \cong \angle 3$	3.	Angles supplementary to the same angle are congruent.
4.	$l_1 \parallel l_2$.	4.	Given two lines intercepted by a transversal, if a pair of corresponding angles are congruent, then the two lines are parallel.

PROBLEM

If line $\overline{AB}$ is parallel to line $\overline{CD}$ and line $\overline{EF}$ is parallel to line $\overline{GH}$, prove that $m\angle 1 = m\angle 2$.

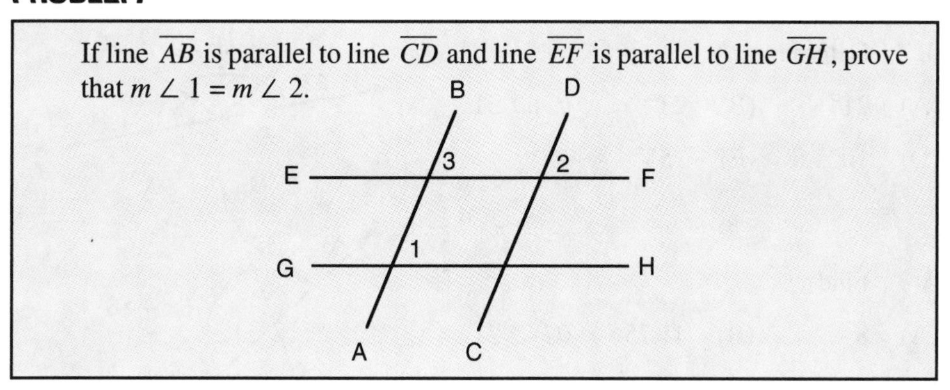

SOLUTION

To show $\angle 1 \cong \angle 2$, we relate both to $\angle 3$. Because $\overline{EF} \parallel \overline{GH}$, corresponding angles 1 and 3 are congruent. Since $\overline{AB} \parallel \overline{CD}$, corresponding angles 3 and 2 are congruent. Because both $\angle 1$ and $\angle 2$ are congruent to the same angle, it follows that $\angle 1 \cong \angle 2$.

	Statement		Reason
1.	$\overline{EF} \parallel \overline{GH}$	1.	Given.
2.	$m\angle 1 = m\angle 3$	2.	If two parallel lines are cut by a transversal, corresponding angles are of equal measure.
3.	$\overline{AB} \parallel \overline{CD}$	3.	Given.
4.	$m\angle 2 = m\angle 3$	4.	If two parallel lines are cut by a transversal, corresponding angles are equal in measure.

5. $m\angle 1 = m\angle 2$ — 5. If two quantities are equal to the same quantity, they are equal to each other.

DRILL: LINES AND ANGLES

Intersection Lines

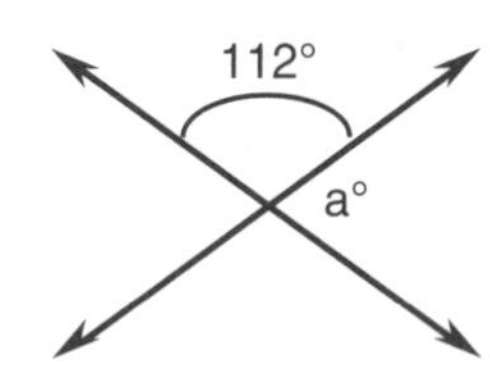

1. Find a.

(A) 38° (B) 68° (C) 78°
(D) 90° (E) 112°

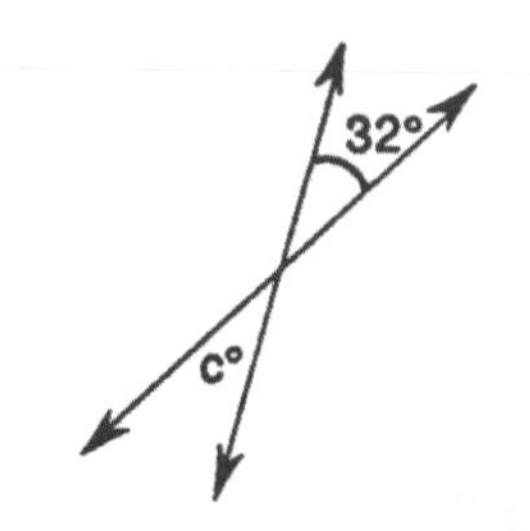

2. Find c.

(A) 32° (B) 48° (C) 58°
(D) 82° (E) 148°

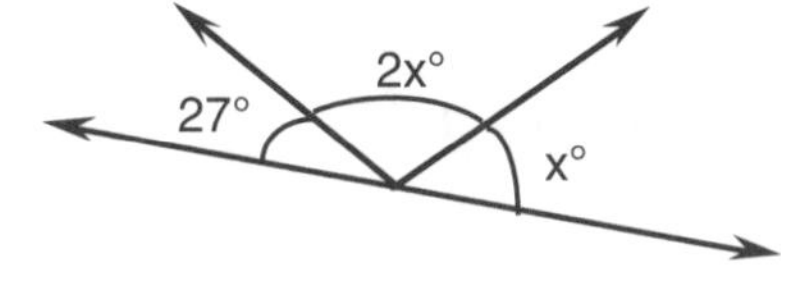

3. Determine x.

(A) 21° (B) 23° (C) 51°
(D) 102° (E) 153°

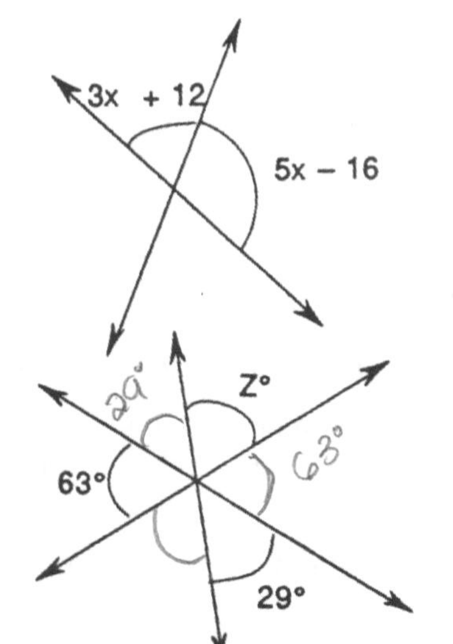

4. Find x.

(A) 8 (B) 11.75 (C) 21
(D) 23 (E) 32

5. Find z.

(A) 29° (B) 54° (C) 61°
(D) 88° (E) 92°

Perpendicular Lines

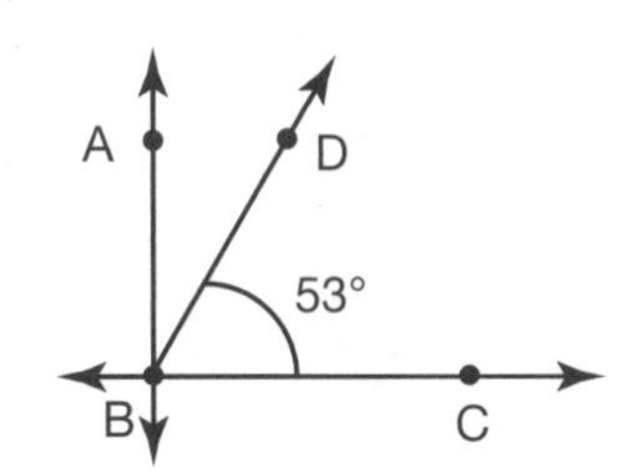

6. $\overline{BA} \perp \overline{BC}$ and $m\angle DBC = 53$. Find $m\angle ABD$.

(A) 27° (B) 33° (C) 37°
(D) 53° (E) 90°

7. $m\angle 1 = 90°$. Find $m\angle 2$.

(A) 80° (B) 90° (C) 100°

(D) 135° (E) 180°

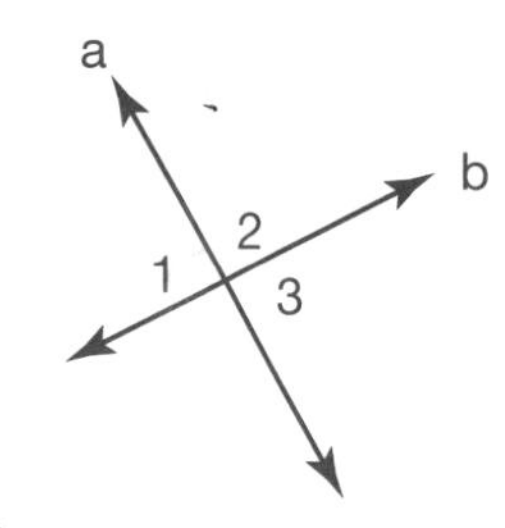

8. If $n \perp p$, which of the following statements is true?

(A) $\angle 1 \cong \angle 2$

(B) $\angle 4 \cong \angle 5$

(C) $m\angle 4 + m\angle 5 > m\angle 1 + m\angle 2$

(D) $m\angle 3 > m\angle 2$

(E) $m\angle 4 = 90°$

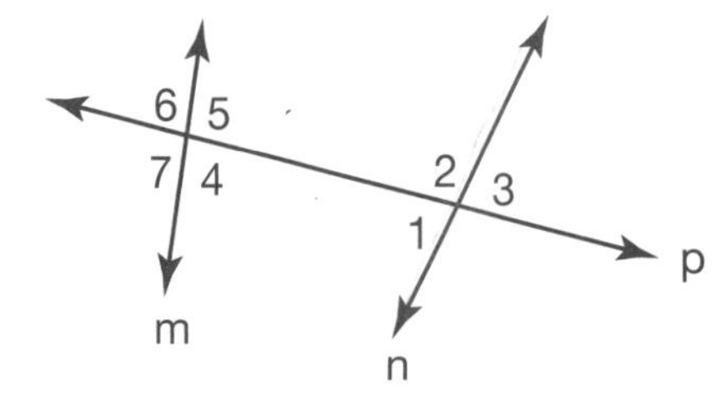

9. $\overline{CD} \perp \overline{EF}$. If $m\angle 1 = 2x$, $m\angle 2 = 30°$, and $m\angle 3 = x$, find x.

(A) 5° (B) 10° (C) 12°

(D) 20° (E) 25°

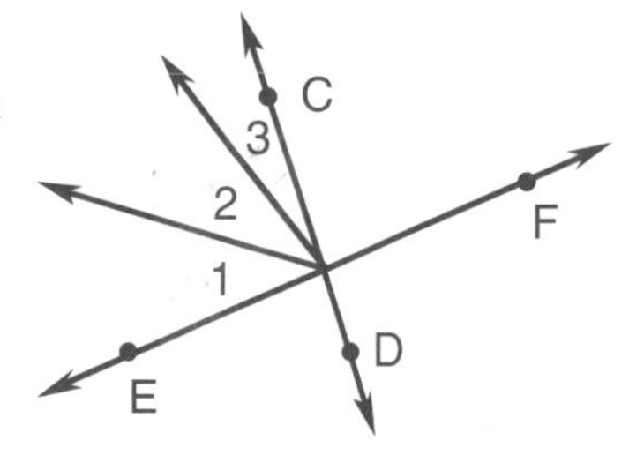

10. In the figure, $p \perp t$ and $q \perp t$. Which of the following statements is false?

(A) $\angle 1 \cong \angle 4$

(B) $\angle 2 \cong \angle 3$

(C) $m\angle 2 + m\angle 3 = m\angle 4 + m\angle 6$

(D) $m\angle 5 + m\angle 6 = 180°$

(E) $m\angle 2 > m\angle 5$

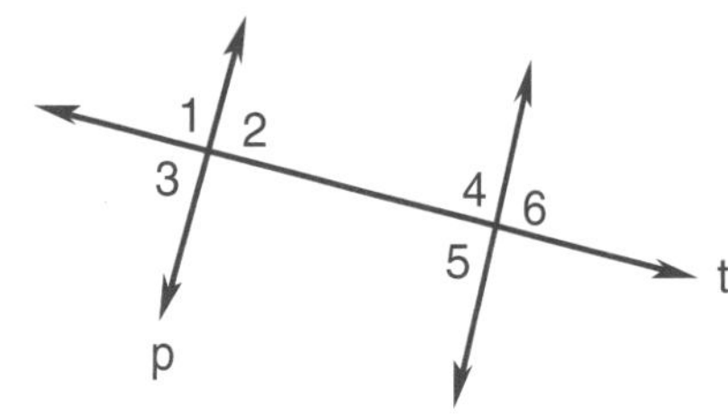

Parallel Lines

11. If $a \parallel b$, find z.

(A) 26° (B) 32° (C) 64°

(D) 86° (E) 116°

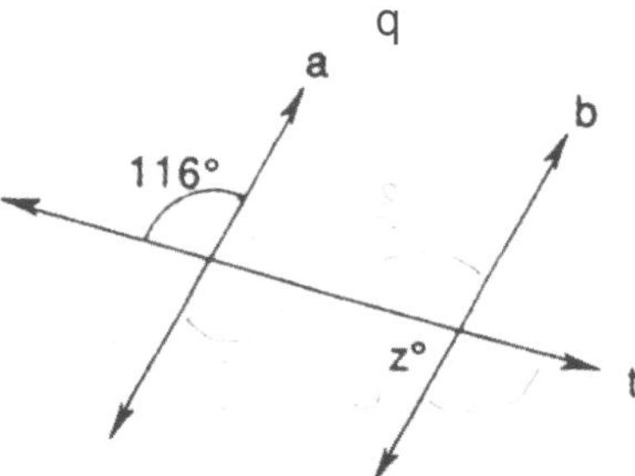

12. In the figure, $p \parallel q \parallel r$. Find $m\angle 7$.

(A) 27° (B) 33° (C) 47°

(D) 57° (E) 64°

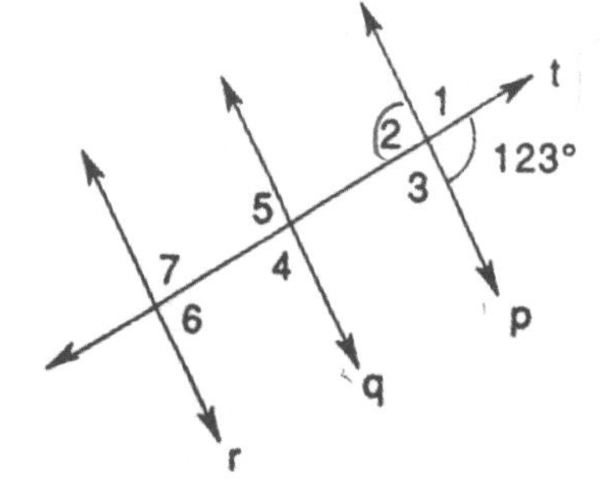

13. If $m \parallel n$, which of the following statements is false?

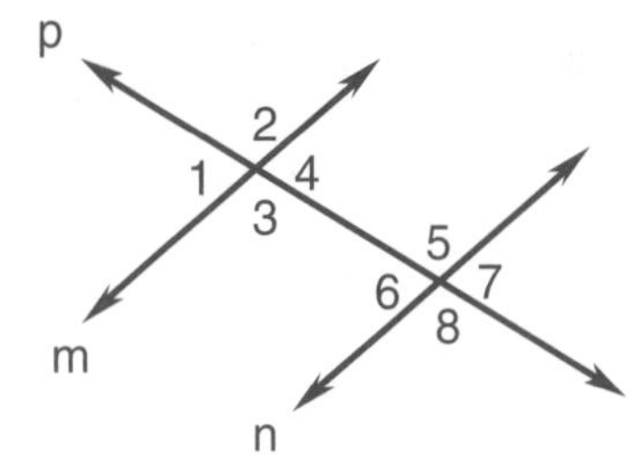

(A) $\angle 2 \cong \angle 5$

(B) $\angle 3 \cong \angle 6$

(C) $m\angle 4 + m\angle 5 = 180°$

(D) $\angle 2 \cong \angle 8$

(E) $m\angle 7 + m\angle 3 = 180°$

14. If $r \parallel s$, find $m\angle 2$.

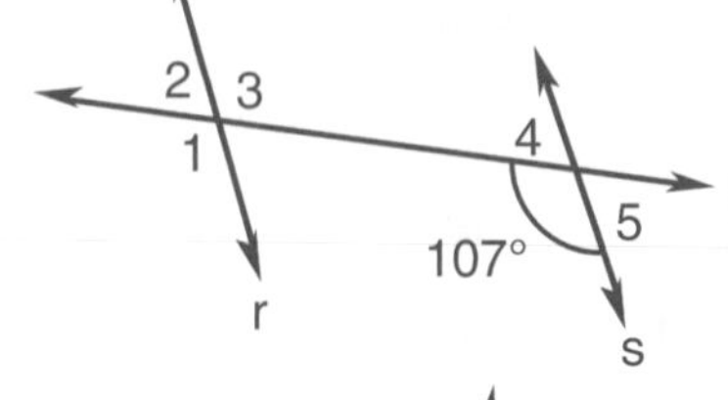

(A) 17° (B) 27° (C) 43°

(D) 67° (E) 73°

15. If $a \parallel b$ and $c \parallel d$, find $m\angle 5$.

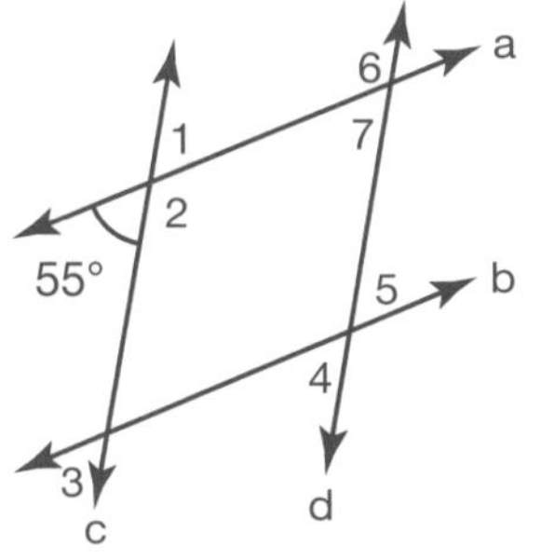

(A) 55° (B) 65° (C) 75°

(D) 95° (E) 125°

2. POLYGONS (CONVEX)

A **polygon** is a figure with the same number of sides as angles.

An **equilateral polygon** is a polygon all of whose sides are of equal length.

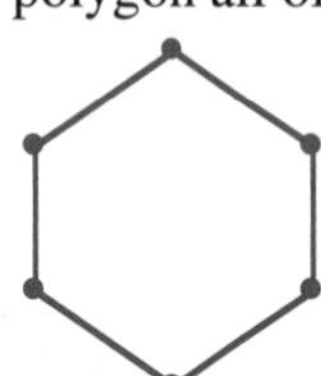

An **equiangular polygon** is a polygon all of whose angles are of equal measure.

A **regular polygon** is a polygon that is both equilateral and equiangular.

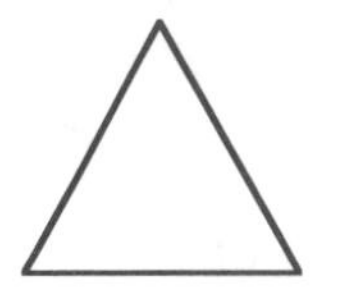 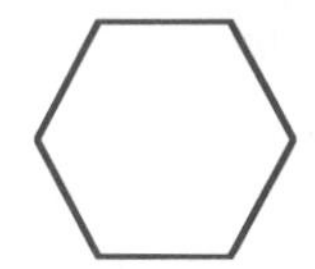 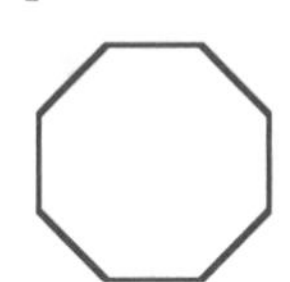

PROBLEM

Each interior angle of a regular polygon contains 120°. How many sides does the polygon have?

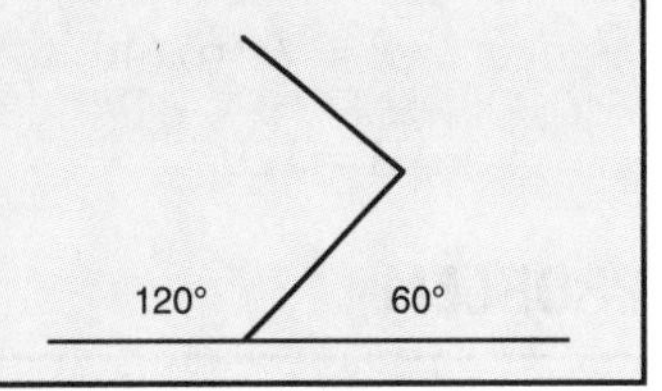

SOLUTION

At each vertex of a polygon, the exterior angle is supplementary to the interior angle, as shown in the diagram.

Since we are told that the interior angles measure 120 degrees, we can deduce that the exterior angle measures 60°.

Each exterior angle of a regular polygon of n sides measure $360°/_n$ degrees. We know that each exterior angle measures 60°, and, therefore, by setting $360°/_n$ equal to 60°, we can determine the number of sides in the polygon. The calculation is as follows:

$$\frac{360°}{n} = 60°$$

$$60°n = 360°$$

$$n = 6.$$

Therefore, the regular polygon, with interior angles of 120°, has 6 sides and is called a hexagon.

The area of a regular polygon can be determined by using the **apothem** and **radius** of the polygon. The apothem (a) of a regular polygon is the segment from the center of the polygon perpendicular to a side of the polygon. The radius (r) of a regular polygon is the segment joining any vertex of a regular polygon with the center of that polygon.

(1) All radii of a regular polygon are congruent.

(2) The radius of a regular polygon is congruent to a side.

(3) All apothems of a regular polygon are congruent.

The **area** of a regular polygon equals one-half the product of the length of the apothem and the perimeter.

$$\text{Area} = \tfrac{1}{2}\, a \times p$$

PROBLEM

Find area of the regular pentagon whose radius is 8 and whose apothem is 6.

SOLUTION

If the radius is 8, the length of a side is also 8. Therefore, the perimeter of the pentagon is 40.

$A = {}^1/_2\, a \times p$

$A = {}^1/_2\, (6)\,(40)$

$A = 120.$

PROBLEM

Find the area of a regular hexagon if one side has length 6.

SOLUTION

Since the length of a side equals 6, the radius also equals 6 and the perimeter equals 36. The base of the right triangle, formed by the radius and apothem, is half the length of a side, or 3. Using the Pythagorean theorem, you can find the length of the apothem.

$a^2 + b^2 = c^2$

$a^2 + (3)^2 = (6)^2$

$a^2 = 36 - 9$

$a^2 = 27$

$a = 3\sqrt{3}$

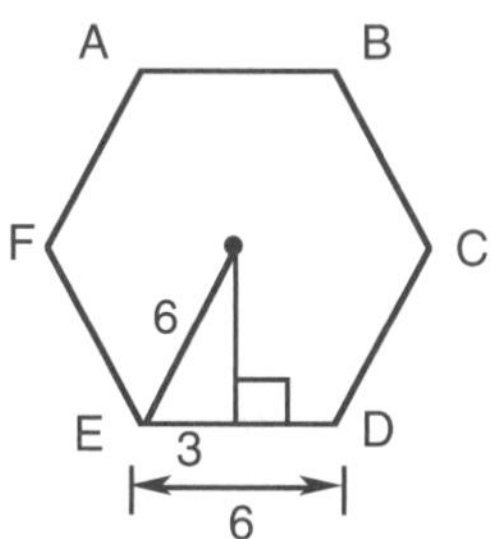

The apothem equals $3\sqrt{3}$. Therefore, the area of the hexagon

$= {}^1/_2\, a \times p$

$= {}^1/_2 (3\sqrt{3})\,(36)$

$= 54\sqrt{3}$

DRILL: REGULAR POLYGONS

1. Find the measure of an interior angle of a regular pentagon.

(A) 55 (B) 72 (C) 90 (D) 108 (E) 540

2. Find the measure of an exterior angle of a regular octagon.

(A) 40 (B) 45 (C) 135 (D) 540 (E) 1080

3. Find the sum of the measures of the exterior angles of a regular triangle.

(A) 90 (B) 115 (C) 180 (D) 250 (E) 360

4. Find the area of a square with a perimeter of 12 cm.

(A) 9 cm^2 (B) 12 cm^2 (C) 48 cm^2 (D) 96 cm^2 (E) 144 cm^2

5. A regular triangle has sides of 24 mm. If the apothem is $4\sqrt{3}$ mm, find the area of the triangle.

(A) 72 mm^2 (B) $96\sqrt{3}$ mm^2 (C) 144 mm^2

(D) $144\sqrt{3}$ mm^2 (E) 576 mm^2

6. Find the area of a regular hexagon with sides of 4 cm.

(A) $12\sqrt{3}$ cm^2 (B) 24 cm^2 (C) $24\sqrt{3}$ cm^2

(D) 48 cm^2 (E) $48\sqrt{3}$ cm^2

7. Find the area of a regular decagon with sides of length 6 cm and an apothem of length 9.2 cm.

(A) 55.2 cm^2 (B) 60 cm^2 (C) 138 cm^2

(D) 138.3 cm^2 (E) 276 cm^2

8. The perimeter of a regular heptagon (7-gon) is 36.4 cm. Find the length of each side.

(A) 4.8 cm (B) 5.2 cm (C) 6.7 cm (D) 7 cm (E) 10.4 cm

9. The apothem of a regular quadrilateral is 4 in. Find the perimeter.

(A) 12 in (B) 16 in (C) 24 in (D) 32 in (E) 64 in

10. A regular triangle has a perimeter of 18 cm; a regular pentagon has a perimeter of 30 cm; a regular hexagon has a perimeter of 33 cm. Which figure (or figures) have sides with the longest measure?

(A) regular triangle

(B) regular triangle and regular pentagon

(C) regular pentagon

(D) regular pentagon and regular hexagon

(E) regular hexagon

3. TRIANGLES

A closed three-sided geometric figure is called a **triangle**. The points of the intersection of the sides of a triangle are called the **vertices** of the triangle.

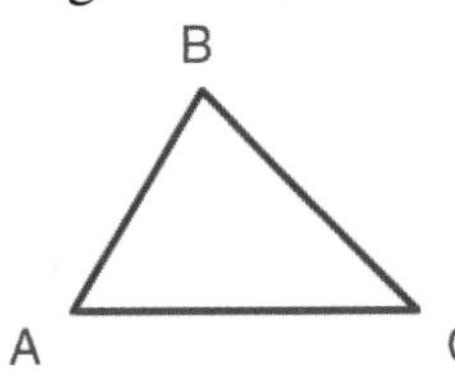

The **perimeter** of a triangle is the sum of the measures of the sides of the triangle.

A triangle with no equal sides is called a **scalene** triangle.

A triangle having at least two equal sides is called an **isosceles** triangle. The third side is called the **base** of the triangle.

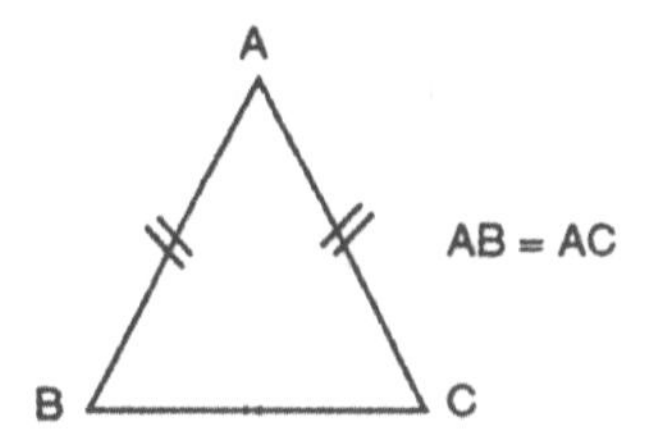

A side of a triangle is a line segment whose endpoints are the vertices of two angles of the triangle.

An interior angle of a triangle is an angle formed by two sides and includes the third side within its collection of points.

An equilateral triangle is a triangle having three equal sides. $AB = AC = BC$

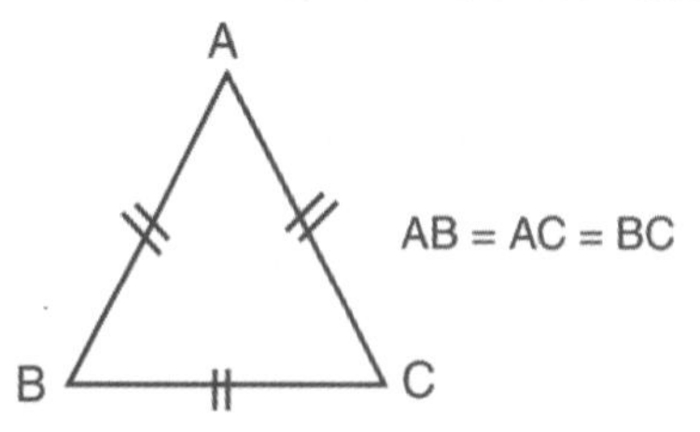

A triangle with an obtuse angle (greater than 90°) is called an **obtuse triangle**.

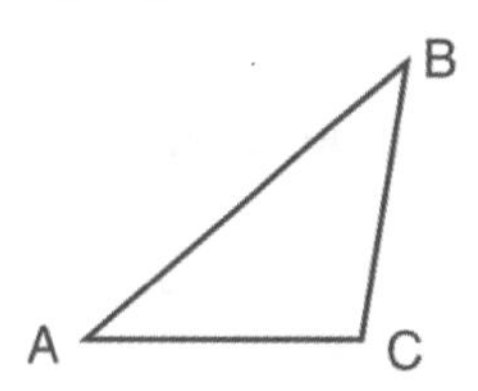

An **acute triangle** is a triangle with three acute angles (less than 90°).

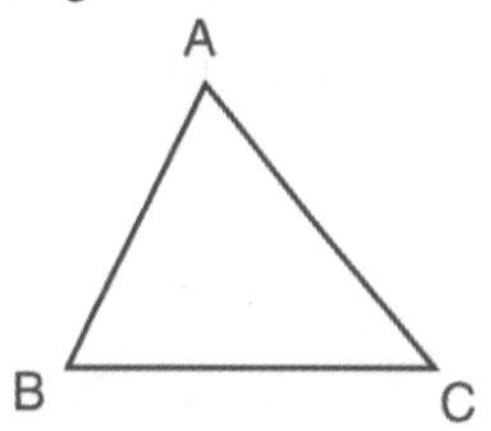

A triangle with a right angle is called a **right triangle**. The side opposite the right angle in a right triangle is called the hypotenuse of the right triangle. The other two sides are called arms or legs of the right triangle.

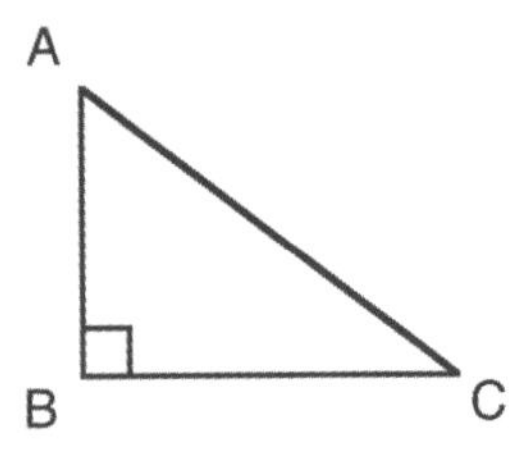

An **altitude** of a triangle is a line segment from a vertex of the triangle perpendicular to the opposite side.

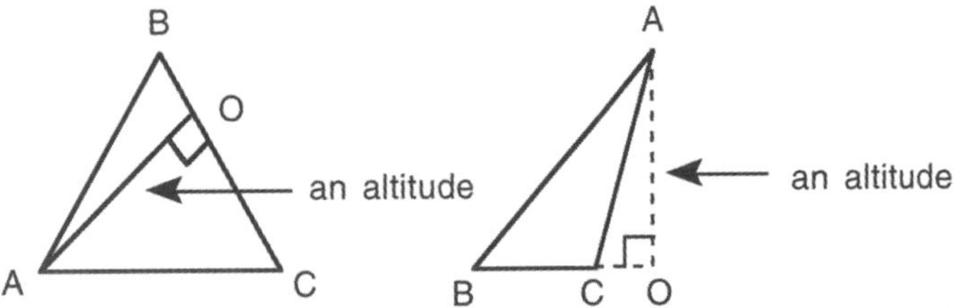

A line segment connecting a vertex of a triangle and the midpoint of the opposite side is called a **median** of the triangle.

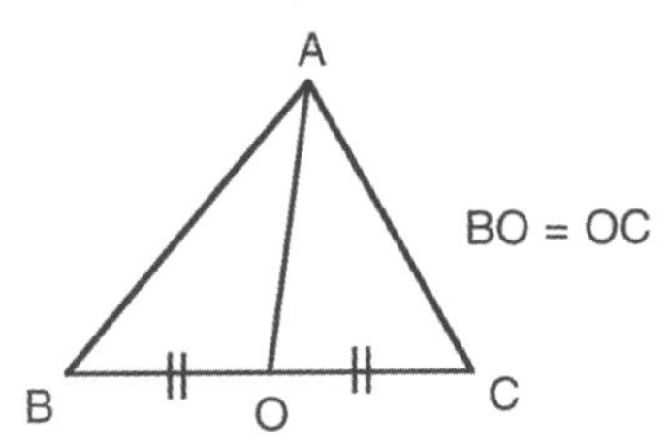

A line that bisects and is perpendicular to a side of a triangle is called a **perpendicular bisector** of that side.

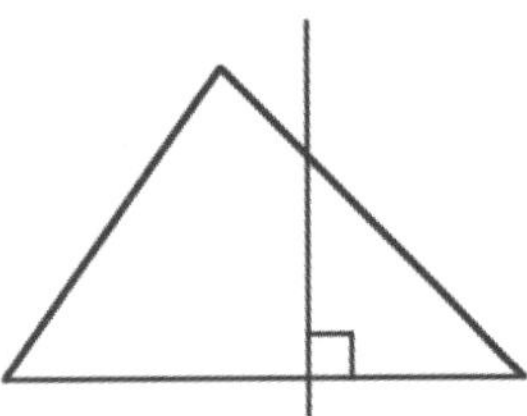

An **angle bisector** of a triangle is a line that bisects an angle and extends to the opposite side of the triangle.

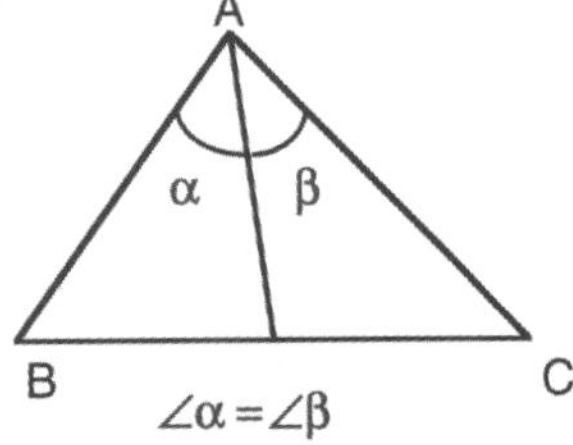

The line segment that joins the midpoints of two sides of a triangle is called a midline of the triangle.

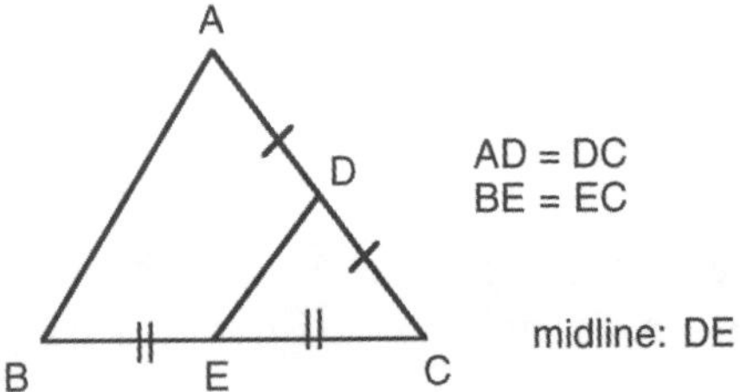

An exterior angle of a triangle is an angle formed outside a triangle by one side of the triangle and the extension of an adjacent side.

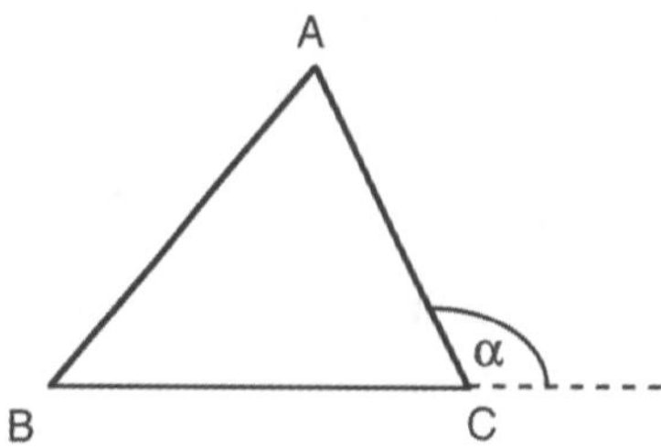

A triangle whose three interior angles have equal measure is said to be equiangular.

Three or more lines (or rays or segments) are concurrent if there exists one point common to all of them, that is, if they all intersect at the same point.

PROBLEM

Determine which of the triangles listed below are right, acute, and obtuse triangles.

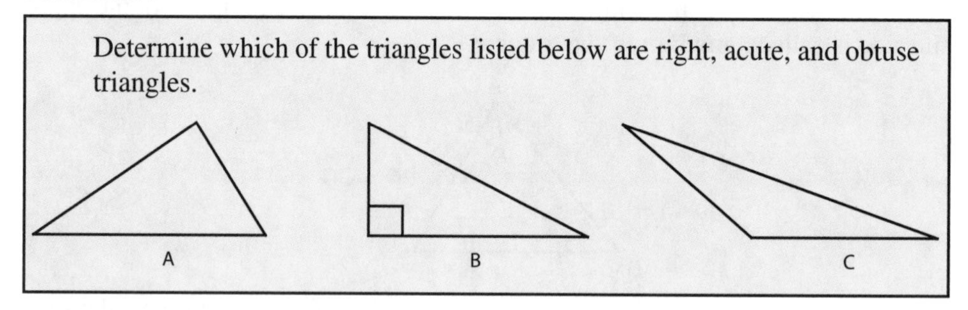

SOLUTION

Triangle *A* is an acute triangle. An acute triangle is a triangle comprised of three acute angles. Triangle *B* is a right triangle. A right triangle is a triangle with a right angle. Triangle *C* is an obtuse triangle. An obtuse triangle is a triangle with one obtuse angle.

PROBLEM

Determine which of the triangles listed below are scalene, isosceles, and equilateral.

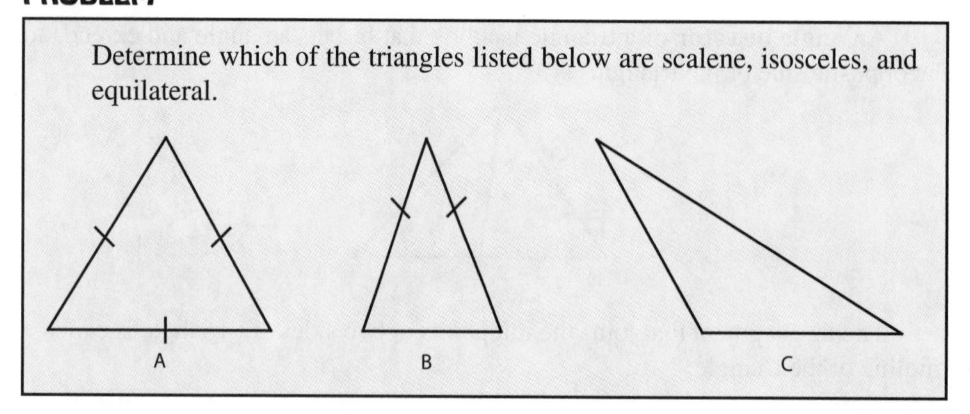

SOLUTION

Triangle *A* is an equilateral triangle. An equilateral triangle is a triangle with three congruent sides. Triangle *B* is an isosceles triangle. An isosceles triangle is a triangle with at least two congruent sides. Therefore, an equilateral triangle is also a type of isosceles triangle. Triangle C is a scalene triangle. A scalene triangle is a triangle with no congruent sides.

PROBLEM

Determine if the sides of the triangle below are equal.

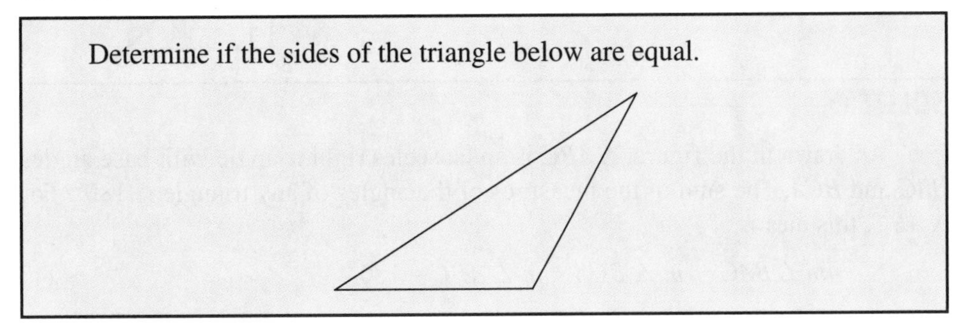

SOLUTION

The triangle is a scalene triangle. Since a scalene triangle does not have any congruent sides, none of the sides of the triangle are equal.

PROBLEM

The measure of the vertex angle of an isosceles triangle exceeds the measurement of each base angle by 30°. Find the value of each angle of the triangle.

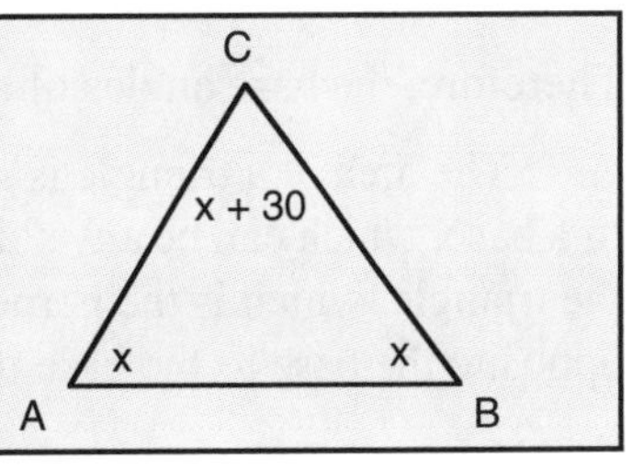

SOLUTION

We know that the sum of the values of the angles of a triangle is 180°. In an isosceles triangle, the angles opposite the congruent sides (the base angles) are, themselves, congruent and of equal value.

Therefore,

(1) Let x = the measure of each base angle.

(2) Then $x + 30$ = the measure of the vertex angle.

We can solve for x algebraically because the sum of all the measures will be 180°.

$$x + x + (x + 30) = 180$$
$$3x + 30 = 180$$
$$3x = 150$$
$$x = 50$$

Therefore, the base angles each measure 50°, and the vertex angle measures 80°.

PROBLEM

Prove that the base angles of an isosceles right triangle measure 45°.

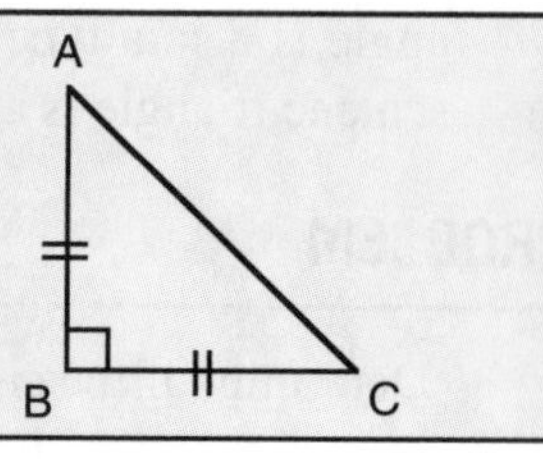

SOLUTION

As drawn in the figure, ΔABC is an isosceles right triangle with base angles BAC and BCA. The sum of the measures of the angles of any triangle is 180°. For ΔABC, this means

$$m\angle BAC + m\angle BCA + m\angle ABC = 180° \qquad (1)$$

But $m\angle ABC = 90°$ because ΔABC is a right triangle. Furthermore, $m\angle BCA = m\angle BAC$, since the base angles of an isosceles triangle are congruent. Using these facts in equation (1)

$$m\angle BAC + m\angle BCA + 90° = 180°$$

or $$2m\angle BAC = 2m\angle BCA = 90°$$

or $$m\angle BAC = m\angle BCA = 45°.$$

Therefore, the base angles of an isosceles right triangle measure 45°.

The area of a triangle is given by the formula $A = {}^1/_2\, bh$, where b is the length of a base, which can be any side of the triangle and h is the corresponding height of the triangle, which is the perpendicular line segment that is drawn from the vertex opposite the base to the base itself.

$$A = {}^1/_2\, bh$$

$$A = {}^1/_2\,(10)\,(3)$$

$$A = 15$$

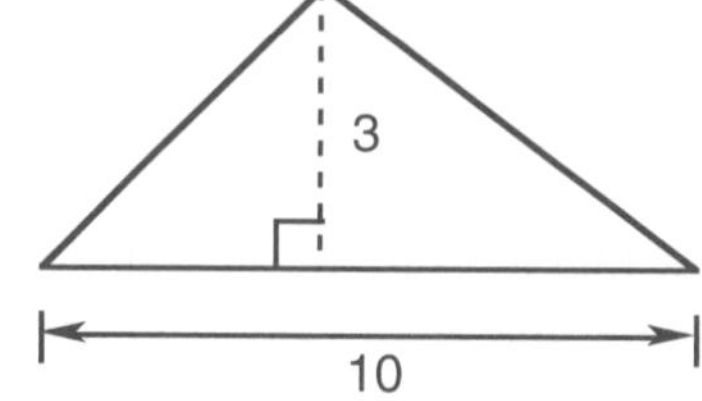

The area of a right triangle is found by taking ${}^1/_2$ the product of the lengths of its two arms.

$$A = {}^1/_2\,(5)\,(12)$$

$$A = 30$$

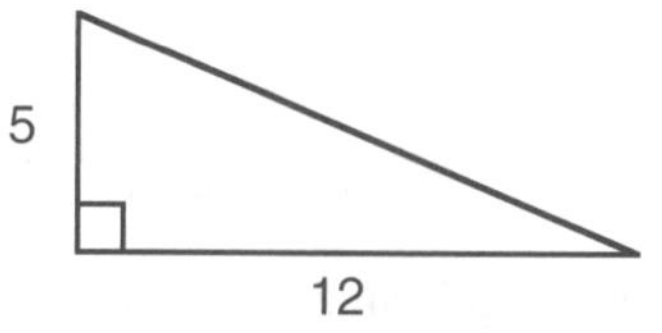

PROBLEM

Find the cosine of the marked angle in the figure below.

SOLUTION

Step 1 is to write the formula for determining the cosine of an angle.

cosine = length of adjacent side/length of hypotenuse

Step 2 is to substitute the values into the formula.

$$\text{cosine} = \frac{2}{8}$$

Step 3 is to calculate the cosine.

cosine = 0.25

The correct answer is 0.25.

PROBLEM

If the sine of $\angle ABC$ in the figure below is $^1/_2$, what is the length of side *AC*?

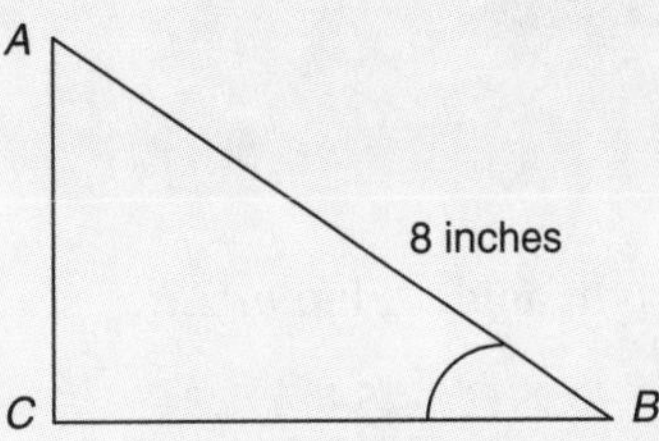

SOLUTION

Step 1 is to write the formula for the sine of an angle.

sine = length of opposite side/length of hypotenuse

Step 2 is to substitute the known values. From the figure it is known that the hypotenuse is 8 inches. It is also known that the sine is $^1/_2$. The length of the opposite side is x.

$$\frac{1}{2} = \frac{x}{8}$$

Step 3 is to solve the equation.

$x = 4$

The correct answer is that side *AC* is 4 inches long.

PROBLEM

If the cosine of $\angle ABC$ in the figure below is 0.75, what is the length of side *BC*?

A

12 meters

C B

SOLUTION

Step 1 is to write the formula for the cosine of an angle.

cosine = length of adjacent side/length of hypotenuse

Step 2 is to substitute the known values. From the figure it is known that the hypotenuse is 12 meters. It is also known that the cosine is 0.75. The length of the adjacent side is x.

$$0.75 = \frac{x}{12}$$

Step 3 is to solve the equation.

$$x = 9$$

The correct answer is that side BC is 9 meters long.

DRILL: TRIANGLES

Angle Measures

1. In ΔPQR, $\angle Q$ is a right angle. Find $m\angle R$.

(A) 27° (B) 33° (C) 54°

(D) 67° (E) 157°

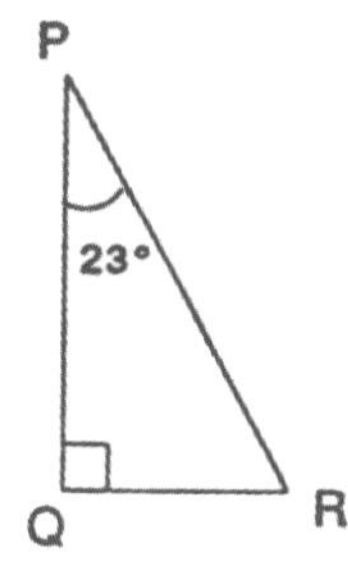

2. ΔMNO is isosceles. If the vertex angle, $\angle N$, has a measure of 96°, find the measure of $\angle M$.

(A) 21° (B) 42° (C) 64°

(D) 84° (E) 96°

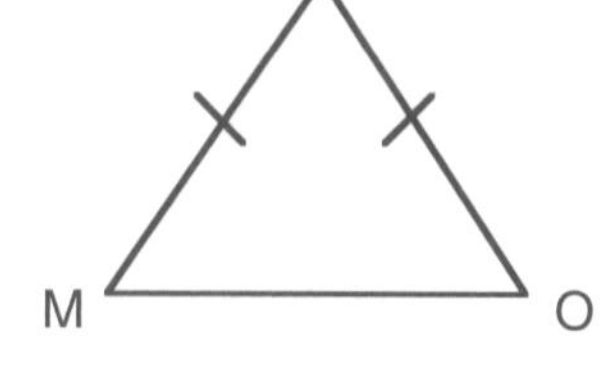

3. Find x.

(A) 15° (B) 25° (C) 30°

(D) 45° (E) 90°

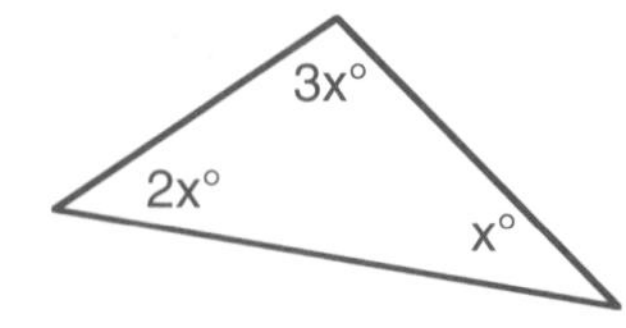

4. Find $m\angle 1$.

(A) 40 (B) 66 (C) 74

(D) 114 (E) 140

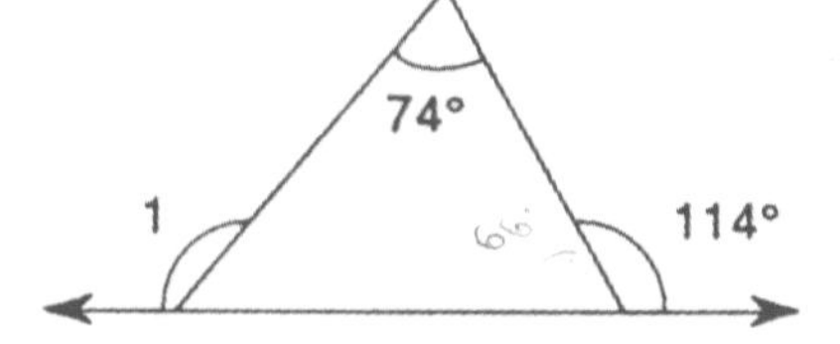

5. ΔABC is a right triangle with a right angle at B. ΔBDC is a right triangle with right angle $\angle BDC$. If $m\angle C = 36$, find $m\angle A$.

(A) 18 (B) 36 (C) 54

(D) 72 (E) 180

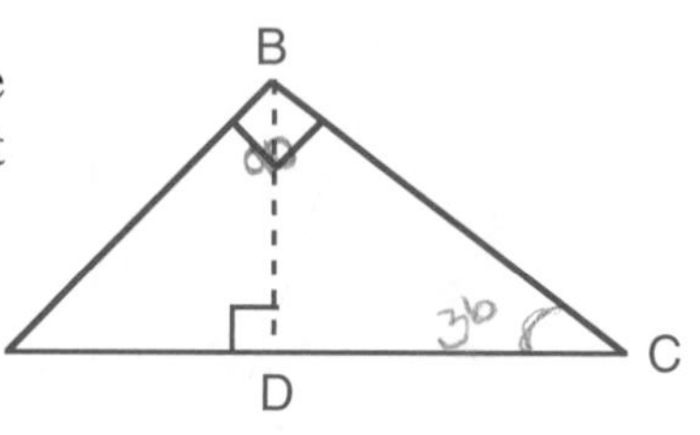

Similar Triangles

6. The two triangles shown are similar. Find b.

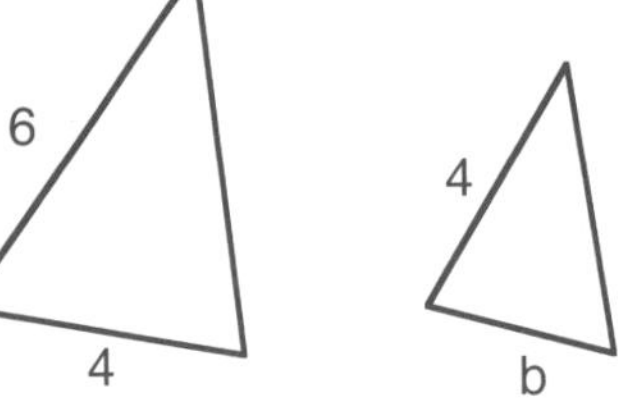

(A) 2 2/3 (B) 3 (C) 4
(D) 16 (E) 24

7. The two triangles shown are similar. Find $m\angle 1$.

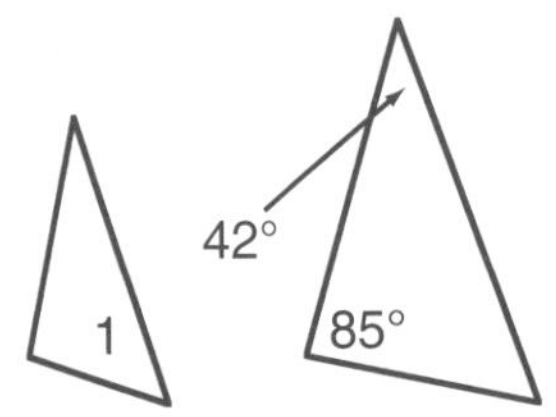

(A) 48 (B) 53 (C) 74
(D) 127 (E) 180

8. The two triangles shown are similar. Find a and b.

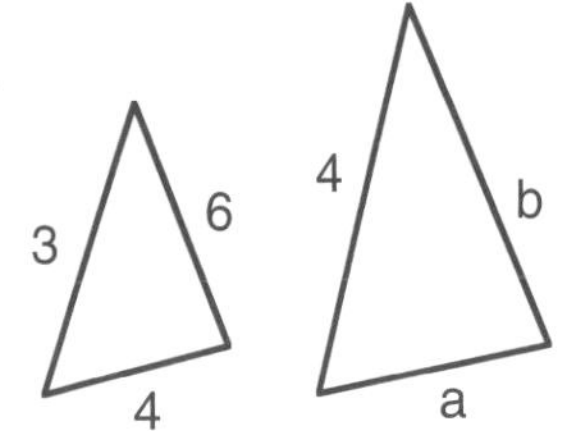

(A) 5 and 10 (B) 4 and 8
(C) 4 2/3 and 7 1/3 (D) 5 and 8
(E) 5 1/3 and 8

9. The perimeter of ΔLXR is 45 and the perimeter of ΔABC is 27. If $LX = 15$, find the length of AB.

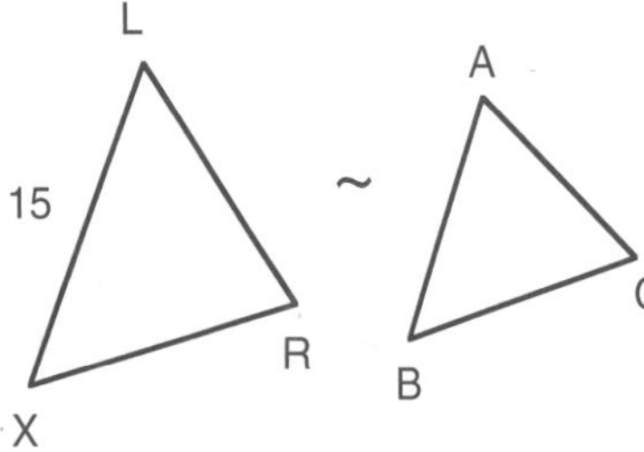

(A) 9 (B) 15 (C) 27
(D) 45 (E) 72

10. Find b.

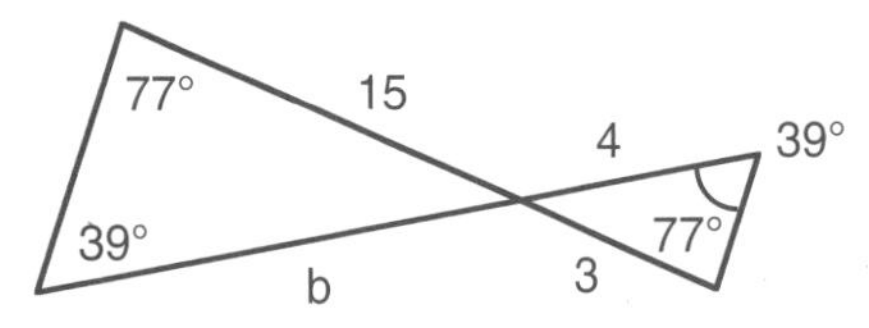

(A) 9 (B) 15 (C) 20
(D) 45 (E) 60

Area

11. Find the area of ΔMNO.

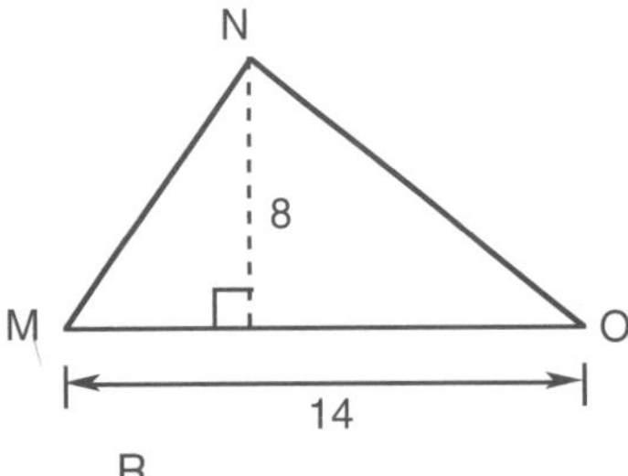

(A) 22 (B) 49 (C) 56
(D) 84 (E) 112

12. Find the area of ΔPQR.

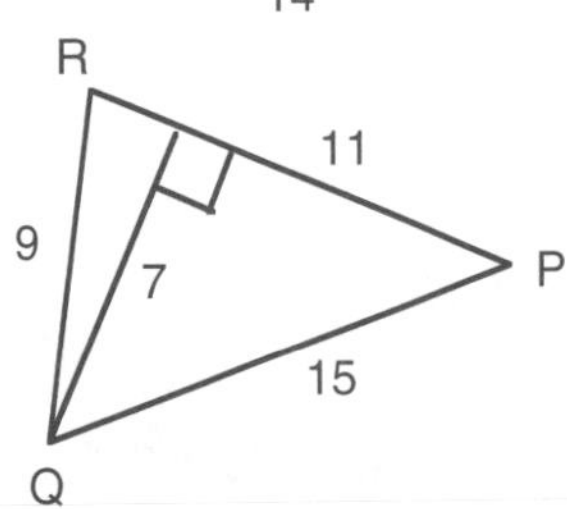

(A) 31.5 (B) 38.5 (C) 53
(D) 77 (E) 82.5

13. Find the area of Δ *STU*.

(A) $4\sqrt{2}$ (B) $8\sqrt{2}$ (C) $12\sqrt{2}$

(D) $16\sqrt{2}$ (E) $32\sqrt{2}$

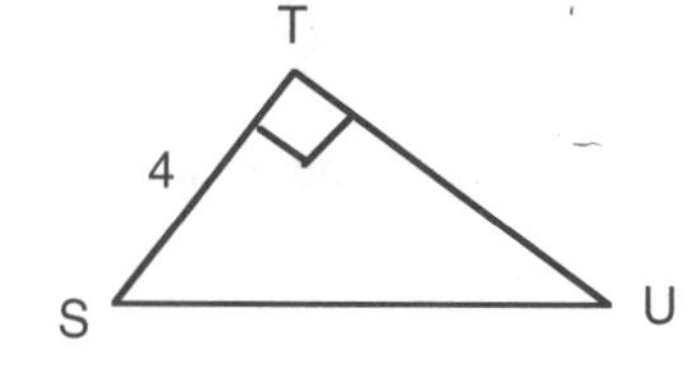

14. Find the area of Δ *ABC*.

(A) 54 cm^2 (B) 81 cm^2 (C) 108 cm^2

(D) 135 cm^2 (E) 180 cm^2

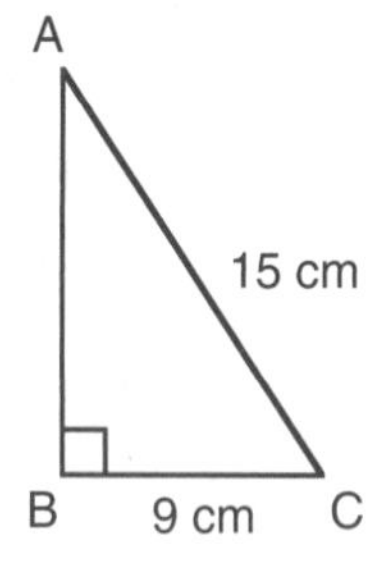

15. Find the area of Δ *XYZ*.

(A) 20 cm^2 (B) 50 cm^2 (C) $50\sqrt{2}$ cm^2

(D) 100 cm^2 (E) 200 cm^2

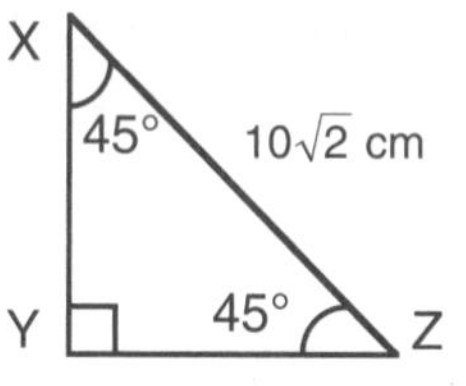

4. CHANGE IN LINEAR DIMENSION

PROBLEM

Given triangle *ABC*, if angle *A* was made smaller, and *AB* and *AC* stay the same length, what would happen to *CB*?

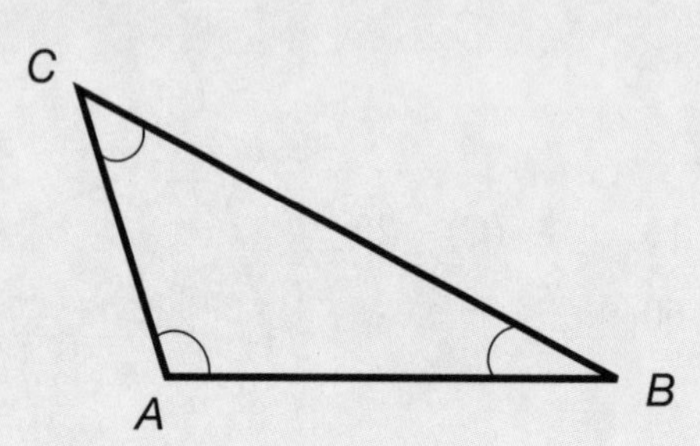

SOLUTION

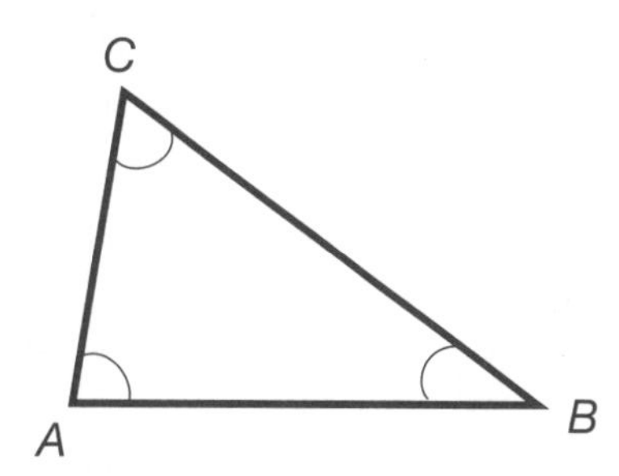

The length of line *CB* gets shorter.

PROBLEM

Given triangle ABC, if angle A was made larger and AB and AC stay the same length, what would happen to CB?

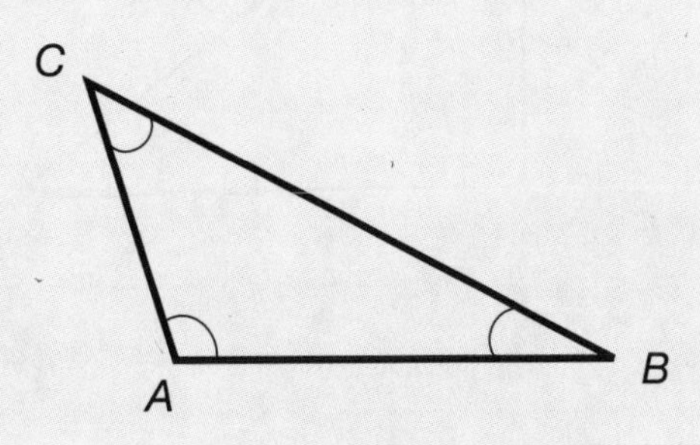

SOLUTION

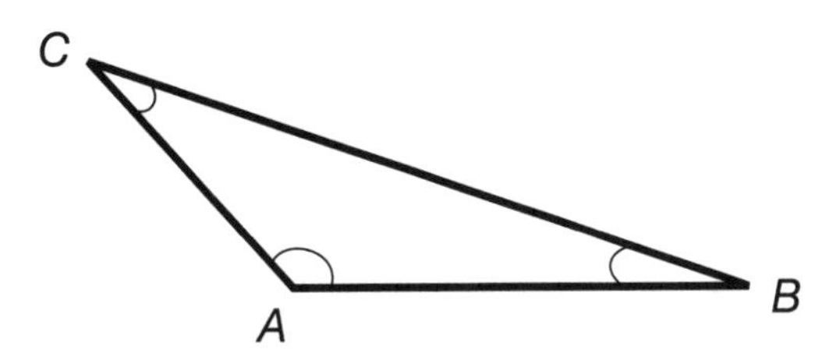

The length of line CB gets longer.

5. THE PYTHAGOREAN THEOREM

The Pythagorean Theorem tells you about right triangles. A right triangle is a triangle that has a right angle in it. A right angle measures 90 degrees. The Pythagorean Theorem tells you that $A^2 + B^2 = C^2$.

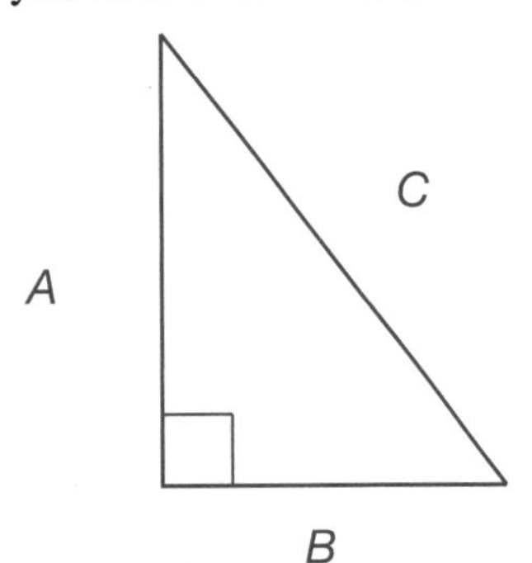

A and B are called legs, and C is called the hypotenuse. The hypotenuse is the side of the triangle that is opposite the right angle. The Pythagorean Theorem is useful in the following way: Given the length of any two sides of a right triangle, you can figure out the length of the third.

PROBLEM

In a right triangle, one leg is 3 inches and the other leg is 4 inches. What is the length of the hypotenuse?

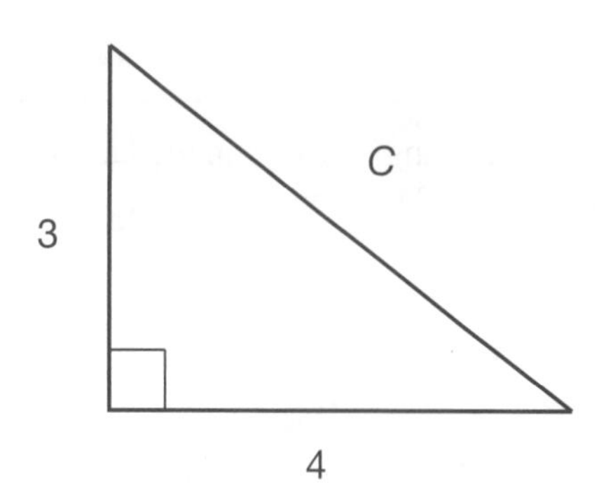

SOLUTION

$A^2 + B^2 = C^2$

$3^2 + 4^2 = C^2$

$9 + 16 = C^2$

$25 = C^2$

$C = \sqrt{25} = 5$ inches

PROBLEM

> If one leg of a right triangle is 6 inches, and the hypotenuse is 10 inches, what is the length of the other leg?

SOLUTION

First, write down the equation for the Pythagorean Theorem. Next, plug the information you are given into the equation. The hypotenuse *C* is equal to 10 and one of the legs *B* is equal to 6. Solve for *A*.

$A^2 + B^2 = C^2$

$A^2 + 6^2 = 10^2$

$A^2 + 36 = 100$

$A^2 = 100 - 36$

$A^2 = 64$

$A = \sqrt{64} = 8$ inches

PROBLEM

> What is the value of *A* in the triangle shown below?
>
> 12 13
>
> *A*

SOLUTION

In order to answer this question, you need to use the Pythagorean Theorem: $A^2 + B^2 = C^2$, where A and B are the legs and C is the hypotenuse. The hypotenuse is the side opposite the right angle. The present problem is asking you to find the value of the missing leg.

$A^2 + B^2 = C^2$

$A^2 + 12^2 = 13^2$

$A^2 + 144 = 169$

$A^2 = 169 - 144$

$A^2 = 25$

$A = \sqrt{25} = 5$

6. QUADRILATERALS

A **quadrilateral** is a polygon with four sides.

PARALLELOGRAMS

A **parallelogram** is a quadrilateral whose opposite sides are parallel.

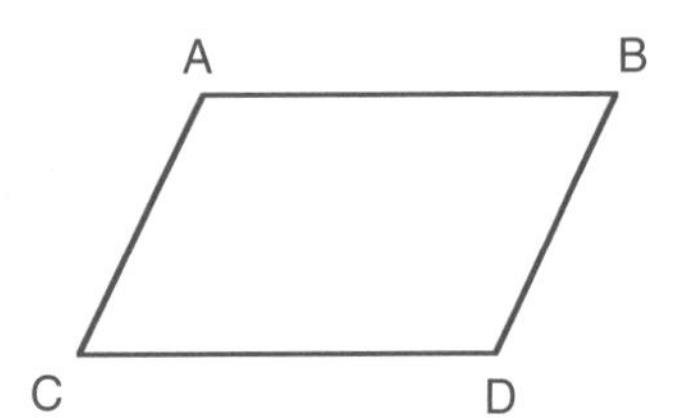

Two angles that have their vertices at the endpoints of the same side of a parallelogram are called **consecutive angles**.

The perpendicular segment connecting any point of a line containing one side of the parallelogram to the line containing the opposite side of the parallelogram is called the **altitude** of the parallelogram.

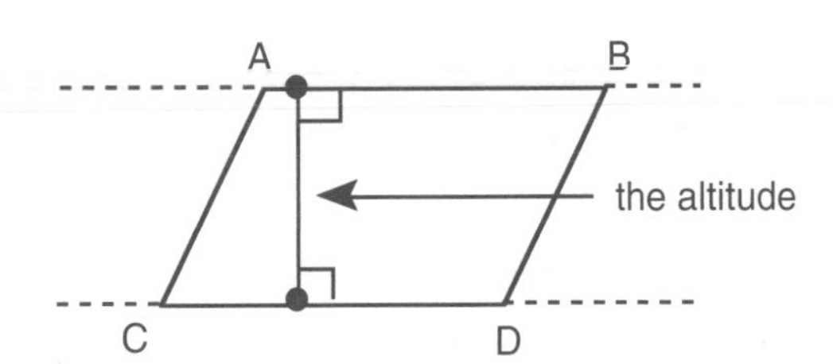

A diagonal of a polygon is a line segment joining any two non-consecutive vertices.

The area of a parallelogram is given by the formula $A = bh$ where b is the base and h is the height drawn perpendicular to that base. Note that the height equals the altitude of the parallelogram

$A = bh$

$A = (10)\ (3)$

$A = 30$

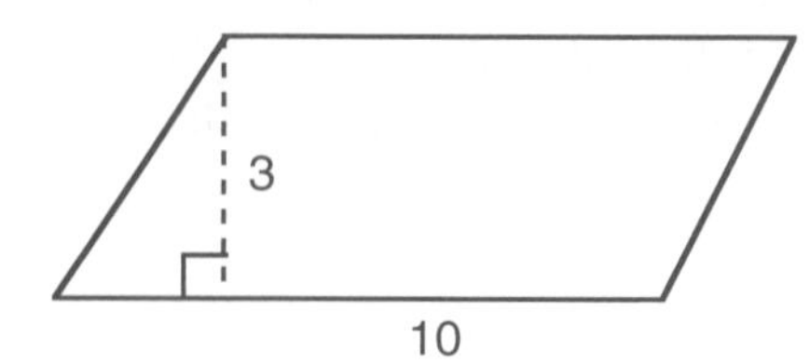

RECTANGLES

A rectangle is a parallelogram with right angles.

The diagonals of a rectangle are equal.

If the diagonals of a parallelogram are equal, the parallelogram is a rectangle.

If a quadrilateral has four right angles, then it is a rectangle.

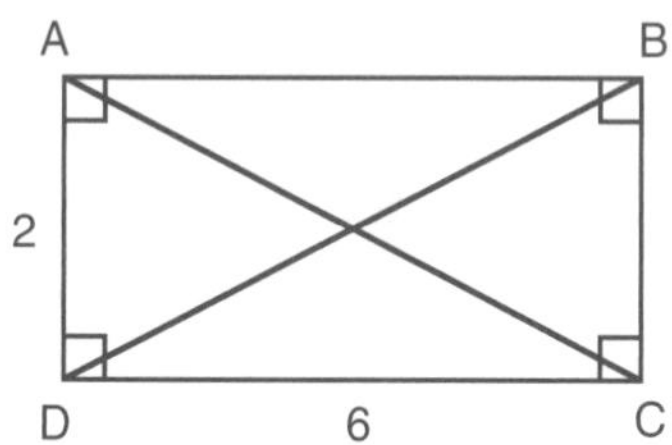

The area of a rectangle is given by the formula $A = lw$ where l is the length and w is the width.

$A = lw$

$A = (3)\ (10)$

$A = 30$

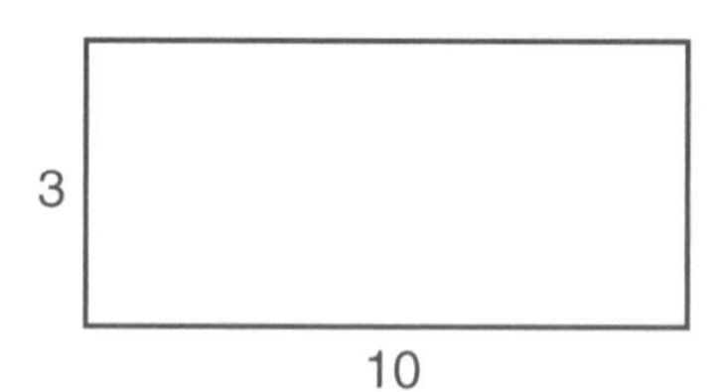

RHOMBI

A rhombus is a parallelogram with all sides equal.

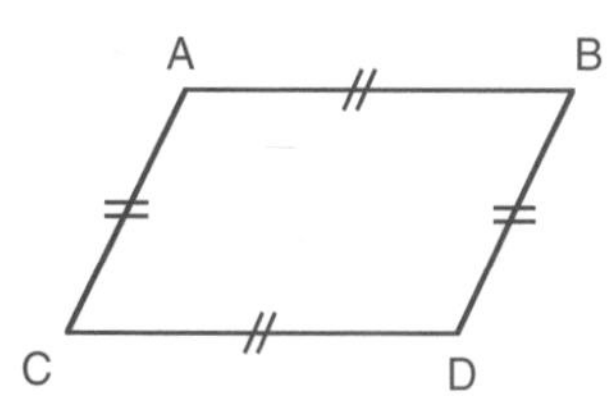

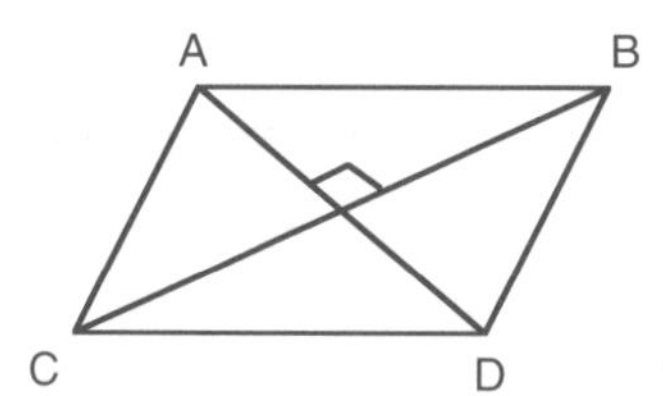

The diagonals of a rhombus are perpendicular to each other.

The diagonals of a rhombus bisect the angles of the rhombus.

If the diagonals of a parallelogram are perpendicular, the parallelogram is a rhombus.

If a quadrilateral has four equal sides, then it is a rhombus.

A parallelogram is a rhombus if either diagonal of the parallelogram bisects the angles of the vertices it joins.

SQUARES

A square is a rhombus with a right angle.

A square is an equilateral quadrilateral.

A square has all the properties of parallelograms, and rectangles.

A rhombus is a square if one of its interior angles is a right angle.

In a square, the measure of either diagonal can be calculated by multiplying the length of any side by the square root of 2.

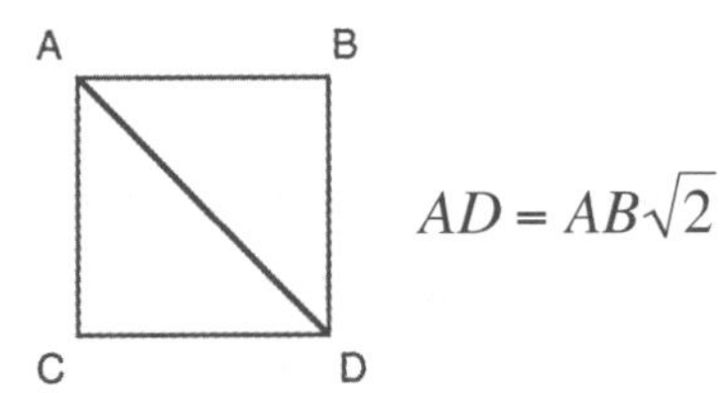

The area of a square is given by the formula $A = s^2$ where s is the side of the square. Since all sides of a square are equal, it does not matter which side is used.

$A = s^2$

$A = 6^2$

$A = 36$

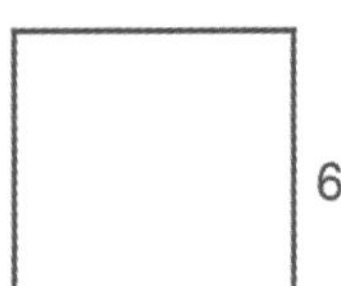

The area of a square can also be found by taking $^1/_2$ the product of the length of the diagonal squared.

$A = {}^1/_2\, d^2$

$A = {}^1/_2\, (8)^2$

$A = 32$

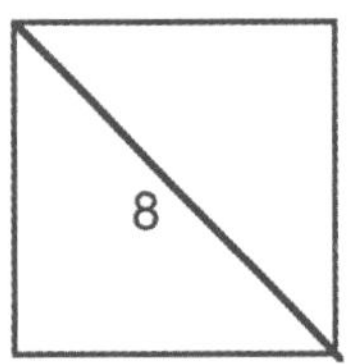

TRAPEZOIDS

A **trapezoid** is a quadrilateral with two and only two sides parallel. The parallel sides of a trapezoid are called **bases**.

The **median** of a trapezoid is the line joining the midpoints of the non-parallel sides.

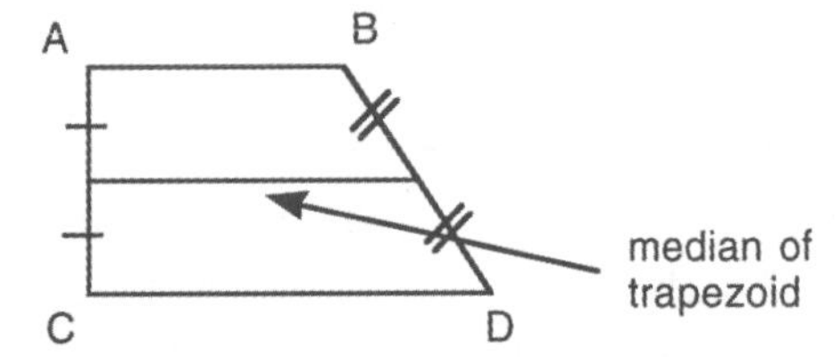

The perpendicular segment connecting any point in the line containing one base of the trapezoid to the line containing the other base is the **altitude** of the trapezoid.

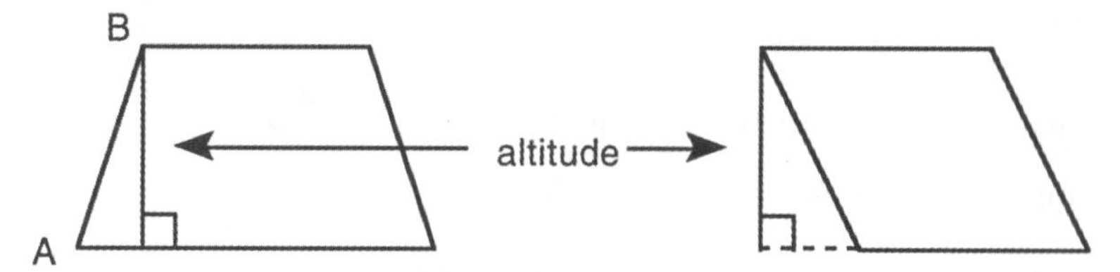

An **isosceles trapezoid** is a trapezoid whose non-parallel sides are equal. A pair of angles including only one of the parallel sides is called **a pair of base angles**.

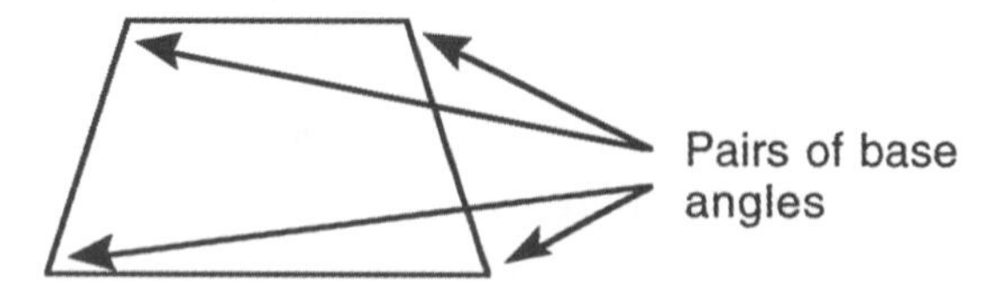

The median of a trapezoid is parallel to the bases and equal to one-half their sum.

The base angles of an isosceles trapezoid are equal.

The diagonals of an isosceles trapezoid are equal.

The opposite angles of an isosceles trapezoid are supplementary.

PROBLEM

Prove that all pairs of consecutive angles of a parallelogram are supplementary.

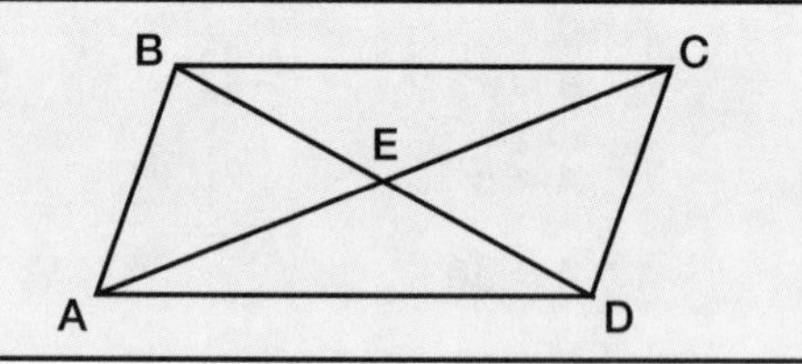

SOLUTION

We must prove that the pairs of angles $\angle BAD$ and $\angle ADC$, $\angle ADC$ and $\angle DCB$, $\angle DCB$ and $\angle CBA$, and $\angle CBA$ and $\angle BAD$ are supplementary. (This means that the sum of their measures is 180°.)

Because *ABCD* is a parallelogram, $\overline{AB} \parallel \overline{CD}$. Angles *BAD* and *ADC* are consecutive interior angles, as are $\angle CBA$ and $\angle DCB$. Since the consecutive interior angles formed by 2 parallel lines and a transversal are supplementary, $\angle BAD$ and $\angle ADC$ are supplementary, as are $\angle CBA$ and $\angle DCB$.

Similarly, $\overline{AD} \parallel \overline{BC}$. Angles *ADC* and *DCB* are consecutive interior angles, as are $\angle CBA$ and $\angle BAD$. Since the consecutive interior angles formed by 2 parallel lines and a transversal are supplementary, $\angle CBA$ and $\angle BAD$ are supplementary, as are $\angle ADC$ and $\angle DCB$.

PROBLEM

In the accompanying figure, ΔABC is given to be an isosceles right triangle with $\angle ABC$ a right angle and $AB \cong BC$. Line segment $\overline{BD}$, which bisects $\overline{CA}$, is extended to *E*, so that $\overline{BD} \cong \overline{DE}$. Prove *BAEC* is a square.

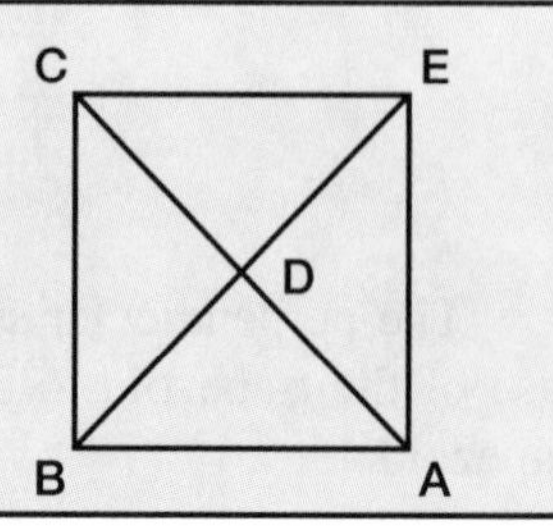

SOLUTION

A square is a rectangle in which two consecutive sides are congruent. This definition will provide the framework for the proof in this problem. We will prove that *BAEC* is a parallelogram that is specifically a rectangle with consecutive sides congruent, namely a square.

Statement	Reason
1. $\overline{BD} \cong \overline{DE}$	1. Given.
2. $\overline{AD} \cong \overline{DC}$	2. $\overline{BD}$ bisects $\overline{CA}$.
3. *BAEC* is a parallelogram	3. If diagonals of a quadrilateral bisect each other, then the quadrilateral is a parallelogram.
4. $\angle ABC$ is a right angle	4. Given.
5. *BAEC* is a rectangle	5. A parallelogram, one of whose angles is a right angle, is a rectangle.
6. $AB \cong BC$	6. Given.
7. *BAEC* is a square	7. If a rectangle has two congruent consecutive sides, then the rectangle is a square.

DRILL: QUADRILATERALS

PARALLELOGRAMS, RECTANGLES, RHOMBI, SQUARES, TRAPEZOIDS

1. In parallelogram *WXYZ*, $WX = 14$, $WZ = 6$, $ZY = 3x + 5$, and $XY = 2y - 4$. Find x and y.

(A) 3 and 5 (B) 4 and 5 (C) 4 and 6

(D) 6 and 10 (E) 6 and 14

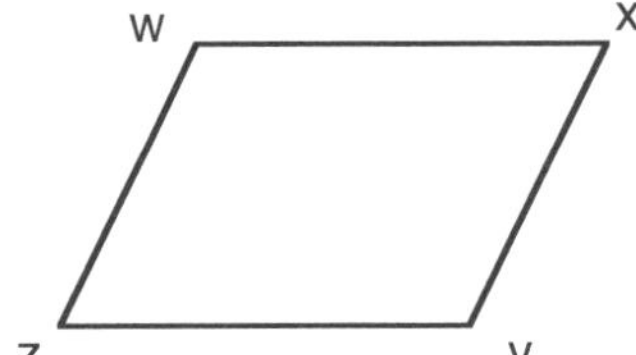
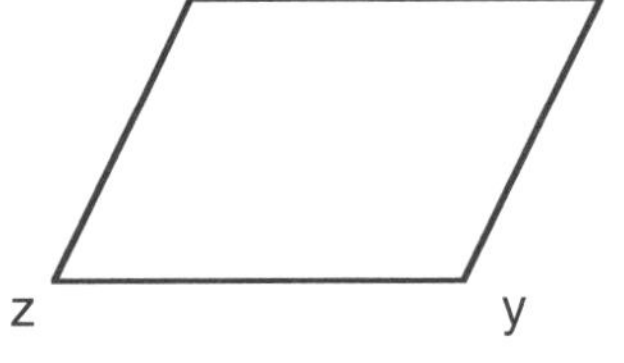

2. Quadrilateral *ABCD* is a parallelogram. If $m \angle B = 6x + 2$ and $m \angle D = 98$, find x.

(A) 12 (B) 16 (C) 16 2/3

(D) 18 (E) 20

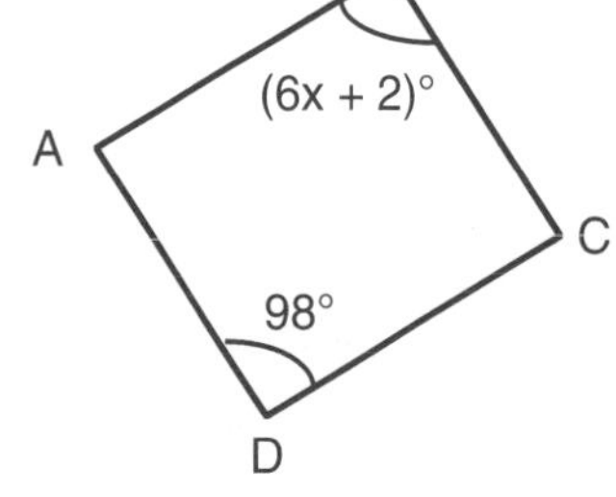

3. Find the area of parallelogram *STUV*.

(A) 56 (B) 90 (C) 108

(D) 162 (E) 180

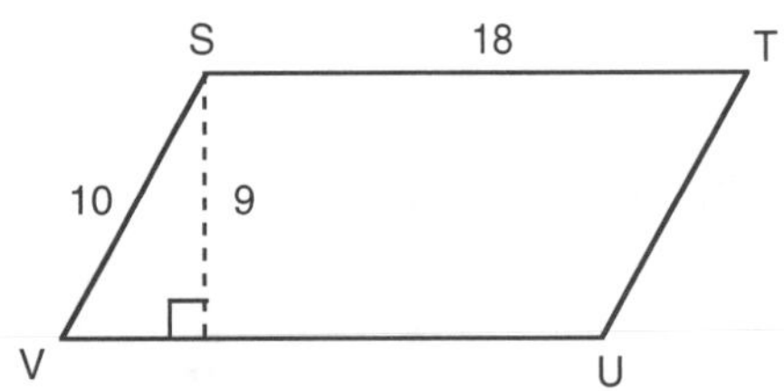

4. Find the area of parallelogram *MNOP*.

(A) 19 (B) 32 (C) $32\sqrt{3}$

(D) 44 (E) $44\sqrt{3}$

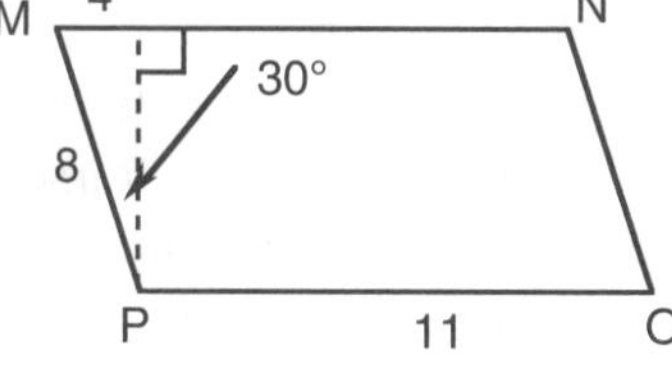

5. Find the perimeter of rectangle *PQRS,* if the area is 99 in².

(A) 31 in (B) 38 in (C) 40 in

(D) 44 in (E) 121 in

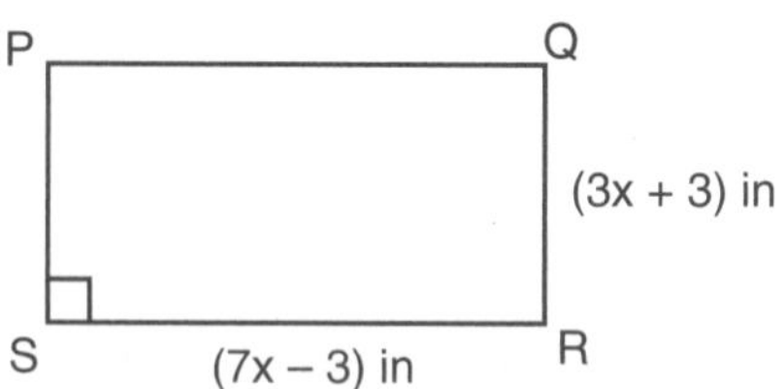

6. In rectangle *ABCD*, *AD* = 6 cm and *DC* = 8 cm. Find the length of the diagonal *AC*.

(A) 10 cm (B) 12 cm (C) 20 cm

(D) 28 cm (E) 48 cm

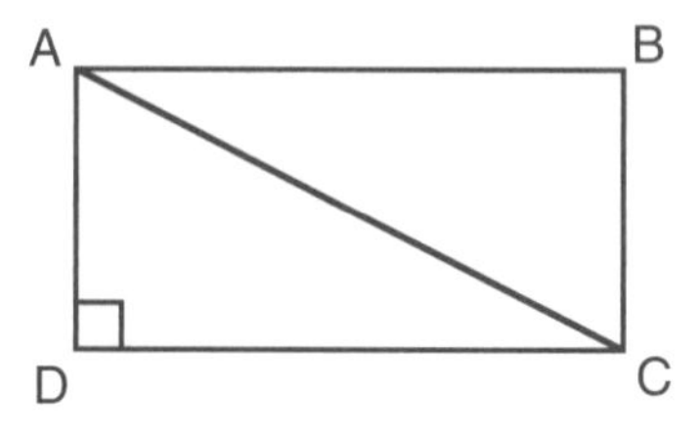

7. Find the area of rectangle *UVXY*.

(A) 17 cm² (B) 34 cm² (C) 35 cm²

(D) 70 cm² (E) 140 cm²

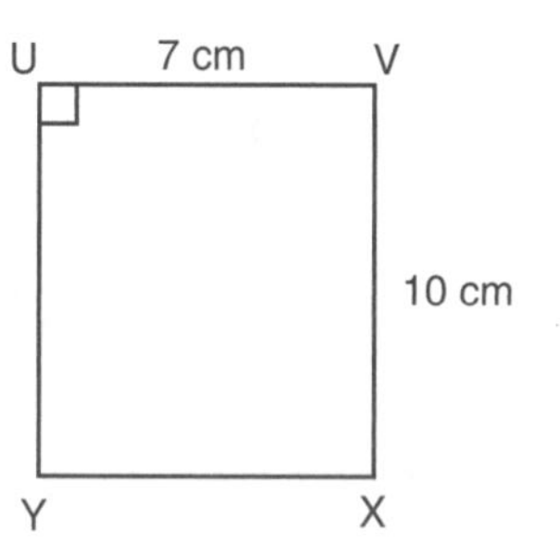

8. Find *x* in rectangle *BCDE* if the diagonal *EC* is 17 mm.

(A) 6.55 mm (B) 8 mm (C) 8.5 mm

(D) 17 mm (E) 34 mm

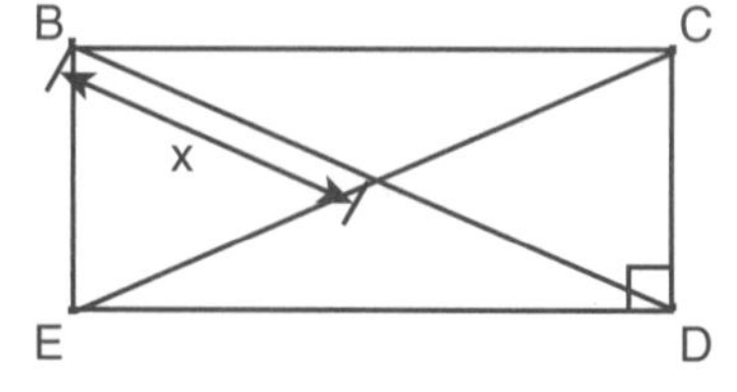

9. In rhombus *DEFG, DE* = 7 cm. Find the perimeter of the rhombus.

(A) 14 cm (B) 28 cm (C) 42 cm

(D) 49 cm (E) 56 cm

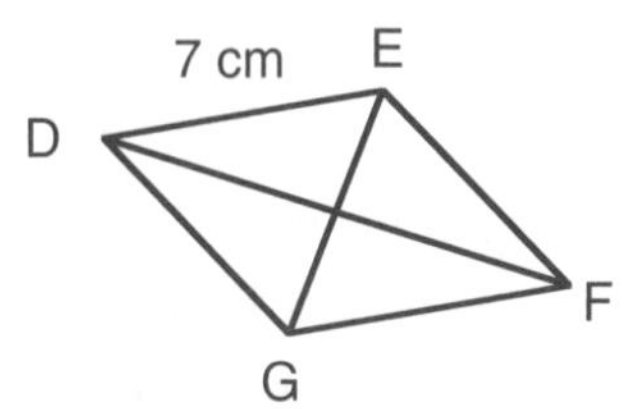

10. In rhombus *RHOM*, the diagonal $\overline{RO}$ is 8 cm and the diagonal $\overline{HM}$ is 12 cm. Find the area of the rhombus.

(A) 20 cm² (B) 40 cm² (C) 48 cm²

(D) 68 cm² (E) 96 cm²

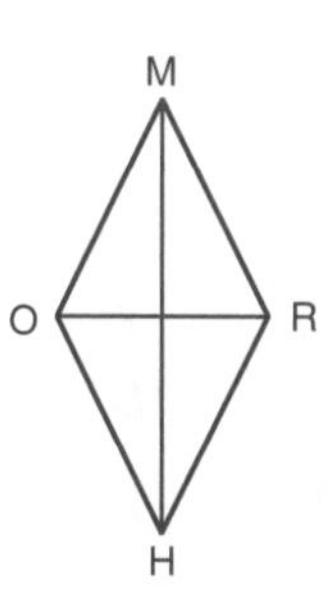

11. In rhombus $GHIJ$, $GI = 6$ cm and $HJ = 8$ cm. Find the length of GH.

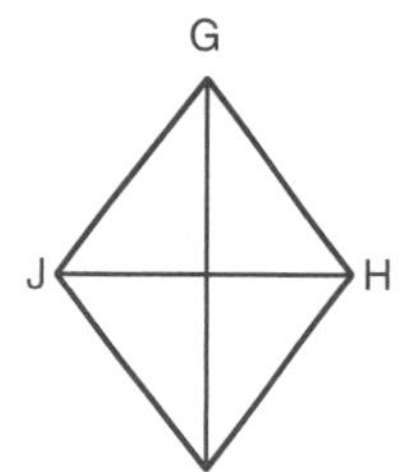

(A) 3 cm (B) 4 cm (C) 5 cm

(D) $4\sqrt{3}$ cm (E) 14 cm

12. In rhombus $CDEF$, CD is 13 mm and DX is 5 mm. Find the area of the rhombus.

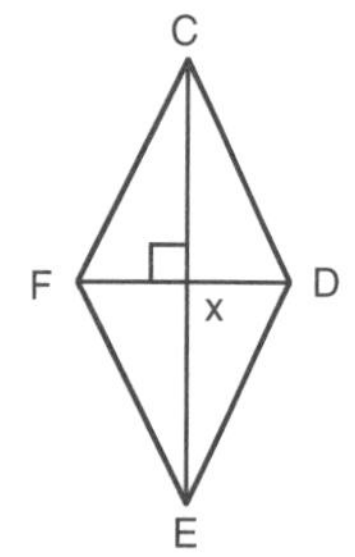

(A) 31 mm² (B) 60 mm² (C) 78 mm²

(D) 120 mm² (E) 260 mm²

13. Quadrilateral $ATUV$ is a square. If the perimeter of the square is 44 cm, find the length of $\overline{AT}$.

(A) 4 cm (B) 11 cm (C) 22 cm (D) 30 cm (E) 40 cm

14. The area of square $XYZW$ is 196 cm². Find the perimeter of the square.

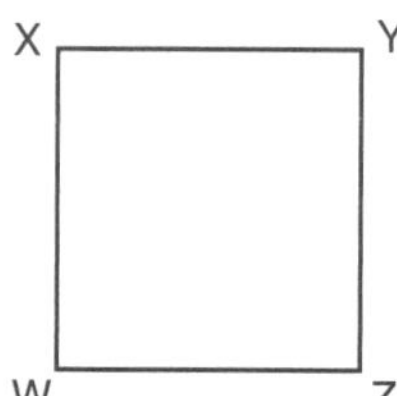
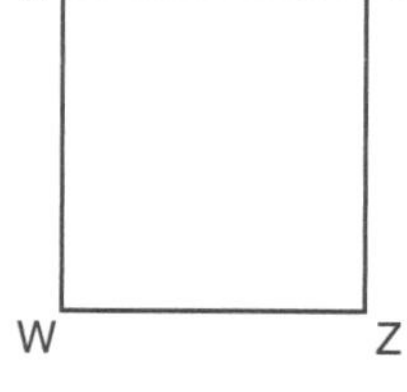

(A) 28 cm (B) 42 cm (C) 56 cm

(D) 98 cm (E) 196 cm.

15. In square $MNOP$, MN is 6 cm. Find the length of diagonal $\overline{MO}$.

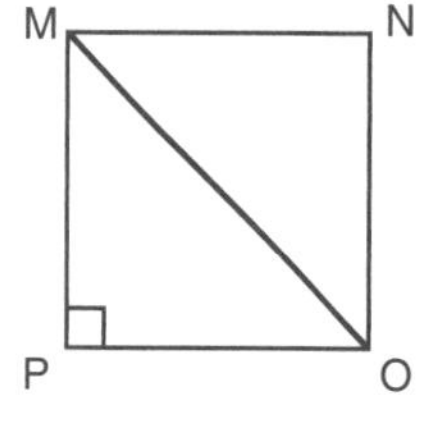

(A) 6 cm (B) $6\sqrt{2}$ cm (C) $6\sqrt{3}$ cm

(D) $6\sqrt{6}$ cm (E) 12 cm

16. In square $ABCD$, $AB = 3$ cm. Find the area of the square.

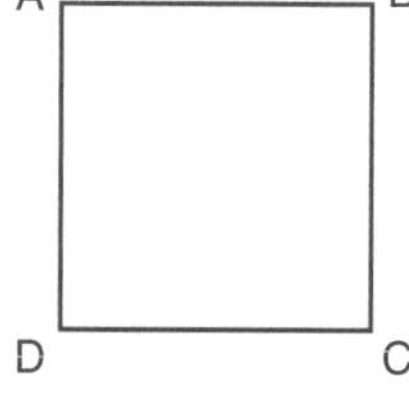
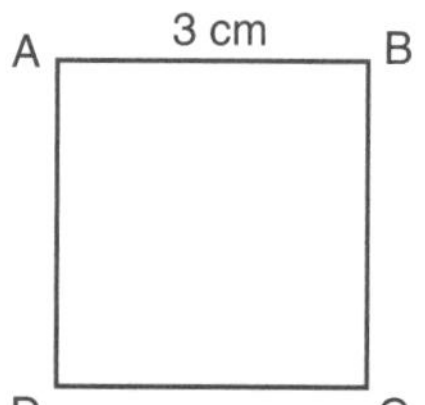

(A) 9 cm² (B) 12 cm² (C) 15 cm²

(D) 18 cm² (E) 21 cm²

17. $ABCD$ is an isosceles trapezoid. Find the perimeter.

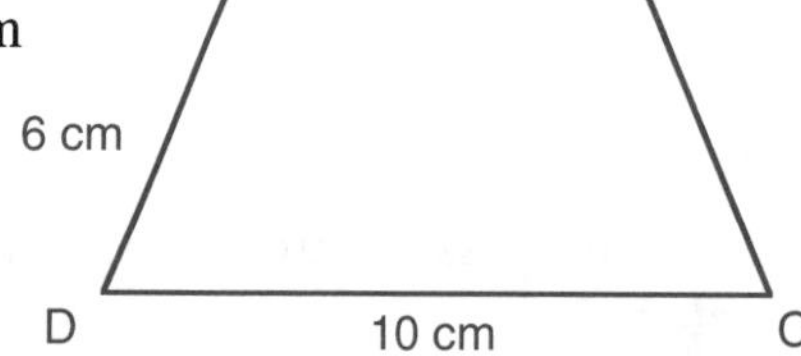

(A) 21 cm (B) 27 cm (C) 30 cm

(D) 50 cm (E) 54 cm

18. Find the area of trapezoid *MNOP*.

M 4 mm N
30°
P 3 mm O

(A) $(17 + 3\sqrt{3})$ mm^2

(B) 33/2 mm^2

(C) $33\sqrt{3}/2$ mm^2

(D) 33 mm^2

(E) $33\sqrt{3}$ mm^2

19. Trapezoid *XYZW* is isosceles. If $m\angle W = 58°$ and $m\angle Z = 4x - 6°$, find x.

(A) 8 (B) 12 (C) 13

(D) 16 (E) 58

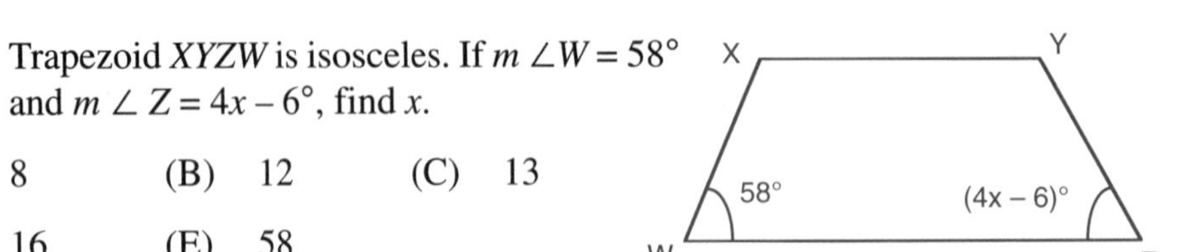

7. CIRCLES

A **circle** is a set of points in the same plane equidistant from a fixed point called its center.

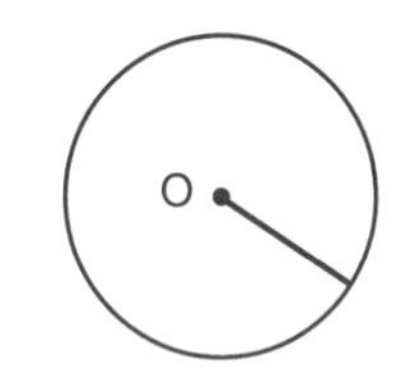

A **radius** of a circle is a line segment drawn from the center of the circle to any point on the circle.

A portion of a circle is called an **arc** of the circle.

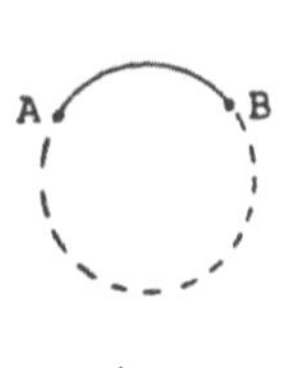

Arc

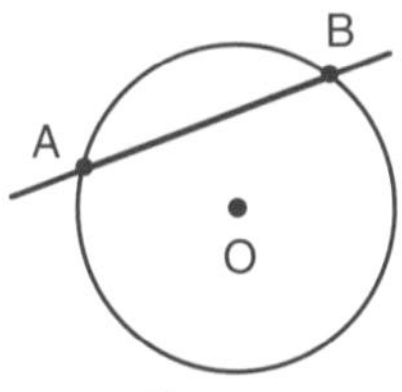

Secant

A line that intersects a circle in two points is called a **secant.**

A line segment joining two points on a circle is called a **chord** of the circle.

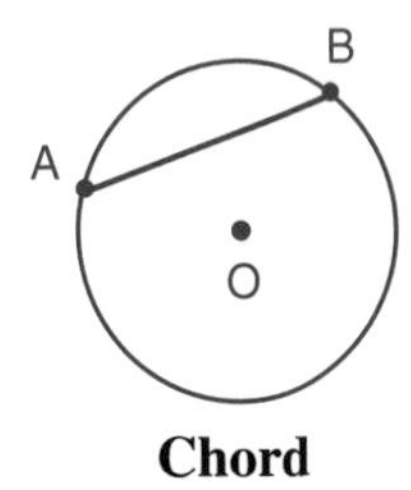

Chord

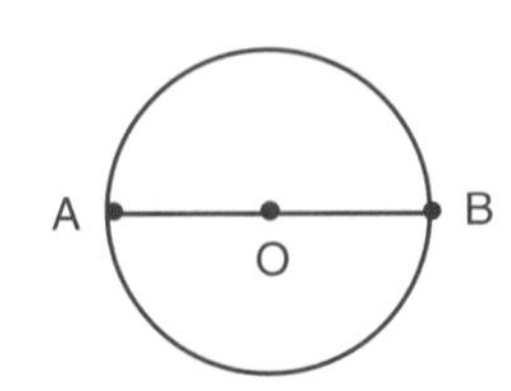

Diameter

A chord that passes through the center of the circle is called a **diameter** of the circle.

The line passing through the centers of two (or more) circles is called the **line of centers**.

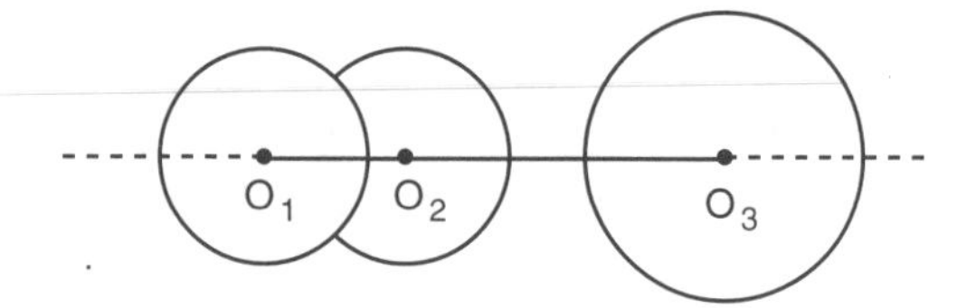

An angle whose vertex is on the circle and whose sides are chords of the circle is called an **inscribed angle**.

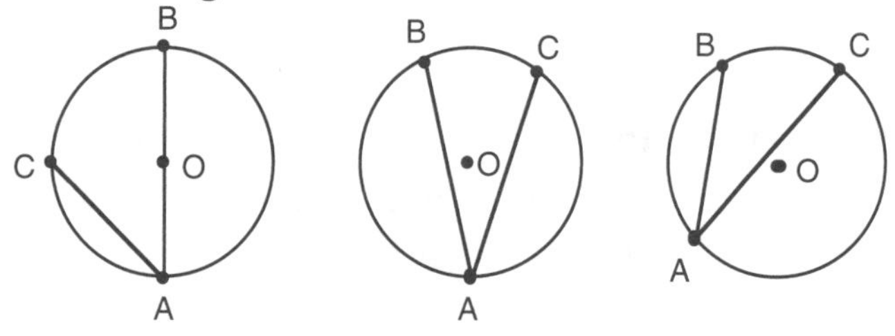

An angle whose vertex is at the center of a circle and whose sides are radii is called a **central angle.**

The measure of a minor arc is the measure of the central angle that intercepts that arc.

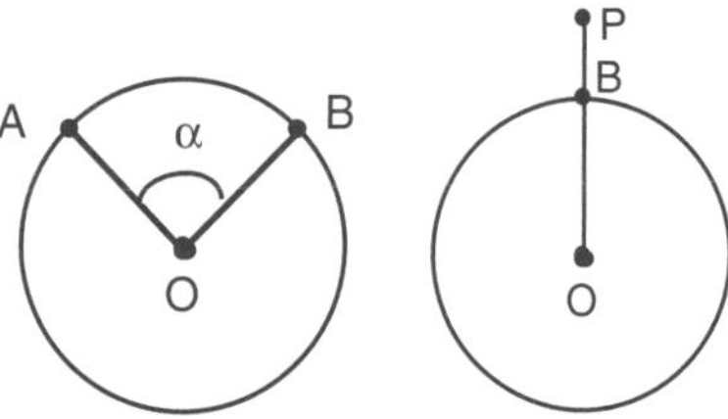

$$m\widehat{AB} = \alpha = m\angle AOB$$

The distance from a point *P* to a given circle is the distance from that point to the point where the circle intersects with a line segment with endpoints at the center of the circle and point *P*. The distance of point *P* to the diagrammed circle (above right) with center *O* is the line segment *PB* of line segment *PO*.

A line that has one and only one point of intersection with a circle is called a tangent to that circle, while their common point is called a **point of tangency**.

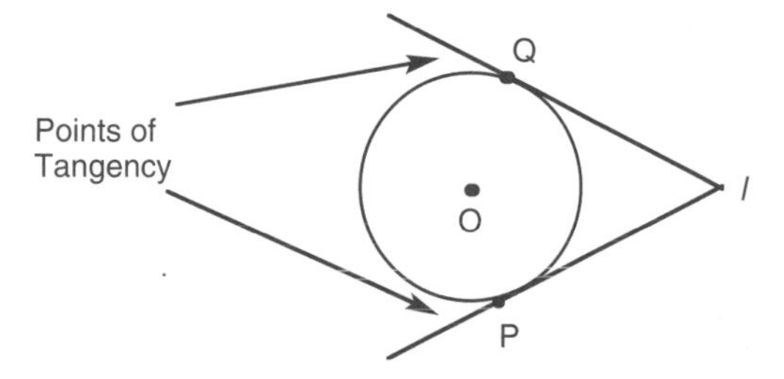

Congruent circles are circles whose radii are congruent.

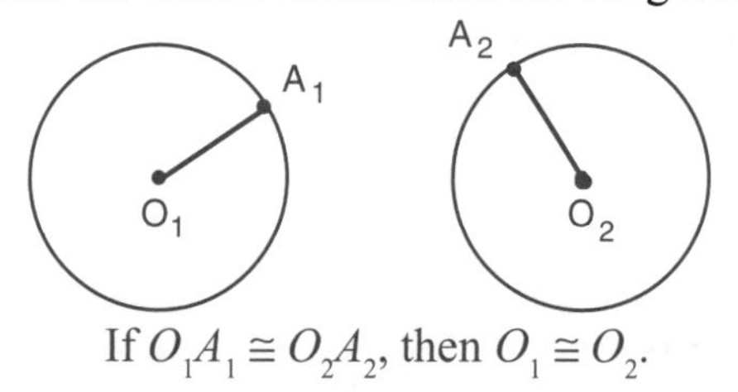

If $O_1A_1 \cong O_2A_2$, then $O_1 \cong O_2$.

The measure of a semicircle is 180°.

A **circumscribed circle** is a circle passing through all the vertices of a polygon.

Circles that have the same center and unequal radii are called **concentric circles**.

Circumscribed Circle **Concentric Circles**

The **circumference** of a circle is given by $2\pi r$, and the **area** of a circle is given by πr^2, where r is the radius and π (*pi*) is 3.14.

PROBLEM

Using the figure below, determine the length of the line segment CB.

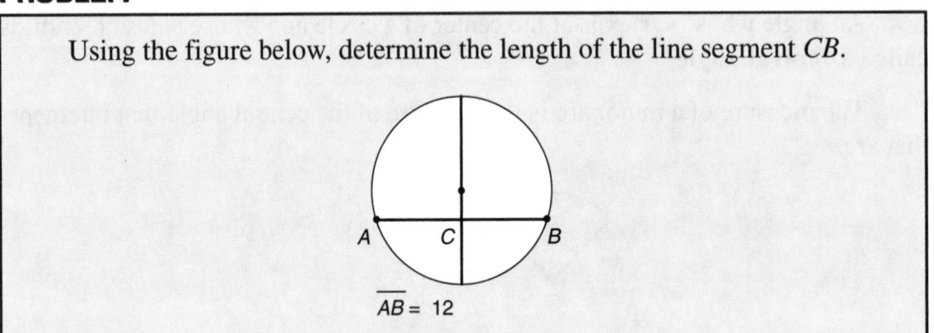

SOLUTION

If a chord is perpendicular to the diameter as in the above figure, then the diameter bisects the chord. Since the length of the chord is 12, the line segment CB is 6.

PROBLEM

Given that the coordinates of the diameter of a circle are (1, 1) and (6, 6), what is the radius of the circle?

SOLUTION

Step 1 is to use the distance formula to calculate the diameter.

$$D = \sqrt{(6-1)^2 + (6-1)^2}$$

$$D = \sqrt{25+25} = \sqrt{50} = 7.07$$

Step 2 is to divide the diameter by 2 to obtain the radius.

$$r = \frac{7.07}{2}$$

$$r = 3.535$$

The correct answer is $r = 3.535$.

PROBLEM

The radius of a circle is $y + 2$. Give the length of the diameter.

SOLUTION

Step 1 is to write the equation for calculating the diameter of a circle.

$D = 2r$, where r is the radius

Step 2 is to substitute the value of the radius into the equation.

$D = 2(y + 2)$

Step 3 is to solve the equation.

$D = 2y + 4$

The correct answer is the diameter $= 2y + 4$.

PROBLEM

A and B are points on circle Q such that ΔAQB is equilateral. If length of side $AB = 12$, find the length of arc AB.

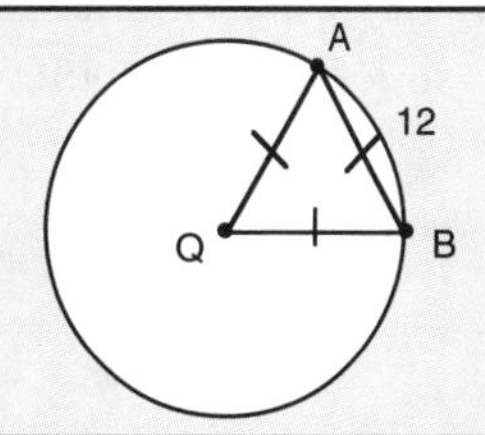

SOLUTION

To find the arc length of $\overset{\frown}{AB}$, we must find the measure of the central angle $\angle AQB$ and the measure of the radius $\overline{QA}$. $\angle AQB$ is an interior angle of the equilateral triangle ΔAQB. Therefore, $m\angle AQB = 60°$. Similarly, in the equilateral ΔAQB, $AQ = AB = QB = 12$. Given the radius, r, and the central angle, n, the arc length is given by $n/360 \times 2\pi r$. Therefore, by substitution, $60/360 \times 2\pi \times 12 = {}^1/_6 \times 2\pi \times 12 = 4\pi$. Therefore, the length of $\overset{\frown}{AB} = 4\pi$.

PROBLEM

In circle O, the measure of $\overset{\frown}{AB}$ is 80°. Find the measure of $\angle A$.

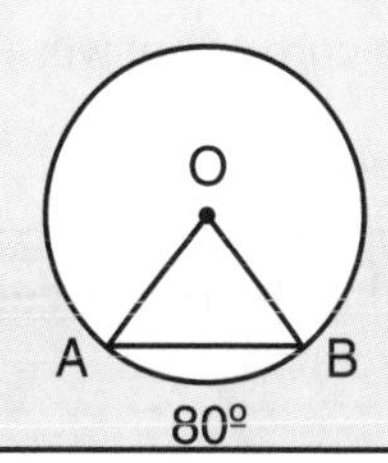

SOLUTION

The accompanying figure shows that $\overset{\frown}{AB}$ is intercepted by central angle AOB. By definition, we know that the measure of the central angle is the measure of its intercepted arc. In this case,

$$m\ \widehat{AB} = m\angle AOB = 80°.$$

Radius $\overline{OA}$ and radius $\overline{OB}$ are congruent and form two sides of ΔOAB. By a theorem, the angles opposite these two congruent sides must, themselves, be congruent. Therefore, $m\angle A = m\angle B$.

The sum of the measures of the angles of a triangle is 180°. Therefore,

$$m\angle A + m\angle B + m\angle AOB = 180°.$$

Since $m\angle A = m\angle B$, we can write

$$m\angle A + m\angle A + 80° = 180°$$

or $2m\angle A = 100°$

or $m\angle A = 50°$.

Therefore, the measure of $\angle A$ is 50°.

PROBLEM

Using the figure below, determine the measure of $\angle ABP$.

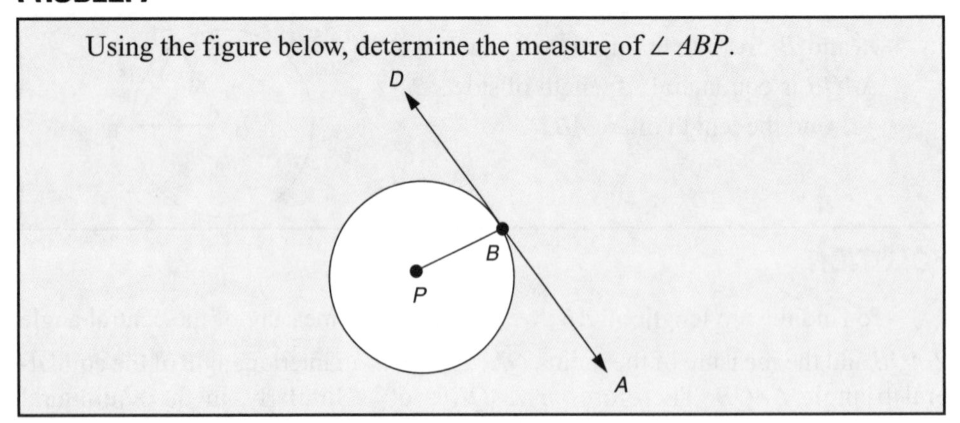

SOLUTION

Step 1 is to recognize that line AD is tangent to Circle P.

In Step 2, since line AD is tangent to Circle P, $\angle ABP$ must be a right angle.

The correct answer is 90 degrees.

PROBLEM

Using the figure below, determine the measure of $\angle ABC$.

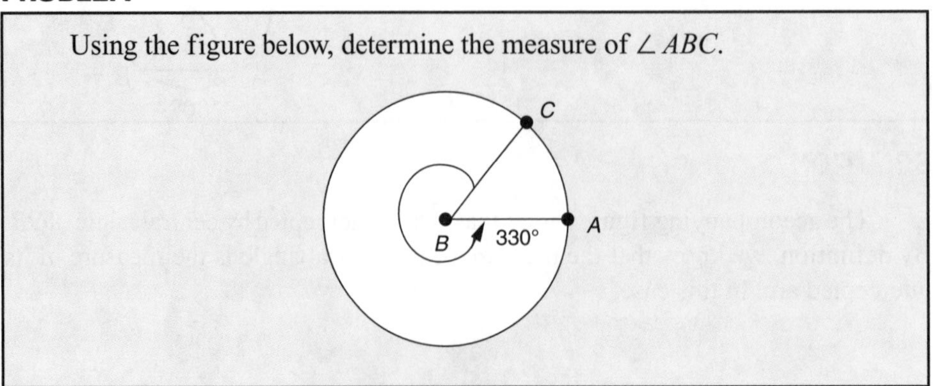

SOLUTION

Since the angle of the arc is given, and a circle is 360 degrees, the measure of $\angle ABC$ must be 360 – the angle of the arc.

$360 - 330 = 30$

The correct answer is 30 degrees.

DRILL: CIRCLES

Circumference, Area, Concentric Circles

1. Find the circumference of circle A if its radius is 3 mm.

(A) 3π mm (B) 6π mm (C) 9π mm (D) 12π mm (E) 15π mm

2. The circumference of circle H is 20π cm. Find the length of the radius.

(A) 10 cm (B) 20 cm (C) 10π cm (D) 15π cm (E) 20π cm

3. The circumference of circle A is how many millimeters larger than the circumference of circle B?

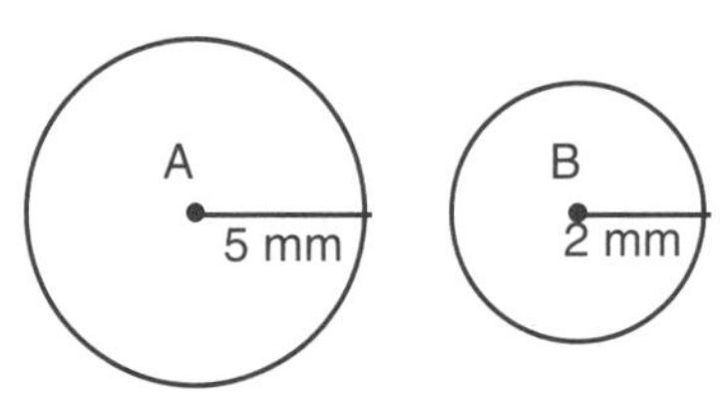

(A) 3 (B) 6 (C) 3π

(D) 6π (E) 7π

4. If the diameter of circle X is 9 cm and if $\pi = 3.14$, find the circumference of the circle to the nearest tenth.

(A) 9 cm (B) 14.1 cm (C) 21.1 cm (D) 24.6 cm (E) 28.3 cm

5. Find the area of circle I.

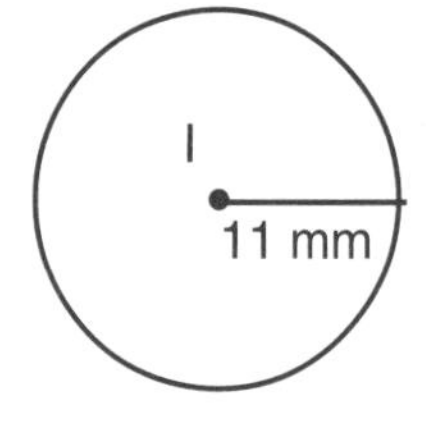

(A) 22 mm^2 (B) 121 mm^2

(C) 121π mm^2 (D) 132 mm^2

(E) 132π mm^2

6. The diameter of circle Z is 27 mm. Find the area of the circle.

(A) 91.125 mm^2 (B) 182.25 mm^2 (C) 191.5π mm^2

(D) 182.25π mm^2 (E) 729 mm^2

7. The area of circle B is 225π cm^2. Find the length of the diameter of the circle.

(A) 15 cm (B) 20 cm (C) 30 cm (D) 20π cm (E) 25π cm

8. The area of circle X is 144π mm^2 while the area of circle Y is 81π mm^2. Write the ratio of the radius of circle X to that of circle Y.

(A) 3 : 4 (B) 4 : 3 (C) 9 : 12 (D) 27 : 12 (E) 18 : 24

9. The circumference of circle M is 18π cm. Find the area of the circle.

(A) 18π cm^2 (B) 81 cm^2 (C) 36 cm^2 (D) 36π cm^2 (E) 81π cm^2

10. In two concentric circles, the smaller circle has a radius of 3 mm while the larger circle has a radius of 5 mm. Find the area of the shaded region.

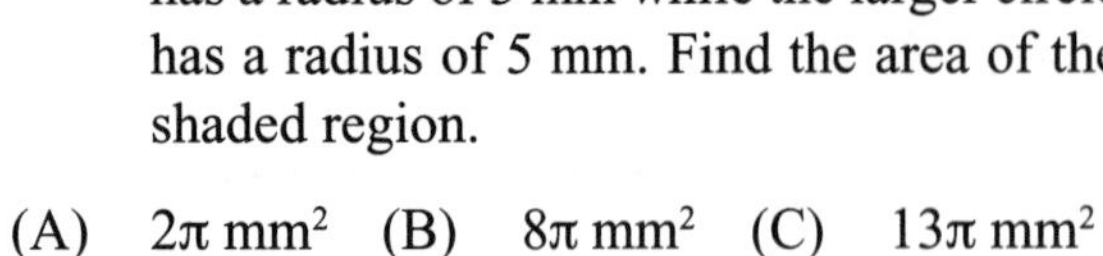

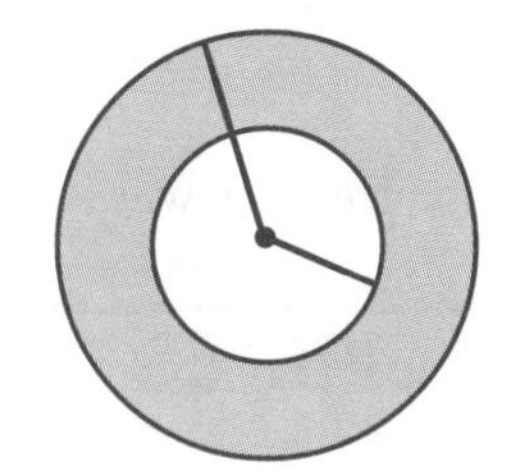

(A) 2π mm^2 (B) 8π mm^2 (C) 13π mm^2

(D) 16π mm^2 (E) 26π mm^2

11. The radius of the smaller of two concentric circles is 5 cm while the radius of the larger circle is 7 cm. Determine the area of the shaded region.

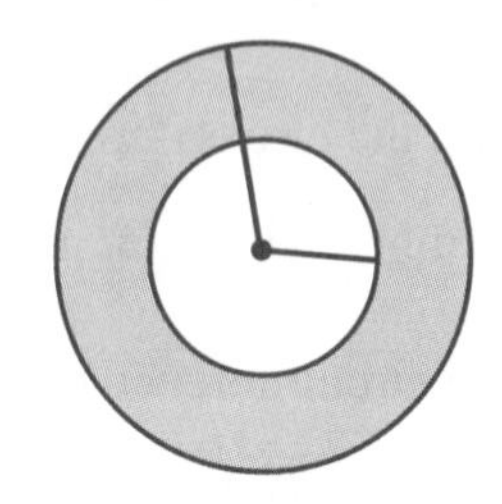

(A) 7π cm^2 (B) 24π cm^2 (C) 25π cm^2

(D) 36π cm^2 (E) 49π cm^2

12. Find the measure of arc MN if $m\angle MON =$ 62°.

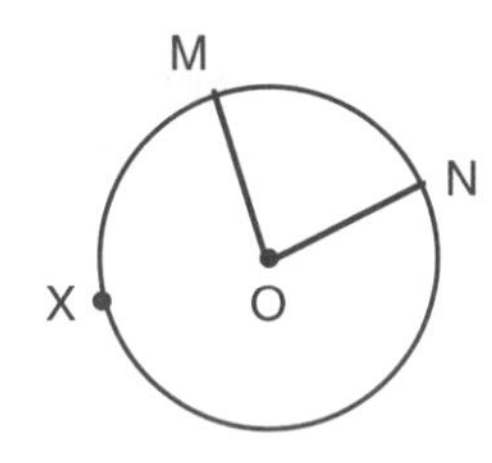

(A) 16° (B) 32° (C) 59°

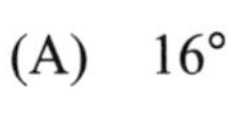

(D) 62° (E) 124°

13. Find the measure of arc AXC.

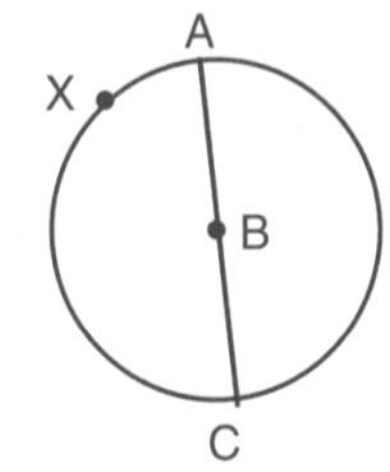

(A) 150° (B) 160° (C) 180°

(D) 270° (E) 360°

14. If arc $MXP = 236°$, find the measure of arc MP.

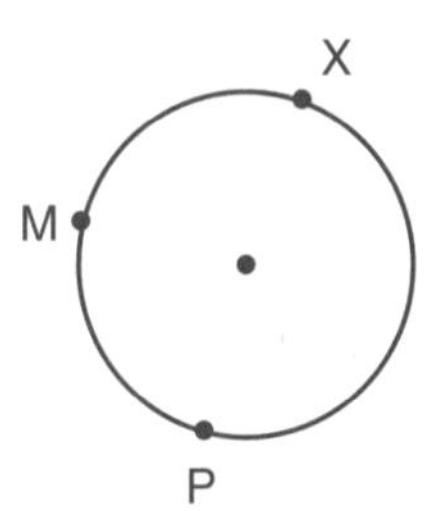

(A) 62° (B) 124° (C) 236°

(D) 270° (E) 360°

15. In circle S, major arc PQR has a measure of 298°. Find the measure of the central angle $\angle PSR$.

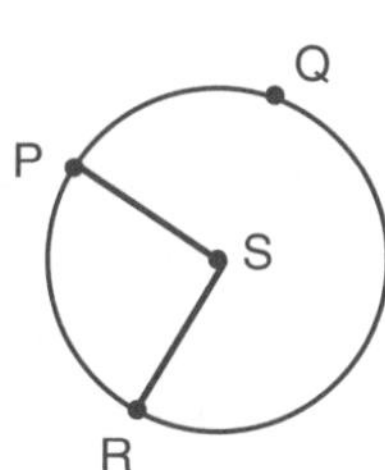

(A) 62° (B) 124° (C) 149°

(D) 298° (E) 360°

16. Find the measure of arc XY in circle W.

(A) 40° (B) 120° (C) 140°

(D) 180° (E) 220°

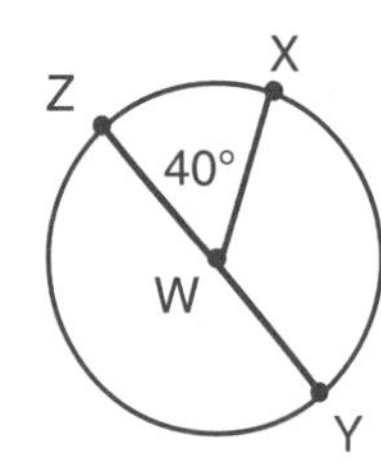

17. Find the area of the sector shown.

(A) 4 cm^2 (B) $2\pi\text{ cm}^2$ (C) 16 cm^2

(D) $8\pi\text{ cm}^2$ (E) $16\pi\text{ cm}^2$

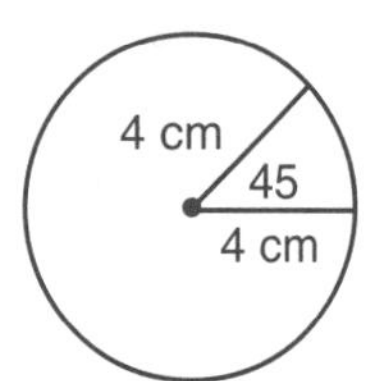

18. Find the area of the shaded region.

(A) 10 (B) 5π (C) 25

(D) 20π (E) 25π

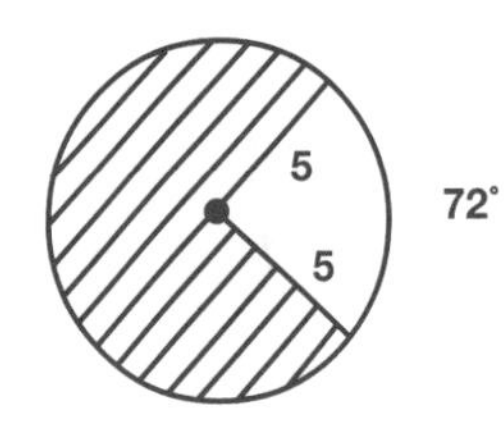

19. Find the area of the sector shown.

(A) $\frac{9\pi\text{ mm}^2}{4}$ (B) $\frac{9\pi\text{ mm}^2}{2}$ (C) 18 mm^2

(D) $6\pi\text{ mm}^2$ (E) $9\pi\text{ mm}^2$

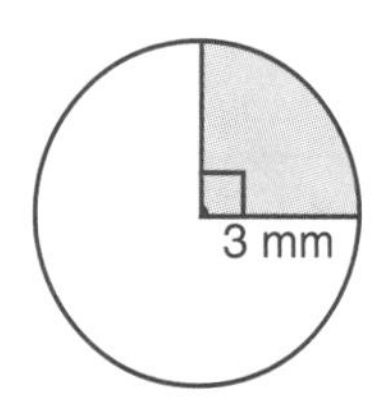

20. If the area of the square is 100 cm^2, find the area of the sector.

(A) $10\pi\text{ cm}^2$ (B) 25 cm^2 (C) $25\pi\text{ cm}^2$

(D) 100 cm^2 (E) $100\pi\text{ cm}^2$

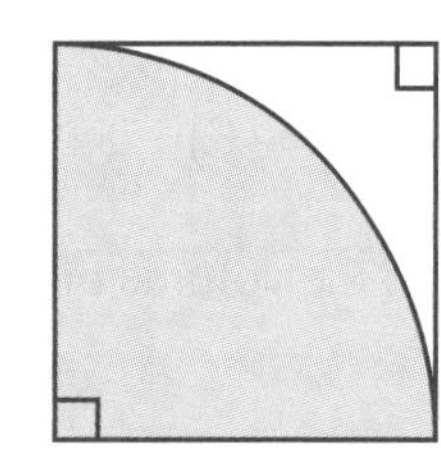

8. SOLIDS

Solid geometry is the study of figures which consist of points not all in the same plane.

RECTANGULAR SOLIDS

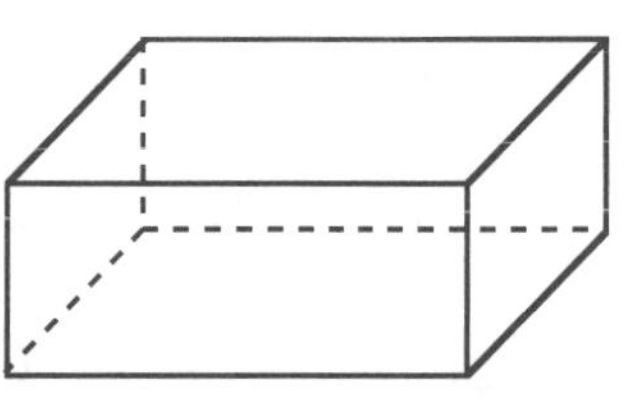

A solid with lateral faces and bases that are rectangles is called a **rectangular solid**.

The surface area of a rectangular solid is the sum of the areas of all the faces.

The volume of a rectangular solid is equal to the product of its length, width and height.

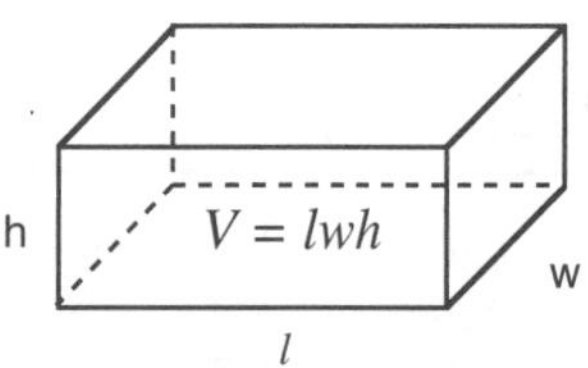

PROBLEM

What are the dimensions of a solid cube whose surface area is numerically equal to its volume?

SOLUTION

The surface area of a cube of edge length a is equal to the sum of the areas of its 6 faces. Since a cube is a regular polygon, all 6 faces are congruent. Each face of a cube is a square of edge length a. Hence, the surface area of a cube of edge length a is

$$S = 6a^2.$$

The volume of a cube of edge length a is

$$V = a^3.$$

We require that $A = V$, or that

$$6a^2 = a^3 \quad \text{or} \quad a = 6$$

Hence, if a cube has edge length 6, its surface area will be numerically equal to its volume.

DRILL: SOLIDS

1. Find the surface area of the rectangular prism shown.

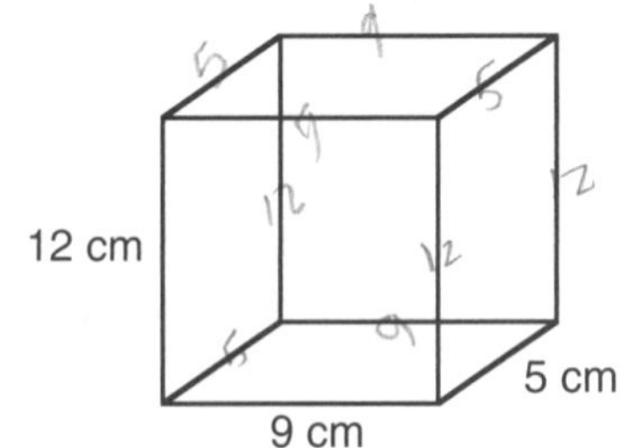

(A) 138 cm² (B) 336 cm² (C) 381 cm²

(D) 426 cm² (E) 540 cm²

2. Find the volume of the rectangular storage tank shown.

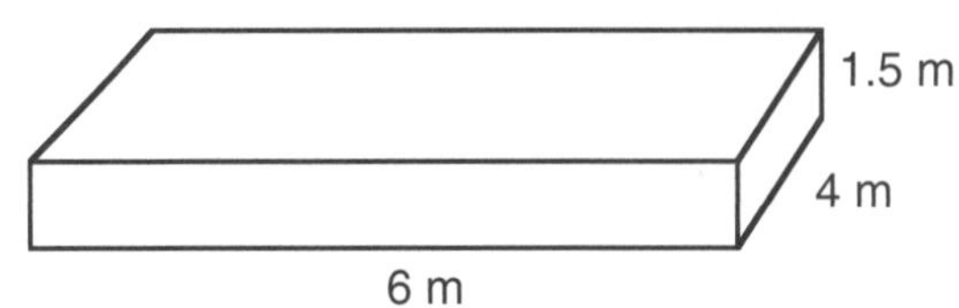

(A) 24 m³ (B) 36 m³ (C) 38 m³ (D) 42 m³ (E) 45 m³

3. The lateral area of a cube is 100 cm². Find the length of an edge of the cube.

(A) 4 cm (B) 5 cm (C) 10 cm (D) 12 cm (E) 15 cm

9. PERIMETERS, AREAS, VOLUMES

PROBLEM

Calculate the circumference of a circle that has a radius of 12 inches.

SOLUTION

Step 1 is to write the formula for the circumference of a circle.

circumference = $2\pi r$

Step 2 is to substitute the known values into the equation. Pi (π) is approximately 3.14. The radius is 12.

circumference = 2(3.14)12

Step 3 is to solve the equation.

circumference = 75.4

The correct answer is that the circumference = 75.4 inches.

PROBLEM

Calculate the area of a circle that has a diameter of 10 meters.

SOLUTION

Step 1 is to write the formula for the area of a circle.

$A = \pi r^2$

Step 2 is to substitute the known values into the equation. Pi is approximately 3.14. The diameter of the circle is 10 meters, so the radius is 5 meters.

$A = \pi(5)^2$

Step 3 is to solve the equation.

$A = 78.5$

The correct answer is that the area of the circle is 78.5 m^2.

PROBLEM

Calculate the volume of a sphere that has a radius of 2 meters.

SOLUTION

Step 1 is to write the formula for the volume of a sphere.

$$V = \frac{4}{3}\pi r^3$$

Step 2 is to substitute the known values into the equation. Pi is approximately 3.14 and the radius is 2 meters.

$$V = \frac{4}{3}\pi(2)^3$$

Step 3 is to solve the equation.

$V = 33.49$ or ~33.5

The correct answer is that the volume of the sphere is 33.5 m^3.

PROBLEM

Calculate the volume of a rectangular solid that has a length of 5, a height of 1.5, and a width of 4.

SOLUTION

Step 1 is to write the formula for the volume of a rectangular solid.

$$V = L \times W \times H$$

Step 2 is to substitute the known values into the formula.

$$V = 5 \times 4 \times 1.5$$

Step 3 is to solve the equation.

$$V = 30$$

The correct answer is that the volume of the rectangular solid = 30.

PROBLEM

Calculate the volume of a pyramid that has a height of 3 feet. The area of the base of the pyramid was calculated to be 15 square feet.

SOLUTION

Step 1 is to write the formula for the volume of a pyramid.

$V = \frac{1}{3}Bh$, where B is the area of the base of the pyramid

Step 2 is to substitute the known values into the formula.

$$V = \frac{1}{3}(15)(3)$$

Step 3 is to solve the equation.

$$V = 15$$

The correct answer is that the volume of the pyramid is 15 cubic feet.

PROBLEM

Calculate the volume of the cone below.

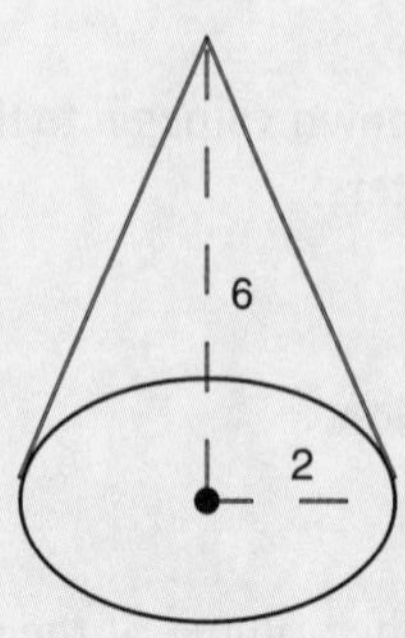

SOLUTION

Step 1 is to write the formula for the volume of a cone.

$$V = \frac{1}{3}\pi r^2 h$$

Step 2 is to substitute the known values into the equation.

$$V = \frac{1}{3}\pi\,(2)^2(6)$$

Step 3 is to solve the equation.

$$V = 25.12$$

The correct answer is that the volume of the cone is 25.12 cubic inches.

PROBLEM

Calculate the volume of the cylinder below.

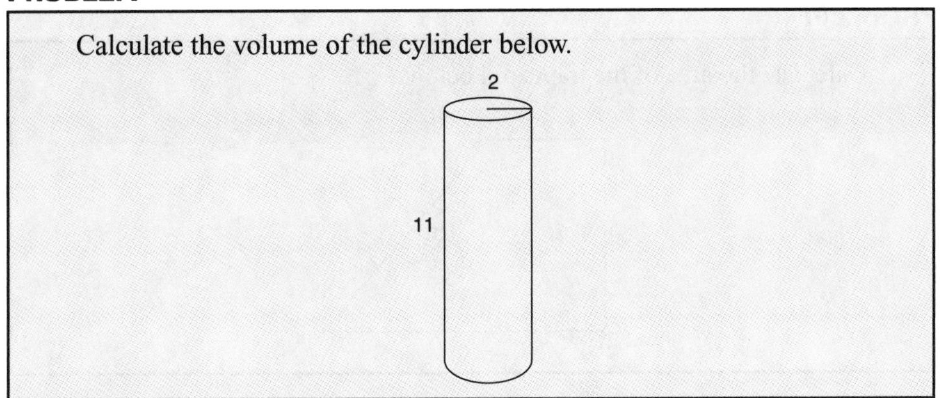

SOLUTION

Step 1 is to write the formula for the volume of a cylinder.

$$V = \pi r^2 h$$

Step 2 is to substitute the known values into the formula.

$$V = \pi(2)^2(11)$$

Step 3 is to solve the equation.

$$V = 138.16$$

The correct answer is that the volume is 138.16 m^3.

PROBLEM

Calculate the total area of the cone below. Round to the nearest meter.

SOLUTION

Step 1 is to write the formula for the total area of a cone.

$A = \pi rs + \pi r^2$, where s is the slant height

Step 2 is to substitute the known values into the formula.

$A = \pi(2)(5) + \pi(2)^2$

Step 3 is to solve the equation.

$A = 10\pi + 4\pi = \pi(10 + 4) = \pi(14)$

$A = 44$

The correct answer is 44 m^3.

PROBLEM

Calculate the area of the trapezoid below.

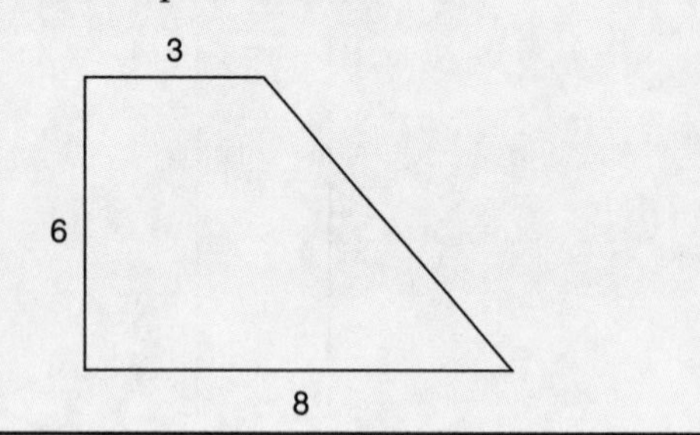

SOLUTION

Step 1 is to write the formula for the area of a trapezoid.

$A = \frac{1}{2}h(b1 + b2)$, where $b1$ and $b2$ are the bases

Step 2 is to substitute the known values into the formula.

$A = \frac{1}{2}(6)(3 + 8)$

Step 3 is to solve the equation.

$A = 33$

The correct answer is that the area of the trapezoid is 33 square feet.

PROBLEM

Calculate the area of the rhombus below.

SOLUTION

Step 1 is to write the formula for the area of a rhombus.

$$A = \frac{1}{2}(d1)(d2),$$ where $d1$ and $d2$ are the diagonals of the rhombus

Step 2 is to substitute the known values into the formula.

$$A = \frac{1}{2}(5)(4)$$

Step 3 is to solve the equation.

$$A = 10$$

The correct answer is that the area of the rhombus is 10 square feet.

10. COORDINATE GEOMETRY, PLOTTING, AND GRAPHING

The rectangular coordinate system, having vertical and horizontal lines meeting each other at right angles and thus forming a rectangular grid, is often called the Cartesian coordinate system. It is named after the French philosopher and mathematician, René Descartes, who invented it. The vertical line is usually labeled with the capital letter Y and called the Y axis. The horizontal line is usually labeled with the capital letter X and called the X axis. The point where the X and Y axes intersect is called the **origin** and is labeled with the letter o.

Above the origin, numbers measured along or parallel to the Y axis are positive; below the origin they are negative. To the right of the origin, numbers measured along or parallel to the X axis are positive; to the left they are negative.

A point anywhere on the graph may be located by two numbers, one showing the distance of the point from the Y axis, and the other showing the distance of the point from the X axis. We call the numbers that indicate the position of a point **coordinates**.

PROBLEM

Plot the point (–3, 2) on the graph below.

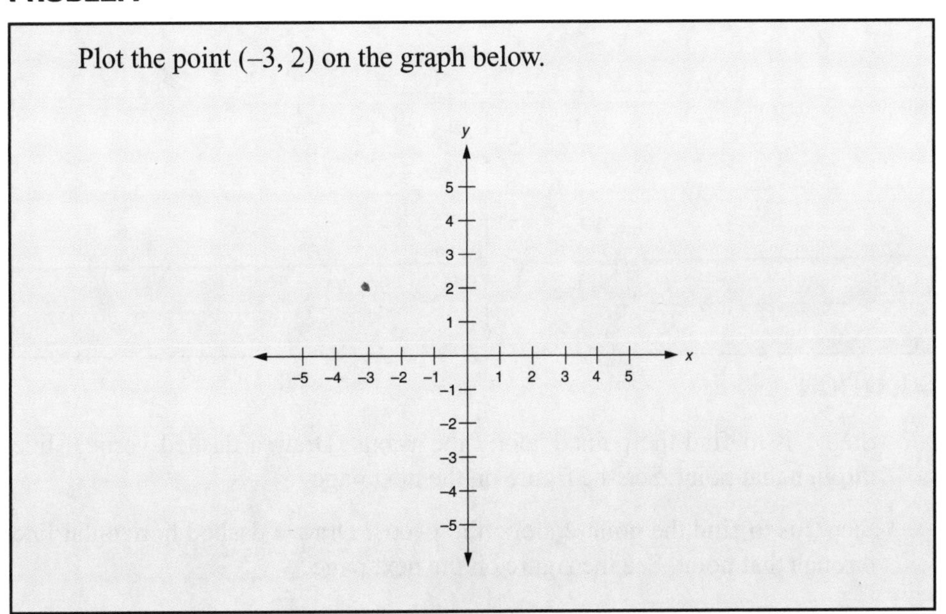

SOLUTION

Step 1 is to find the point –3 along the x-axis. Draw a dashed vertical line through that point. See the figure below.

Step 2 is to find the point 2 along the y-axis. Draw a dashed horizontal line through that point. See the figure below.

Step 3 is to plot the point where the two dashed lines intersect with a solid circle. This is the point (–3, 2). See the figure below.

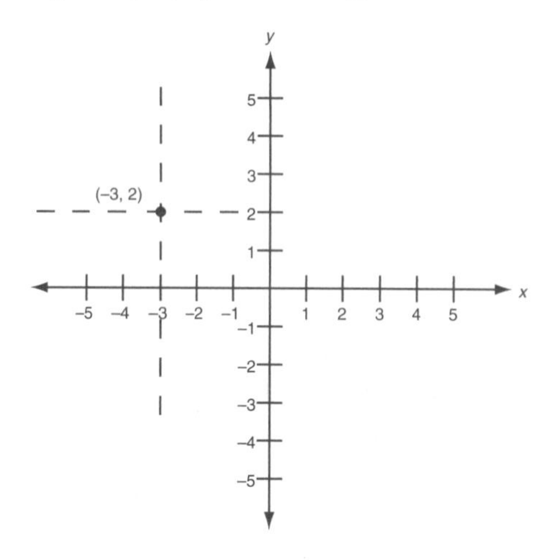

PROBLEM

Plot the points (1, 2) and (–4, –5) on the graph below. Draw a line connecting the points.

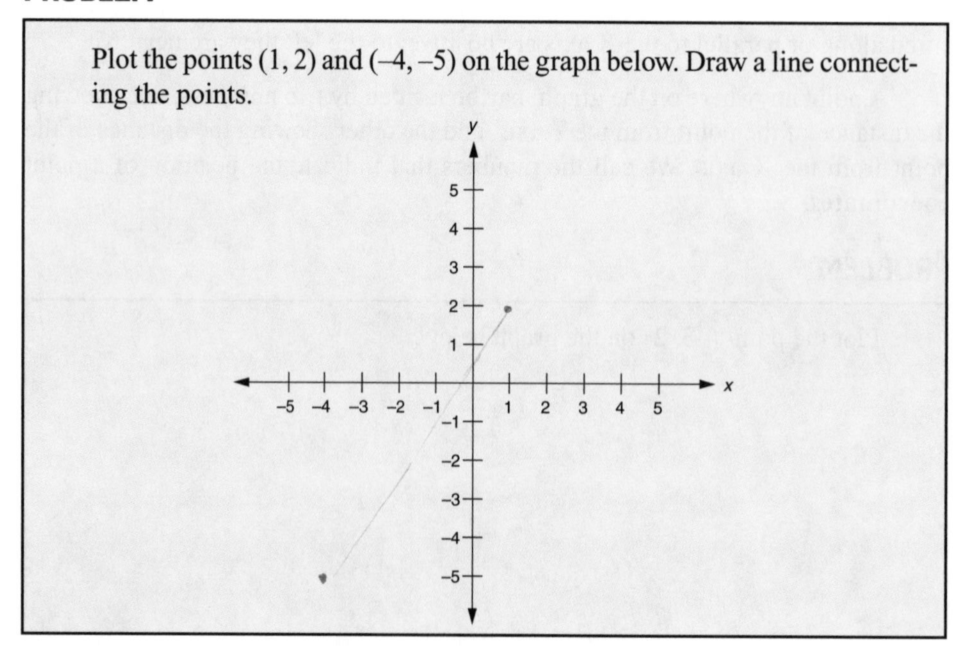

SOLUTION

Step 1 is to find the point 1 along the x-axis. Draw a dashed vertical line through that point. See the figure on the next page.

Step 2 is to find the point 2 along the y-axis. Draw a dashed horizontal line through that point. See the figure on the next page.

Step 3 is to plot the point where the two dashed lines intersect with a solid circle. This is the point (1, 2). See the figure below.

Step 4 is to find the point –4 along the x-axis. Draw a dashed vertical line through that point. See the figure below.

Step 5 is to find the point –5 along the y-axis. Draw a dashed horizontal line through that point. See the figure below.

Step 6 is to plot the point where the two dashed lines intersect with a solid circle. This is the point (–4, –5). See the figure below.

Step 7 is to draw a straight line connecting the two points.

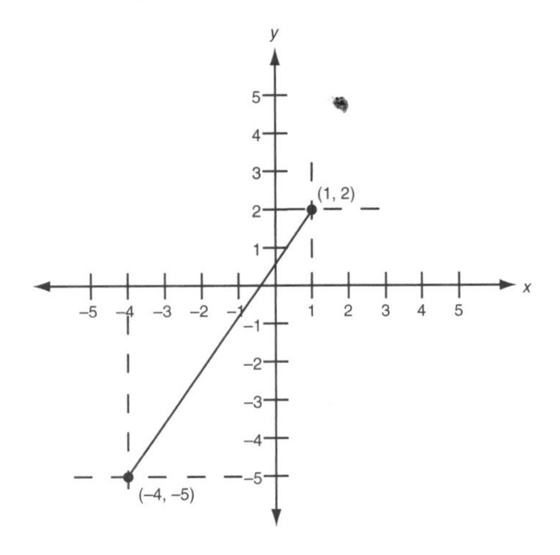

PROBLEM

Examine the figure below for this problem. What is Point A? Does Line a pass through Point A? Where does Line a intercept the x-axis?

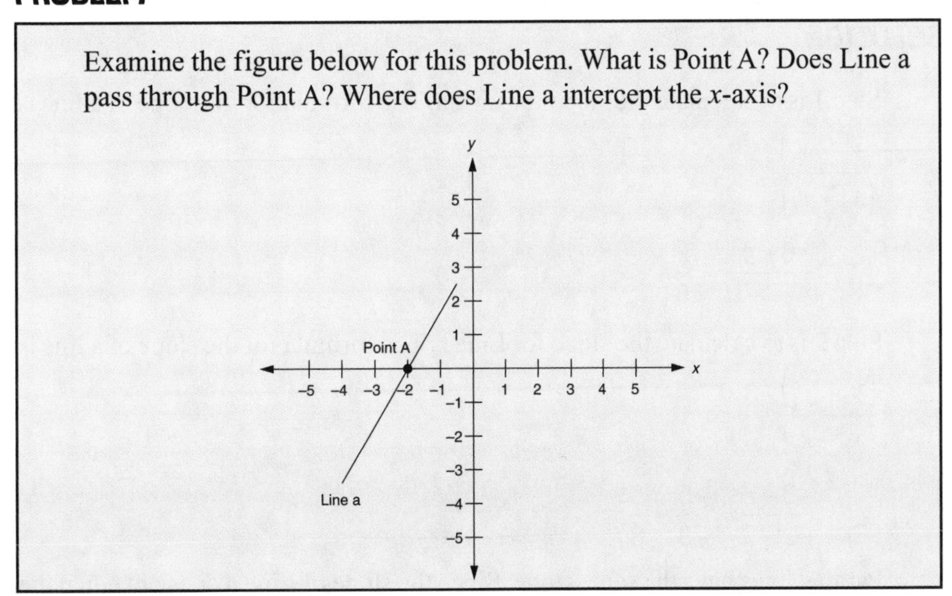

SOLUTION

Point A is the point (–2, 0). The graph illustrates that the coordinates for Point A are (–2, 0).

Line a does pass through Point A. This can be seen by looking at the graph. Since Line a passes through the point (–2, 0), it must pass through Point A.

Use the graph to see where the line crosses the x-axis. This will be where Line a intercepts the x-axis, which is at the point (–2, 0). The y-coordinate will always be 0 when a line intercepts the x-axis.

PROBLEM

A line contains the points (–6, 3) and (–3, 6). What is the slope of the line?

SOLUTION

Step 1 is to use the coordinates in the formula for the slope of a line. The formula is $\frac{(y_2 - y_1)}{(x_2 - x_1)}$.

$$\text{slope} = \frac{6-3}{-3-(-6)}$$

Step 2 is to solve the formula to obtain the slope.

$$\text{slope} = \frac{6-3}{-3+6} = \frac{3}{3} = 1$$

The slope of the line is 1.

PROBLEM

Line a contains the points (–1, 4) and (4, 6). Line b contains the points (5, 2) and (–3, –3). Are the two lines parallel?

SOLUTION

Step 1 is to calculate the slope for Line a. The formula for the slope of a line is $\frac{(y_2 - y_1)}{(x_2 - x_1)}$.

$$\frac{6-4}{4-(-1)} = \frac{2}{5}$$

Step 2 is to calculate the slope for Line b. The formula for the slope of a line is $\frac{(y_2 - y_1)}{(x_2 - x_1)}$.

$$\frac{-3-2}{-3-5} = \frac{-5}{-8} = \frac{5}{8}$$

Parallel lines have the same slope. Since the slope of Line a does not equal the slope of Line b, the lines cannot be parallel.

PROBLEM

Line a contains the points (–2, –2) and (6, 4). Line b contains the points (0, 4) and (3, 0). Are the two lines perpendicular?

SOLUTION

Step 1 is to calculate the slope for Line a. The formula for the slope of a line is $\frac{(y_2 - y_1)}{(x_2 - x_1)}$.

$$\frac{4-(-2)}{6-(-2)} = \frac{4+2}{6+2} = \frac{6}{8} = \frac{3}{4}$$

Step 2 is to calculate the slope for Line b. The formula for the slope of a line is $\frac{(y_2 - y_1)}{(x_2 - x_1)}$.

$$\frac{0-4}{3-0} = -\frac{4}{3}$$

Two lines are perpendicular if the slope of one line (m) equals the negative reciprocal of the other $\left(\frac{-1}{m}\right)$. Since the slope of Line a is $\frac{3}{4}$ and the slope of Line b is $\frac{-4}{3}$, the lines are perpendicular.

PROBLEM

Use the graph below to calculate the slope of Line a.

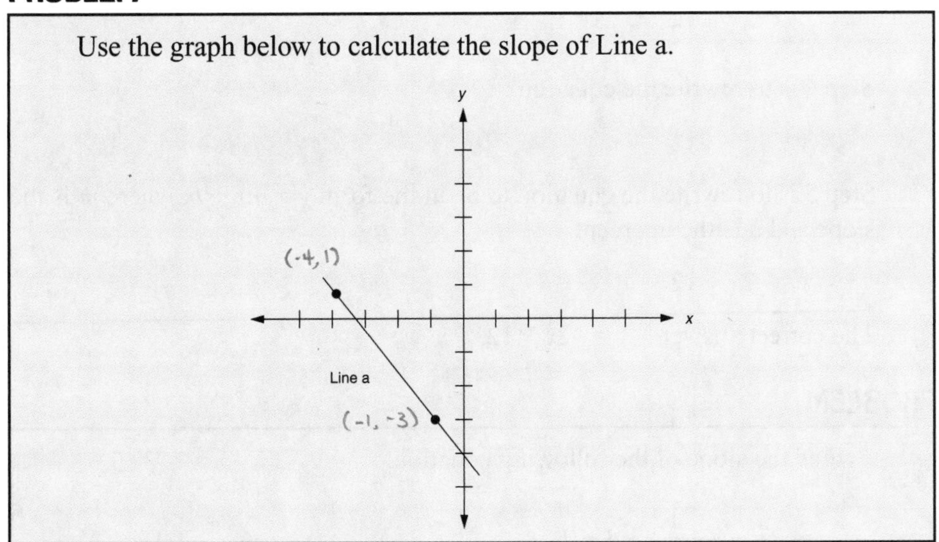

SOLUTION

Step 1 is to find two points that are intersected by Line a.

(–4, 1) and (–1, –3)

Step 2 is to use the formula for the slope of a line to calculate the slope of Line a. The formula is $\frac{(y_2 - y_1)}{(x_2 - x_1)}$.

$$\text{slope} = \frac{-3-1}{-1-(-4)}$$

Step 3 is to solve the formula to obtain the slope.

$$\text{slope} = \frac{-3-1}{-1-(-4)} = -\frac{4}{3}$$

The slope of the line is $-\frac{4}{3}$.

PROBLEM

Put the following equation in y slope–intercept form.

$$y + 2x = 12$$

SOLUTION

Step 1 is to isolate the variable y. To do this, subtract $2x$ from both sides of the equation.

$$y + 2x - 2x = 12 - 2x$$

Step 2 is to simplify the left side of the equation.

$$y + 2x - 2x = y$$

Step 3 is to simplify the right side of the equation.

$$12 - 2x = 12 - 2x$$

Step 4 is to rewrite the equation.

$$y = 12 - 2x$$

Step 5 is to rewrite the equation to be in the form $y = mx + b$, where m is the slope and b is the intercept.

$$y = -2x + 12$$

The correct answer is $y = -2x + 12$.

PROBLEM

Find the slope of the following equation.

$$\frac{1}{2}y - \frac{1}{4}x - 4 = 0$$

SOLUTION

Step 1 is isolate the variable $\frac{1}{2}y$.

$$\frac{1}{2}y - \frac{1}{4}x - 4 = 0$$

$$\frac{1}{2}y = \frac{1}{4}x + 4$$

Step 2 is to multiply both sides of the equation by 2. This will isolate the variable y and put the equation in y slope–intercept form.

$$y = \frac{1}{2}x + 8$$

Step 3 is to find the slope of the line. The coefficient of x is the slope.

$$m = \frac{1}{2}$$

The correct answer is $m = \frac{1}{2}$.

PROBLEM

Graph the equation $\frac{1}{2}y = x - 1$.

SOLUTION

Step 1 is to put the equation in y slope–intercept form.

$$\frac{1}{2}y = x - 1$$

$$y = 2x - 2$$

Step 2 is to choose two arbitrary values of x. Log the values in a chart. Use the values of x to obtain values for y.

x	y
0	–2
3	4

Step 3 is to use the x and y values to form coordinates. Note that there are many different possible coordinates. However, the graph should look the same.

(0, –2) and (3, 4)

Step 4 is to use the coordinates to graph the line.

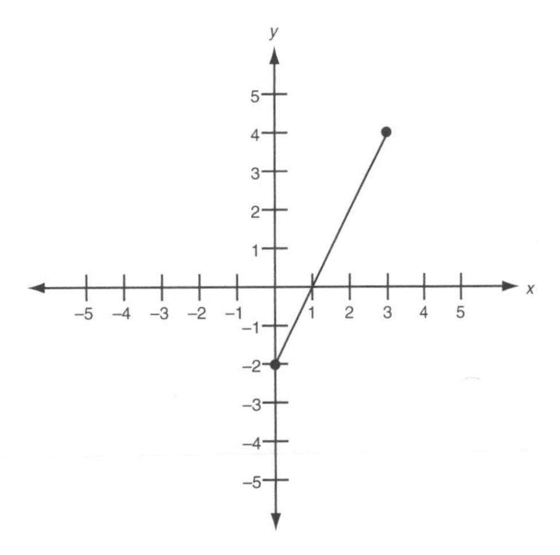

PROBLEM

Graph the equation $\frac{1}{4}y = \frac{1}{2}x - 1$.

SOLUTION

Step 1 is to put the equation in y slope–intercept form.

$$\frac{1}{4}y = \frac{1}{2}x - 1$$

$$y = 2x - 4$$

Step 2 is to choose two arbitrary values of x. Log the values in a chart. Use the values of x to obtain values for y.

x	y
0	–4
2	0

Step 3 is to use the x and y values to form coordinates. Note that there are many different possible coordinates. However, the graph should look the same.

(0, –4) and (2, 0)

Step 4 is to use the coordinates to graph the line.

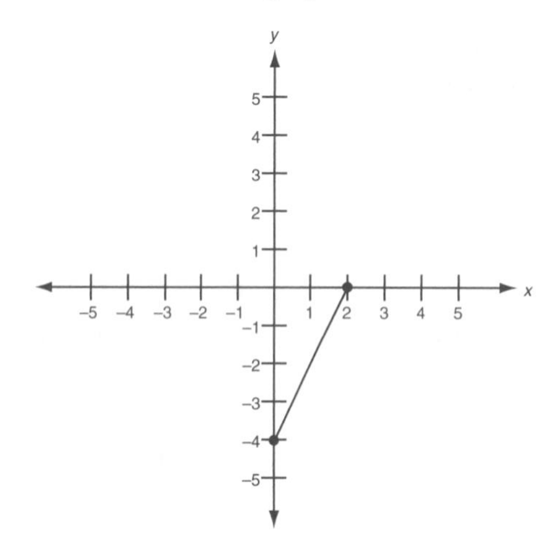

PROBLEM

> Put the following inequality in y slope–intercept form.
>
> $y + 5x > 22$

SOLUTION

Step 1 is to isolate the variable y. To do this, subtract $5x$ from both sides of the inequality.

$$y + 5x - 5x > 22 - 5x$$

Step 2 is to simplify the left side of the inequality.

$$y + 5x - 5x = y$$

Step 3 is to simplify the right side of the inequality.

$$22 - 5x = 22 - 5x$$

Step 4 is to rewrite the inequality.

$$y > 22 - 5x$$

Step 5 is to rewrite the inequality in the form $y = mx + b$, where m is the slope and b is the y–intercept.

$$y > -5x + 22$$

The correct answer is $y > -5x + 12$.

PROBLEM

Graph the inequality $-y > x + 5$.

SOLUTION

Step 1 is to put the inequality in y slope–intercept form.

$$-y > x + 5$$

$$y < -x - 5$$

Step 2 is to choose two arbitrary values of x. Log the values in a chart. Use the values of x to obtain values for y. This will solve the inequality $y < -x - 5$.

x	y
0	–5
–5	0

Step 3 is to use the x and y values to form coordinates. Note that there are many different possible coordinates. However, the graph should look the same.

(0, –5) and (–5, 0)

Step 4 is to use the coordinates to graph the line. Notice that a dashed line connects the points. The points along the line do not satisfy the inequality. Only the points in the shaded region satisfy the inequality.

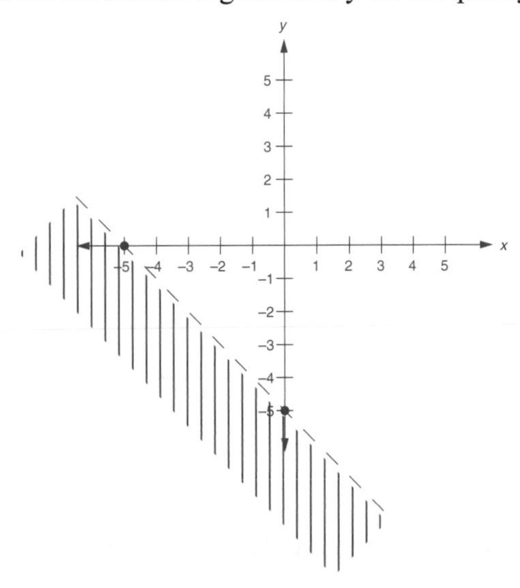

PROBLEM

Graph the inequality $4y \le 4x - 4$.

SOLUTION

Step 1 is to put the inequality in y slope–intercept form.

$$4y \le 4x - 4$$
$$y \le x - 1$$

Step 2 is to choose two arbitrary values of x. Log the values in a chart. Use the values of x to obtain values for y. These points solve the inequality $y \le x - 1$.

x	y
0	−1
2	1

Step 3 is to use the x and y values to form coordinates. Note that there are many different possible coordinates. However, the graph should look the same.

(0, −1) and (2, 1)

Step 4 is to use the coordinates to graph the line. Notice that a solid line connects the points. The points along the line satisfy the inequality. The points in the shaded region also satisfy the inequality.

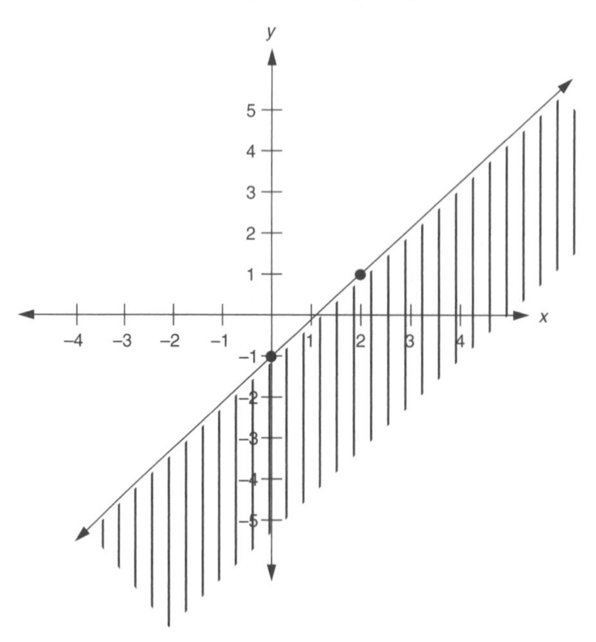

PROBLEM

Graph the inequality $y + 3x \ge 0$.

SOLUTION

Step 1 is to put the inequality in y slope–intercept form.

$y + 3x \geq 0$

$y \geq -3x$

Step 2 is to choose two arbitrary values of x. Log the values in a chart. Use the values of x to obtain values for y. These points solve the inequality $y \geq -3x$.

x	y
0	0
–3	9

Step 3 is to use the x and y values to form coordinates. Note that there are many different possible coordinates. However, the graph should look the same.

(0, 0) and (–3, 9)

Step 4 is to use the coordinates to graph the line. Notice that a solid line connects the points. The points along the line satisfy the inequality and the points in the shaded region satisfy the inequality.

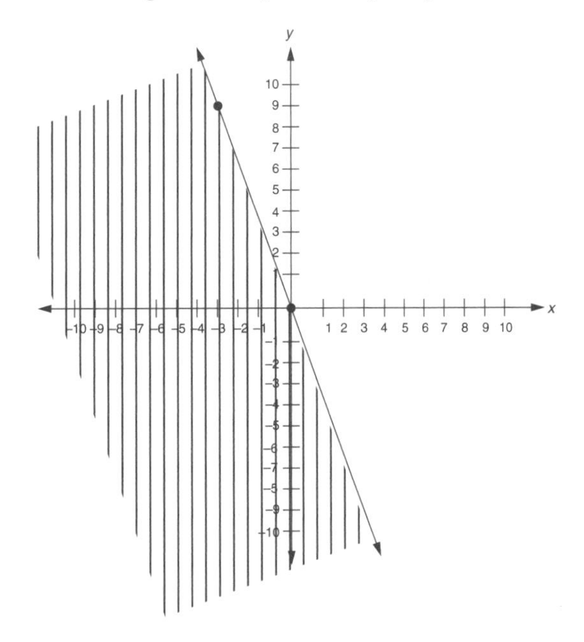

DRILL: COORDINATE GEOMETRY

1. Which point shown has the coordinates (– 3, 2)?

(A) A (B) B (C) C

(D) D (E) E

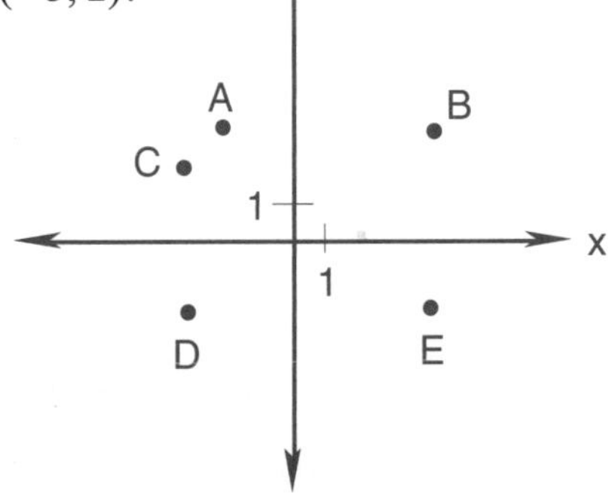

2. Name the coordinates of point A.

(A) (4, 3) (B) (3, – 4) (C) (3, 4)

(D) (– 4, 3) (E) (4, – 3)

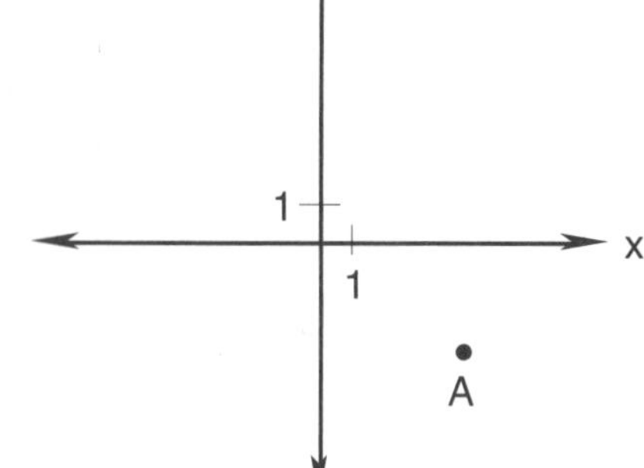

3. Which point shown has the coordinates (2.5, – 1)?

(A) M (B) N (C) P

(D) Q (E) R

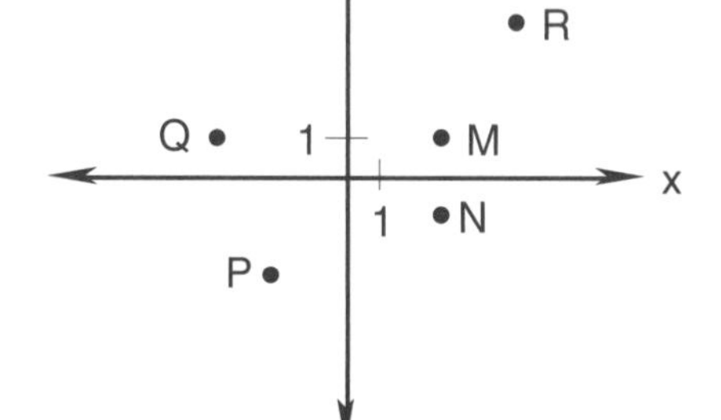

4. The correct *x*-coordinate for point *H* is what number?

(A) 3 (B) 4 (C) – 3

(D) – 4 (E) – 5

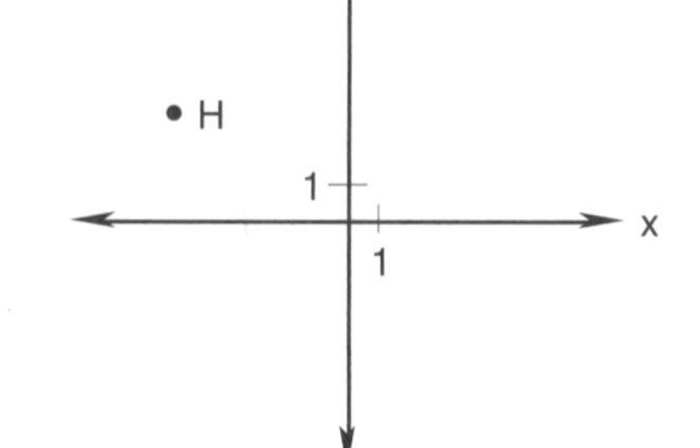

5. The correct *y*-coordinate for point *R* is what number?

(A) –7 (B) 2 (C) – 2

(D) 7 (E) 8

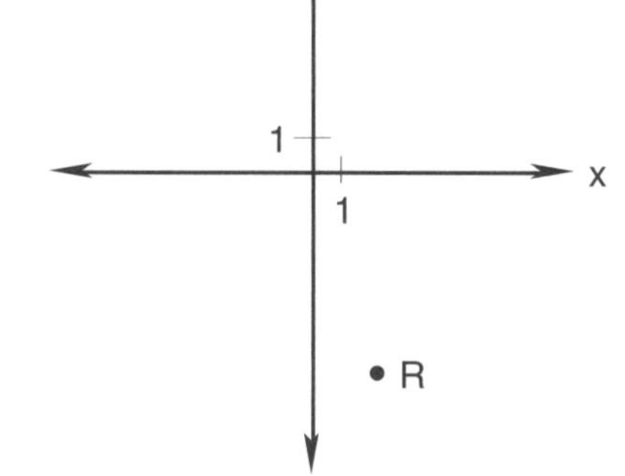

6. Find the distance between (4, – 7) and (– 2, – 7).

(A) 4 (B) 6 (C) 7 (D) 14 (E) 15

7. Find the distance between (3, 8) and (5, 11).

(A) 2 (B) 3 (C) $\sqrt{13}$ (D) $\sqrt{15}$ (E) $3\sqrt{3}$

8. How far from the origin is the point (3, 4)?

(A) 3 (B) 4 (C) 5 (D) $5\sqrt{3}$ (E) $4\sqrt{5}$

9. Find the distance between the point (– 4, 2) and (3, – 5).

(A) 3 (B) $3\sqrt{3}$ (C) 7 (D) $7\sqrt{2}$ (E) $7\sqrt{3}$

10. The distance between points A and B is 10 units. If A has coordinates (4, – 6) and B has coordinates (– 2, y), determine the value of y.

(A) – 6 (B) – 2 (C) 0 (D) 1 (E) 2

11. Find the midpoint between the points (– 2, 6) and (4, 8).

(A) (3, 7) (B) (1, 7) (C) (3, 1) (D) (1, 1) (E) (– 3, 7)

12. Find the coordinates of the midpoint between the points (– 5, 7) and (3, – 1).

(A) (– 4, 4) (B) (3, – 1) (C) (1, – 3) (D) (– 1, 3) (E) (4, – 4).

13. What is the y-coordinate of the midpoint of segment $\overline{AB}$ if A has coordinates (– 3, 7) and B has coordinates (– 3, – 2)?

(A) 5/2 (B) 3 (C) 7/2 (D) 5 (E) 15/2

14. One endpoint of a line segment is (5, – 3). The midpoint is (– 1, 6). What is the other endpoint?

(A) (7, 3) (B) (2, 1.5) (C) (– 7, 15)

(D) (– 2, 1.5) (E) (– 7, 12)

15. The point (– 2, 6) is the midpoint for which of the following pair of points?

(A) (1, 4) and (– 3, 8) (B) (– 1, – 3) and (5, 9)

(C) (1, 4) and (5, 9) (D) (– 1, 4) and (3, – 8)

(E) (1, 3) and (– 5, 9)

11. FRACTALS

Fractals are geometric shapes or patterns made up of identical parts, which are in turn identical to the overall pattern. Fractal objects have a high degree of "self-similarity."

EXAMPLE

Here is an example of a fractal pattern starting with a triangle.

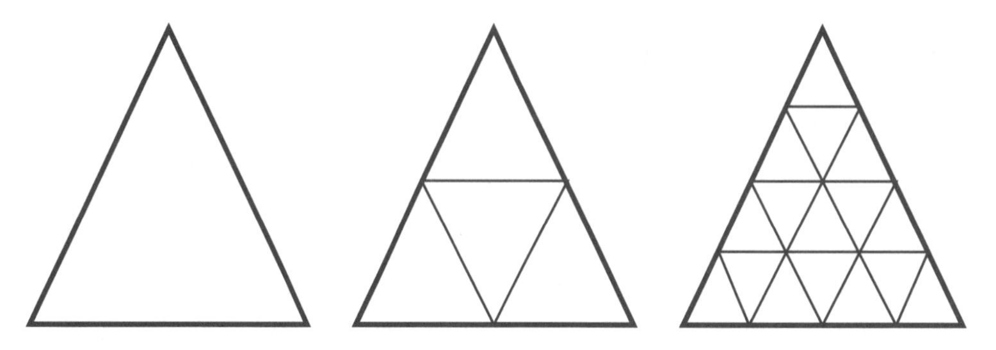

PROBLEM

The figures below show the first 3 stages of a fractal tree. Each of the branches is a smaller version of the main trunk of the tree. What does the fourth stage look like?

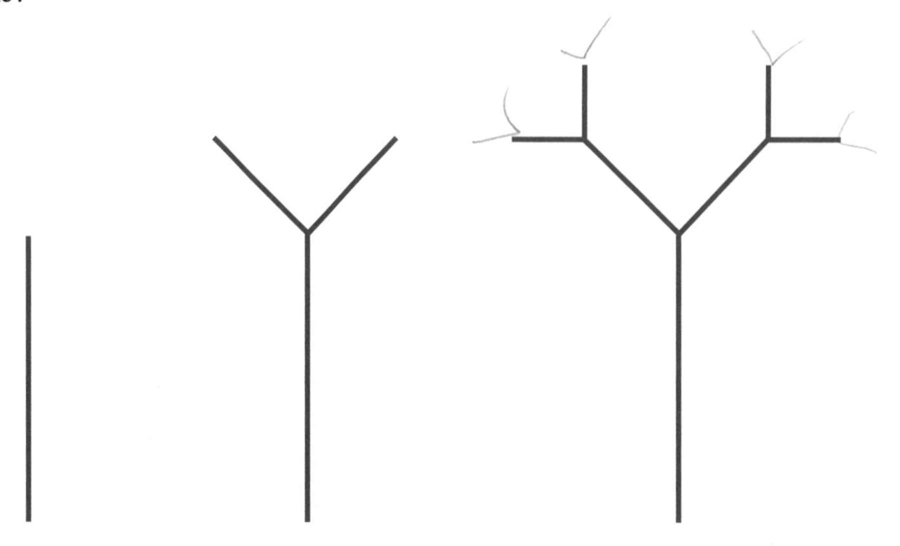

SOLUTION

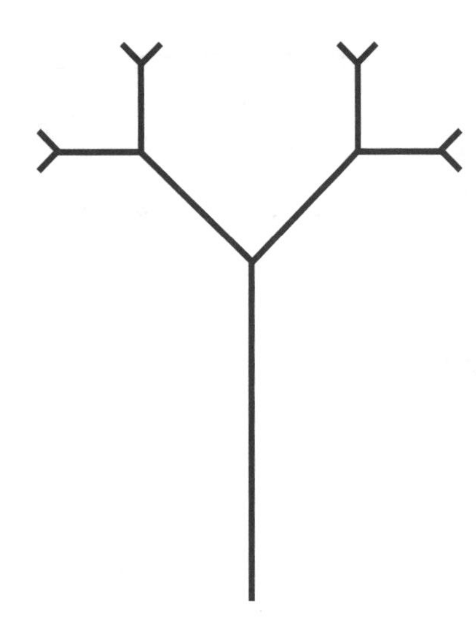

GEOMETRY DRILLS

DRILL–LINES AND ANGLES

1.	(B)	5.	(D)	9.	(D)	13.	(B)
2.	(A)	6.	(C)	10.	(E)	14.	(E)
3.	(C)	7.	(B)	11.	(C)	15.	(A)
4.	(D)	8.	(A)	12.	(D)		

DRILL–REGULAR POLYGONS

1.	(D)	4.	(A)	7.	(E)	10.	(B)
2.	(B)	5.	(D)	8.	(B)		
3.	(E)	6.	(C)	9.	(D)		

DRILL–TRIANGLES

1.	(D)	5.	(C)	9.	(A)	13.	(D)
2.	(B)	6.	(A)	10.	(C)	14.	(A)
3.	(C)	7.	(B)	11.	(C)	15.	(B)
4.	(E)	8.	(E)	12.	(B)		

DRILL–QUADRILATERALS

1.	(A)	6.	(A)	11.	(C)	16.	(A)
2.	(B)	7.	(D)	12.	(D)	17.	(B)
3.	(D)	8.	(C)	13.	(B)	18.	(C)
4.	(E)	9.	(B)	14.	(C)	19.	(D)
5.	(C)	10.	(C)	15.	(B)		

DRILL –CIRCLES

1.	(B)	6.	(D)	11.	(B)	16.	(C)
2.	(A)	7.	(C)	12.	(D)	17.	(B)
3.	(D)	8.	(B)	13.	(C)	18.	(D)
4.	(E)	9.	(E)	14.	(B)	19.	(A)
5.	(C)	10.	(D)	15.	(A)	20.	(C)

DRILL–SOLIDS

1.	(D)	2.	(B)	3.	(C)

DRILL–COORDINATE GEOMETRY

1.	(C)	5.	(A)	9.	(D)	13.	(A)
2.	(E)	6.	(B)	10.	(E)	14.	(C)
3.	(B)	7.	(C)	11.	(B)	15.	(E)
4.	(D)	8.	(C)	12.	(D)		

WORD PROBLEMS REVIEW

One of the main problems students have in mathematics involves solving word problems. The secret to solving these problems is being able to convert words into numbers and variables in the form of an algebraic equation.

The easiest way to approach a word problem is to read the question and ask yourself what you are trying to find. This unknown quantity can be represented by a variable.

Next, determine how the variable relates to the other quantities in the problem. More than likely, these quantities can be explained in terms of the original variable. If not, a separate variable may have to be used to represent a quantity.

Using these variables and the relationships determined among them, an equation can be written. Solve for a particular variable and then plug this number in for each relationship that involves this variable in order to find any unknown quantities.

Lastly, reread the problem to be sure that you have answered the questions correctly and fully.

1. ALGEBRAIC

The following illustrates how to formulate an equation and solve the problem.

EXAMPLE

Find two consecutive odd integers whose sum is 36.

Let x = the first odd integer

Let $x + 2$ = the second odd integer

The sum of the two numbers is 36. Therefore,

$$x + (x + 2) = 36$$

Simplifying,

$$2x + 2 = 36$$

$$2x = 34$$

$$x = 17$$

Substituting 17 for x, we find the second odd integer $= (x + 2) = (17 + 2) = 19$. Therefore, we find that the two consecutive odd integers whose sum is 36 are 17 and 19 respectively.

PROBLEM

Write the algebraic expression for the following (do not solve):

A number multiplied by 3 equals 18. What is the number?

SOLUTION

Step 1 is to represent the unknown quantity. In this problem, the unknown quantity is the missing number.

Let x = the missing number

Step 2 is to write the left side of the equation. It is known that the missing number multiplied by 3 will equal the product. Therefore, the left side of the equation will be:

$$3x$$

Step 3 is to write the right side of the equation. It is known that the product will be 18. Therefore, the right side of the equation will be:

$$18$$

Step 4 is to write both sides of the equation together.

$$3x = 18$$

The correct answer is $3x = 18$.

PROBLEM

Write the algebraic expression for the following. Solve the expression.

Ten is subtracted from a number. It is then divided by 5. The quotient equals 2. What is the number?

SOLUTION

Step 1 is to represent the unknown quantity. In this problem, the unknown quantity is the missing number.

Let x = the missing number

Step 2 is to write the left side of the equation. It is known that the missing number minus 10, then divided by 5, will equal the given quotient. Therefore, the left side of the equation will be:

$$\frac{(x-10)}{5}$$

Step 3 is to write the right side of the equation. It is known that the quotient is 2. Therefore, the right side of the equation will be:

$$2$$

Step 4 is to write both sides of the equation together.

$$\frac{(x-10)}{5} = 2$$

Step 5 is to solve the equation.

$$\frac{(x-10)}{5} = 2$$

$$x - 10 = 10$$

$$x = 20$$

The correct answer is $x = 20$.

PROBLEM

Write the algebraic expression for the following. Solve the expression.

A number multiplied by 4 equals 44. What is the number?

SOLUTION

Step 1 is to represent the unknown quantity. In this problem, the unknown quantity is the missing number.

Let x = the missing number

Step 2 is to write the left side of the equation. It is known that the missing number multiplied by 4 will equal the product. Therefore, the left side of the equation will be:

$4x$

Step 3 is to write the right side of the equation. It is known the product will be 44. Therefore, the right side of the equation will be:

44

Step 4 is to write both sides of the equation together.

$4x = 44$

Step 5 is to solve the equation.

$$4x = 44$$

$$x = \frac{44}{4}$$

$$x = 11$$

The correct answer is $x = 11$.

PROBLEM

Write the algebraic expression for the following. Solve the expression.

Kelly gave $25 to her cousin Kayde for her birthday. Combined with her savings, Kayde now has $700 toward college. How much does Kayde have in savings?

SOLUTION

Step 1 is to represent the unknown quantity. In this problem, the unknown quantity is the amount of money that Kayde has in savings.

Let x = the amount of money in Kayde's savings

Step 2 is to write the left side of the equation. It is known that the amount Kayde received for her birthday plus the amount she has saved will equal the total amount. Therefore, the left side of the equation will be:

$\$25 + x$

Step 3 is to write the right side of the equation. It is known that the total amount is \$700. Therefore, the right side of the equation will be:

\$700

Step 4 is to write both sides of the equation together.

$\$25 + x = \$700$

Step 5 is to solve the equation.

$\$25 + x = \$700 \qquad x = \$700 - \$25 \qquad x = \$675$

The correct answer is $x = \$675$.

PROBLEM

Write the algebraic expression for the following. Solve the expression.

Island Beach, Inc., is a real estate developer. This year's house sales will be $\frac{1}{2}$ as much as last year's sales plus 5% (of last year's sales). If this year's sales totaled 4,100 homes, how much were last year's sales?

SOLUTION

Step 1 is to represent the unknown quantity. In this problem, the unknown quantity is last year's sales.

Let x = last year's sales

Step 2 is to write the left side of the equation. It is known that last year's sales times $\frac{1}{2}$ plus 5% will equal this year's sales. Therefore, the left side of the equation will be:

$\frac{1}{2}x + 5\%x$

Step 3 is to write the right side of the equation. It is known that this year's sales totaled 4,100 homes. Therefore, the right side of the equation will be:

4,100

Step 4 is to write both sides of the equation together.

$$\frac{1}{2}x + 5\%x = 4{,}100$$

Step 5 is to solve the equation.

$$\frac{1}{2}x + 5\%x = 4{,}100 \qquad \frac{1}{2}x + .05x = 4{,}100 \qquad .55x = 4{,}100$$

$$x = 7{,}455$$

The correct answer is $x = 7{,}455$ homes.

PROBLEM

Write the algebraic expression for the following. Solve the expression.

A land surveyor measured the distance between the Price County Court House and Lake Silver. The distance was measured to be 6,320 meters. However, he needs to convert the distance to kilometers (1 km = 1,000 m). He then must subtract a flood factor (the lake rises during the rainy season). If the distance in meters divided by 1,000 minus the flood factor equals 5.25 kilometers, what is the flood factor?

SOLUTION

Step 1 is to represent the unknown quantity. In this problem, the unknown quantity is the flood factor in kilometers.

Let x = the flood factor in kilometers

Step 2 is to write the left side of the equation. It is known that the distance was measured to be 6,320 meters minus the flood factor. Therefore, the left side of the equation will be:

$$\frac{6{,}320}{1{,}000} - x$$

Step 3 is to write the right side of the equation. It is known that the total distance is 5.25 kilometers. Therefore, the right side of the equation will be:

5.25 kilometers

Step 4 is to write both sides of the equation together.

$$\frac{6{,}320}{1{,}000} - x = 5.25$$

Step 5 is to solve the equation.

$$6.32 - x = 5.25$$
$$-x = -1.07$$
$$x = 1.07$$

The correct answer is $x = 1.07$ kilometers.

PROBLEM

Write the algebraic expression for the following. Solve the expression.

A number squared plus 6 times the number equals –8.

SOLUTION

Step 1 is to represent the unknown quantity. In this problem, the unknown quantity is the missing number.

Let x = the missing number

Step 2 is to write the left side of the equation. It is known that the missing number squared, plus 6 times the missing number, equals –8. Therefore, the left side of the equation will be:

$$x^2 + 6x$$

Step 3 is to write the right side of the equation. It is known that the result will be –8. Therefore, the right side of the equation will be:

$$-8$$

Step 4 is to write both sides of the equation together.

$$x^2 + 6x = -8$$

Step 5 is to put the equation into the standard quadratic equation, by adding 8 to both sides.

$$x^2 + 6x + 8 = -8 + 8$$

Step 6 is to simplify the equation.

$$x^2 + 6x + 8 = 0$$

Step 7 is to solve the equation.

$$x^2 + 6x + 8 = 0$$

$$(x + 4)(x + 2) = 0$$

$$x = -4,\ x = -2$$

The correct answers are $x = -4$ and $x = -2$.

PROBLEM

Write the algebraic expression for the following. Solve the expression.

An engineer is developing a theory on nuclear reactions. The theory states that 4 times the amount of plutonium (in grams) needed plus 5.00 equals the amount of plutonium squared. How much plutonium is needed?

SOLUTION

Step 1 is to represent the unknown quantity. In this problem, the unknown quantity is the amount of plutonium.

Let x = the amount of plutonium

Step 2 is to write the left side of the equation. It is known that 4 times the amount of plutonium (in grams) plus 5.00 equals the amount of plutonium squared. Therefore, the left side of the equation will be:

$4x + 5.00$

Step 3 is to write the right side of the equation. It is known that the result will be the amount of plutonium squared. Therefore, the right side of the equation will be:

x^2

Step 4 is to write both sides of the equation together.

$4x + 5 = x^2$

Step 5 is to put the equation into the standard quadratic equation, by subtracting $4x$ and 5 from both sides.

$4x + 5 - 4x - 5 = x^2 - 4x - 5$

Step 6 is to simplify the equation.

$x^2 - 4x - 5 = 0$

Step 7 is to solve the equation.

$$x^2 - 4x - 5 = 0$$

$$(x - 5)(x + 1) = 0$$

$$x = 5, x = -1$$

Since –1 grams is not a valid value, the correct answer is 5 grams of plutonium.

PROBLEM

Write the algebraic expression for the following. Solve the expression.

Farmer Ted's cows produce 1 gallon of milk for every 6 pounds of feed they consume. If the average weight of a cow is 1,500 pounds and produces 10 gallons of milk, how much feed will the cow consume?

SOLUTION

Step 1 is to represent the unknown quantity. In this problem, the unknown quantity is the amount of feed consumed by the cow.

Let x = the amount of feed consumed by the cow

Step 2 is to write the left side of the equation. It is known that the cows produce 1 gallon of milk for every 6 pounds of feed they consume. It is also known that the average weight of a cow is 1,500 pounds. This is extraneous information and is not used in the equation. Therefore, the left side of the equation will be:

$$\frac{x}{6}$$

Step 3 is to write the right side of the equation. It is known that 10 gallons of milk are produced. Therefore, the right side of the equation will be:

$$10$$

Step 4 is to write both sides of the equation together.

$$\frac{x}{6} = 10$$

Step 5 is to solve the expression.

$$\frac{x}{6} = 10$$

$$x = 10(6)$$

$$x = 60$$

The correct answer is $x = 60$ pounds of feed.

PROBLEM

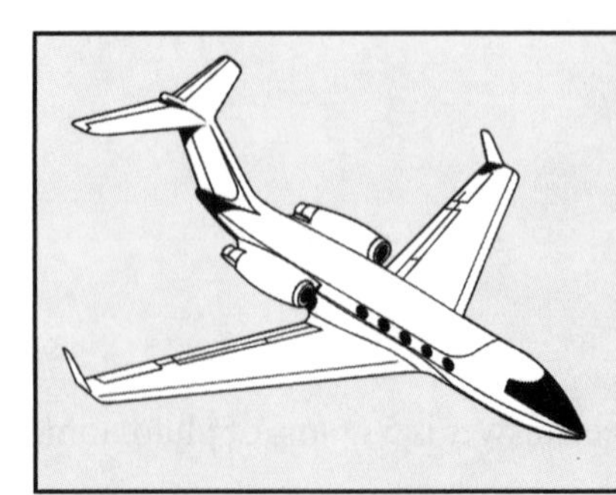

Write the algebraic expression for the following. Solve the expression.

A jet's average speed is 320 miles per hour. If the jet travels for 2 hours and the distance it covers is 2 times the distance a propeller plane could cover, what is the average speed of the propeller plane?

SOLUTION

Step 1 is to represent the unknown quantity. In this problem, the unknown quantity is the average speed of the propeller plane.

Let x = the average speed of the propeller plane

Step 2 is to write the left side of the equation. It is known that the jet's average speed is 320 miles per hour and it travels for 2 hours. Therefore, the left side of the equation will be:

$$\left(\frac{2 \text{ hours} \times 320 \text{ miles}}{\text{hour}}\right)$$

Step 3 is to write the right side of the equation. It is known that the distance it covers is 2 times the distance of a propeller plane. Therefore, the right side of the equation will be:

$2x$

Step 4 is to write both sides of the equation together.

$$\left(\frac{2 \text{ hours} \times 320 \text{ miles}}{\text{hour}}\right) = 2x$$

Step 5 is to solve the expression.

$640 \text{ miles} = 2x \qquad x = 320 \text{ miles}$

In step 6, since the propeller plane travels 320 miles in 2 hours, its speed is:

$$\frac{320}{2} \text{ hours} = 160 \text{ miles per hour}$$

The correct answer is 160 miles per hour.

PROBLEM

Write the algebraic expression for the following. Solve the expression.

Jake's dad will be 4 times Jake's current age when he turns 60. If Jake will be 20 when his dad is 60, what is Jake's current age?

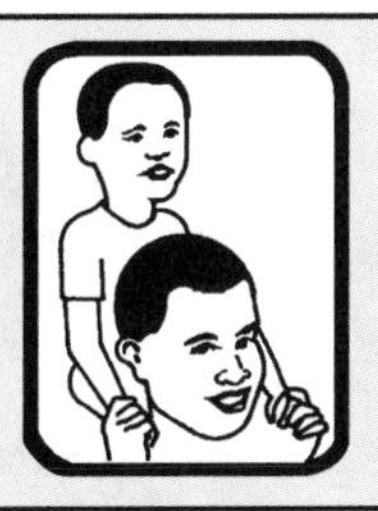

SOLUTION

Step 1 is to represent the unknown quantity. In this problem, the unknown quantity is Jake's current age.

Let x = Jake's current age

Step 2 is to write the left side of the equation. It is known that Jake's current age times 4 will equal his dad's age in the future. Therefore, the left side of the equation will be:

$4x$

Step 3 is to write the right side of the equation. It is known that his dad's age will be 60. Therefore, the right side of the equation will be:

60

Step 4 is to write both sides of the equation together.

$4(x) = 60$

Step 5 is to solve the equation.

$4x = 60 \qquad x = 15$

The correct answer is that Jake's current age is 15.

PROBLEM

Write the algebraic expression for the following. Do not solve.

A number subtracted from 8 is less than 20. What is the number?

SOLUTION

Step 1 is to represent the unknown quantity.

Let x = the unknown number

Step 2 is to set up the left side of the inequality. It is known that the unknown number is subtracted from 8. Therefore, the left side of the inequality will be:

$8 - x$

Step 3 is to set up the right side of the inequality. It is known that the unknown number, when subtracted from 8, will be less than 20. Therefore, the right side of the inequality will be:

20

Step 4 is to write both sides of the inequality together. Remember to use the "less than" sign.

$8 - x < 20$

The correct answer is $8 - x < 20$.

PROBLEM

Write the algebraic expression for the following. Do not solve.

A number plus 6, then divided by 4, is greater than or equal to 90. What is the number?

SOLUTION

Step 1 is to represent the unknown quantity.

Let x = the unknown number

Step 2 is to set up the left side of the inequality. It is known that 6 is added to the unknown number. The result is then divided by 4. Therefore, the left side of the inequality will be:

$$\frac{(x+6)}{4}$$

Step 3 is to set up the right side of the inequality. It is known that when 6 is added to the unknown number, then divided by 4, it will be greater than or equal to 90. Therefore, the right side of the inequality will be:

90

Step 4 is to write both sides of the inequality together. Remember to use the "greater than or equal to" sign.

$$\frac{(x+6)}{4} \geq 90$$

The correct answer is $\frac{(x+6)}{4} \geq 90$.

PROBLEM

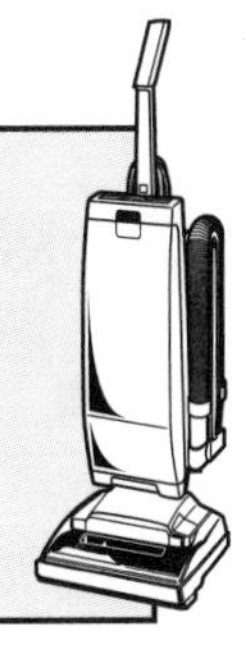

Write the algebraic expression for the following. Do not solve.

> Ms. Taylor is a travelling salesperson. She must sell at least \$200 per week to stay employed with her company, Vacuums R Us. If each product she sells is \$58, how many products must she sell to stay employed?

SOLUTION

Step 1 is to represent the unknown quantity. In this problem, the unknown quantity is the number of products sold.

Let x = the number of products sold

Step 2 is to set up the left side of the inequality. It is known that each product sold is \$58. Therefore, the left side of the inequality will be:

$\$58x$

Step 3 is to set up the right side of the inequality. It is known that she must sell at least \$200 to stay employed. Therefore, the right side of the inequality will be:

\$200

Step 4 is to write both sides of the inequality together. Remember to use the "greater than or equal to" sign.

$\$58x \geq \$200$

The correct answer is $\$58x \geq \$200$.

PROBLEM

Write the algebraic expression for the following. Solve the expression.

> A number, when divided by 2, is less than 2 times the number plus 2. What is the number?

SOLUTION

Step 1 is to represent the unknown quantity.

Let x = the unknown number

Step 2 is to set up the left side of the inequality. It is known that the unknown number is divided by 2. Therefore, the left side of the inequality will be:

$$\frac{x}{2}$$

Step 3 is to set up the right side of the inequality. It is known that the unknown number, when divided by 2, will be less than 2 times the number plus 2. Therefore, the right side of the inequality will be:

$$2x + 2$$

Step 4 is to write both sides of the inequality together. Remember to use the "less than" sign.

$$\frac{x}{2} < 2x + 2$$

Step 5 is to solve the expression.

$$\frac{x}{2} < 2x + 2$$

$$x < 4x + 4$$

$$-3x < 4$$

$$-x < \frac{4}{3}$$

$$x > -\frac{4}{3}$$

The correct answer is $x > -\frac{4}{3}$.

PROBLEM

> Write the algebraic expression for the following. Solve the expression.
>
> A number plus 5, then multiplied by 2, is less than or equal to 46. What is the number?

SOLUTION

Step 1 is to represent the unknown quantity.

Let x = the unknown number

Step 2 is to set up the left side of the inequality. It is known that 5 is added to the unknown number. The result is then multiplied by 2. Therefore, the left side of the inequality will be:

$$2(x + 5)$$

Step 3 is to set up the right side of the inequality. It is known that when 5 is added to the unknown number, then multiplied by 2, it will be less than or equal to 46. Therefore, the right side of the inequality will be:

$$46$$

Step 4 is to write both sides of the inequality together. Remember to use the "less than or equal to" sign.

$2(x + 5) \leq 46$

Step 5 is to solve the expression.

$$2(x+5) \leq 46$$
$$2x+10 \leq 46$$
$$2x \leq 36$$
$$x \leq 18$$

The correct answer is $x \leq 18$.

PROBLEM

Write the algebraic expression for the following. Solve the expression.

Profit

Shares

A stock broker has a formula for making the most profit for his customer. For each share of the Desert Technology Corporation, his customer will make \$3 profit plus \$.50 for each share from dividends. If his customer wants to make at least \$3,500 profit, how many shares are needed?

SOLUTION

Step 1 is to represent the unknown quantity. In this problem, the unknown quantity is the number of shares of stock.

Let x = the number of shares of stock

Step 2 is to set up the left side of the inequality. It is known that each share brings in a profit of \$3 plus an extra \$.50 from dividends. Therefore, the left side of the inequality will be:

$\$3x + \$0.50x$

Step 3 is to set up the right side of the inequality. It is known that he must make at least \$3,500 profit. Therefore, the right side of the inequality will be:

\$3,500

Step 4 is to write both sides of the inequality together. Remember to use the "greater than or equal to" sign.

$\$3x + \$.50x \geq \$3{,}500$

Step 5 is to solve the expression.

$$\$3x + \$.50x \geq \$3{,}500$$
$$\$3.5x \geq \$3{,}500$$
$$x \geq 1{,}000$$

The correct answer is $x \geq 1{,}000$. The stock broker has to buy at least 1,000 shares of the stock to make \$3,500 profit for his customer.

PROBLEM

Write the algebraic expression for the following. Solve the expression.

A number squared plus 10 is less than or equal to 154. What is the number?

SOLUTION

Step 1 is to represent the unknown quantity.

Let x = the unknown number

Step 2 is to set up the left side of the inequality. It is known that the unknown number is squared and then 10 is added. Therefore, the left side of the inequality will be:

$x^2 + 10$

Step 3 is to set up the right side of the inequality. It is known that the unknown number, when squared and then added to 10, will be less than or equal to 154. Therefore, the right side of the inequality will be:

154

Step 4 is to write both sides of the inequality together. Remember to use the "less than or equal to" sign.

$x^2 + 10 \leq 154$

Step 5 is to solve the expression.

$$x^2 + 10 \leq 154$$
$$x^2 \leq 144$$
$$x \leq \pm 12$$

The correct answer is $-12 \leq x \leq 12$.

DRILL: ALGEBRAIC

1. The sum of two numbers is 41. One number is one less than twice the other. Find the larger of the two numbers.

(A) 13 (B) 14 (C) 21 (D) 27 (E) 41

2. The sum of two consecutive integers is 111. Three times the larger integer less two times the smaller integer is 58. Find the value of the smaller integer.

(A) 55 (B) 56 (C) 58 (D) 111 (E) 112

3. The difference between two integers is 12. The sum of the two integers is 2. Find both integers.

(A) 7 and 5 (B) 7 and – 5 (C) – 7 and 5
(D) 2 and 12 (E) – 2 and 12

2. RATE

One of the formulas you will use for rate problems will be:

Rate × Time = Distance

PROBLEM

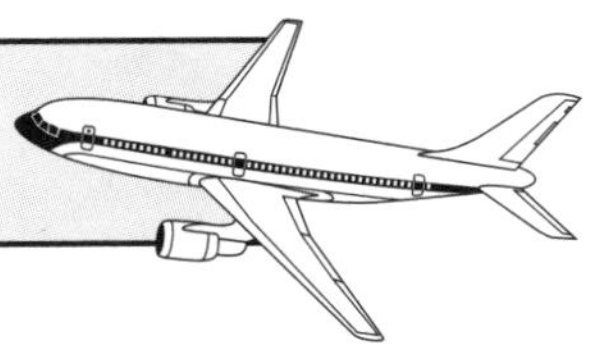

If a plane travels five hours from New York to California at a speed of 600 miles per hour, how many miles does the plane travel?

SOLUTION

Using the formula rate × time = distance, multiply 600 mph × 5 hours = 3000 miles.

The average rate at which an object travels can be solved by dividing the total distance traveled by the total amount of time.

PROBLEM

On a 40-mile bicycle trip, Cathy rode half the distance at 20 mph and the other half at 10 mph. What was Cathy's average speed on the bike trip?

SOLUTION

First you need to break down the problem. On half of the trip, which would be 20 miles, Cathy rode 20 mph. Using the rate formula, $^{\text{distance}}/_{\text{rate}}$ = time, you would compute,

$$\frac{\text{20 miles}}{\text{20 miles per hour}} = 1 \text{ hour}$$

to travel the first 20 miles. During the second 20 miles, Cathy traveled at 10 miles per hour, which would be

$$\frac{\text{20 miles}}{\text{10 miles per hour}} = 2 \text{ hours}$$

Thus, the average speed Cathy traveled would be $^{40}/_{3}$ = 13.33 miles per hour.

In solving for some rate problems you can use cross multiplication involving ratios to solve for x.

PROBLEM

If 2 pairs of shoes cost \$52, then what is the cost of 10 pairs of shoes at this rate?

SOLUTION

$$\frac{2}{52} = \frac{10}{x}, 2x = 52 \times 10, x = \frac{520}{2}, x = \$260.$$

DRILL: RATE

1. Two towns are 420 miles apart. A car leaves the first town traveling toward the second town at 55 mph. At the same time, a second car leaves the other town and heads toward the first town at 65 mph. How long will it take for the two cars to meet?

(A) 2 hr (B) 3 hr (C) 3.5 hr (D) 4 hr (E) 4.25 hr

2. A camper leaves the campsite walking due east at a rate of 3.5 mph. Another camper leaves the campsite at the same time but travels due west. In two hours the two campers will be 15 miles apart. What is the walking rate of the second camper?

(A) 2.5 mph (B) 3 mph (C) 3.25 mph
(D) 3.5 mph (E) 4 mph

3. A bicycle racer covers a 75-mile training route to prepare for an upcoming race. If the racer could increase his speed by 5 mph, he could complete the same course in 3/4 of the time. Find his average rate of speed.

(A) 15 mph (B) 15.5 mph (C) 16 mph
(D) 18 mph (E) 20 mph

3. WORK

In work problems, one of the basic formulas is

$$\frac{1}{x}+\frac{1}{y}=\frac{1}{z}$$

where x and y represent the number of hours it takes two objects or people to complete the work and z is the total number of hours when both are working together.

PROBLEM

Otis can seal and stamp 400 envelopes in 2 hours while Elizabeth seals and stamps 400 envelopes in 1 hour. In how many hours can Otis and Elizabeth, working together, complete a 400-piece mailing at these rates?

SOLUTION

$$\frac{1}{2}+\frac{1}{1}=\frac{1}{z},\ \frac{1}{2}+\frac{2}{2}=\frac{3}{2},\ \frac{3}{2}=\frac{1}{z},\quad 3z=2$$

$z = {}^{2}/_{3}$ of an hour or 40 minutes. Working together, Otis and Elizabeth can seal and stamp 400 envelopes in 40 minutes.

DRILL: WORK

1. It takes Marty 3 hours to type the address labels for his club's newsletter. It only takes Pat $2^1/_4$ hours to type the same amount of labels. How long would it take them working together to complete the address labels?

(A) $^7/_9$ hr (B) $1\,^2/_7$ hr (C) $1\,^4/_5$ hour

(D) $2\,^5/_8$ hr (E) $5\,^1/_4$ hr

2. It takes Troy 3 hours to mow his family's large lawn. With his little brother's help, he can finish the job in only 2 hours. How long would it take the little brother to mow the entire lawn alone?

(A) 4 hr (B) 5 hr (C) 5.5 hr (D) 6 hr (E) 6.75 hr

3. A tank can be filled by one inlet pipe in 15 minutes. It takes an outlet pipe 75 minutes to drain the tank. If the outlet pipe is left open by accident, how long would it take to fill the tank?

(A) 15.5 min (B) 15.9 min (C) 16.8 min

(D) 18.75 min (E) 19.3 min

4. MIXTURE

Mixture problems present the combination of different products and ask you to solve for different parts of the mixture.

PROBLEM

A chemist has an 18% solution and a 45% solution of a disinfectant. How many ounces of each should be used to make 12 ounces of a 36% solution?

SOLUTION

Let x = Number of ounces from the 18% solution, and

y = Number of ounces from the 45% solution.

$$x + y = 12 \quad (1)$$

$$.18x + .45y = .36(12) \quad (2)$$

Note that .18 of the first solution is pure disinfectant and that .45 of the second solution is pure disinfectant. When the proper quantities are drawn from each mixture the result is 12 ounces of mixture which is .36 pure disinfectant.

The second equation cannot be solved with two unknowns. Therefore, write one variable in terms of the other and plug it into the second equation.

$$x = 12 - y \quad (1)$$

$$.18(12 - y) + .45y = .36(12) \quad (2)$$

Simplifying,

$$2.16 - .18y + .45y = 4.32$$

$$.27y = 4.32 - 2.16$$

$$.27y = 2.16$$

$$y = 8$$

Substituting for y in the first equation,

$$x + 8 = 12$$

$$x = 4$$

Therefore, 4 ounces of the first and 8 ounces of the second solution should be used.

PROBLEM

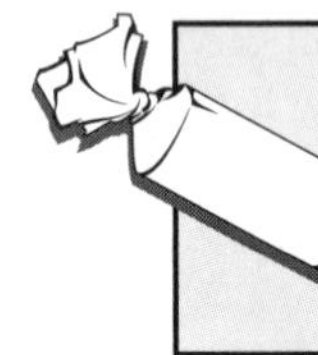

Clark pays $2.00 per pound for 3 pounds of peanut butter chocolates and then decides to buy 2 pounds of chocolate covered raisins at $2.50 per pound. If Clark mixes both together, what is the cost per pound of the mixture?

SOLUTION

The total mixture is 5 pounds and the total value of the chocolates is

$$3(\$2.00) + 2(\$2.50) = \$11.00$$

The price per pound of the chocolates is $^{\$11.00}/_{5\text{ pounds}} = \$2.20$.

DRILL: MIXTURE

1. How many liters of a 20% alcohol solution must be added to 80 liters of a 50% alcohol solution to form a 45% solution?

(A) 4 (B) 8 (C) 16 (D) 20 (E) 32

2. How many kilograms of water must be evaporated from 50 kg of a 10% salt solution to obtain a 15% salt solution?

(A) 15 (B) 15.75 (C) 16 (D) $16.\overline{66}$ (E) 16.75

3. How many pounds of coffee *A* at $3.00 a pound should be mixed with 2.5 pounds of coffee *B* at $4.20 a pound to form a mixture selling for $3.75 a pound?

(A) 1 (B) 1.5 (C) 1.75 (D) 2 (E) 2.25

5. INTEREST

If the problem calls for computing simple interest, the interest is computed on the principal alone. If the problem involves compounded interest, then the interest on the principal is taken into account in addition to the interest earned before.

PROBLEM

> How much interest will Jerry pay on his loan of \$400 for 61 days at 6% per year?

SOLUTION

Use the formula:

Interest = Principal × Rate × Time ($I = P \times R \times T$).

$\$400 \times 6\%/\text{year} \times 61 \text{ days} = \$400 \times .06 \times {}^{1}/_{6}$

$= \$400 \times {}^{1}/_{100} = \$4.00.$

Jerry will pay \$4.00.

PROBLEM

> Mr. Smith wishes to find out how much interest he will receive on \$300 if the rate is 3% compounded annually for three years.

SOLUTION

Compound interest is interest computed on both the principal and the interest it has previously earned. The interest is added to the principal at the end of every year. The interest on the first year is found by multiplying the rate by the principal. Hence, the interest for the first year is

$$3\% \times \$300 = .03 \times \$300 = \$9.00.$$

The principal for the second year is now \$309, the old principal (\$300) plus the interest (\$9). The interest for the second year is found by multiplying the rate by the new principal. Hence, the interest for the second year is

$$3\% \times \$309 = .03 \times \$309 = \$9.27.$$

The principal now becomes \$309 + \$9.27 = \$318.27.

The interest for the third year is found using this new principal. It is

$$3\% \times \$318.27 = .03 \times \$318.27 = \$9.55.$$

At the end of the third year his principal is \$318.27 + 9.55 = \$327.82. To find how much interest was earned, we subtract his starting principal (\$300) from his ending principal (\$327.82), to obtain

$$\$327.82 - \$300.00 = \$27.82.$$

DRILL: INTEREST

1. A man invests \$3,000, part in a 12-month certificate of deposit paying 8% and the rest in municipal bonds that pay 7% a year. If the yearly return from both investments is \$220, how much was invested in bonds?

(A) \$80 (B) \$140 (C) \$220 (D) \$1000 (E) \$2000

2. A sum of money was invested at 11% a year. Four times that amount was invested at 7.5%. How much was invested at 11% if the total annual return was \$1,025?

(A) \$112.75 (B) \$1025 (C) \$2500
(D) \$3400 (E) \$10,000

3. One bank pays 6.5% a year simple interest on a savings account while a credit union pays 7.2% a year. If you had \$1500 to invest for three years, how much more would you earn by putting the money in the credit union?

(A) \$10.50 (B) \$31.50 (C) \$97.50 (D) \$108 (E) \$1500

6. DISCOUNT

If the discount problem asks to find the final price after the discount, first multiply the original price by the percent of discount. Then subtract this result from the original price.

If the problem asks to find the original price when only the percent of discount and the discounted price are given, simply subtract the percent of discount from 100% and divide this percent into the sale price. This will give you the original price.

PROBLEM

A popular bookstore gives 10% discount to students. What does a student actually pay for a book costing \$24.00?

SOLUTION

10% of \$24 is \$2.40 and hence the student pays \$24 – \$2.40 = \$21.60.

PROBLEM

Eugene paid \$100 for a business suit. The suit's price included a 25% discount. What was the original price of the suit?

SOLUTION

Let x represent the original price of the suit and take the complement of .25 (discount price) which is .75.

$$.75x = \$100 \text{ or } x = 133.33$$

So, the original price of the suit is \$133.33.

DRILL: DISCOUNT

1. A man bought a coat marked 20% off for \$156. How much had the coat cost originally?

(A) \$136 (B) \$156 (C) \$175 (D) \$195 (E) \$205

2. A woman saved \$225 on the new sofa, which was on sale for 30% off. What was the original price of the sofa?

(A) \$25 (B) \$200 (C) \$225 (D) \$525 (E) \$750

3. At an office supply store, customers are given a discount if they pay in cash. If a customer is given a discount of \$9.66 on a total order of \$276, what is the percent of discount?

(A) 2% (B) 3.5% (C) 4.5% (D) 9.66% (E) 276%

7. PROFIT

The formula used for the profit problems is

Profit = Revenue – Cost or

Profit = Selling Price – Expenses.

PROBLEM

Four high school and college friends started a business of remodeling and selling old automobiles during the summer. For this purpose they paid \$600 to rent an empty barn for the summer. They obtained the cars from a dealer for \$250 each, and it takes an average of \$410 in materials to remodel each car. How many automobiles must the students sell at \$1,440 each to obtain a gross profit of \$7,000?

SOLUTION

Total Revenues – Total Cost = Gross Profit

Revenue – [Variable Cost + Fixed Cost] = Gross Profit

Let a = number of cars

Revenue = \$1,440a

Variable Cost = (\$250 + 410)a

Fixed Cost = \$600

The desired gross profit is \$7,000.

Using the equation for the gross profit,

$$1{,}440a - [660a + 600] = 7{,}000$$

$$1{,}440a - 660a - 600 = 7{,}000$$
$$780a = 7{,}600$$
$$a = 9.74$$

or to the nearest car, $a = 10$.

PROBLEM

A glass vase sells for $25.00. The net profit is 7%, and the operating expenses are 39%. Find the gross profit on the vase.

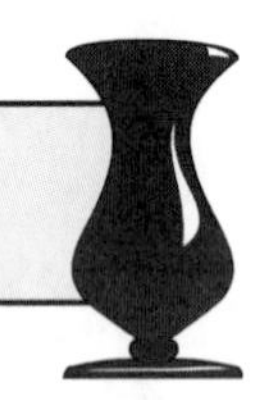

SOLUTION

The gross profit is equal to the net profit plus the operating expenses. The net profit is 7% of the selling cost; thus it is equal to 7% × $25.00 = .07 × $25 = $1.75. The operating expenses are 39% of the selling price, thus equal to 39% × $25 = .39 × $25 = $9.75.

$1.75	net profit
+ $9.75	operating expenses
$11.50	gross profit

DRILL: PROFIT

1. An item cost a store owner $50. She marked it up 40% and advertised it at that price. How much profit did she make if she later sold it at 15% off the advertised price?

(A) $7.50 (B) $9.50 (C) $10.50 (D) $39.50 (E) $50

2. An antique dealer made a profit of 115% on the sale of an oak desk. If the desk cost her $200, how much profit did she make on the sale?

(A) $230 (B) $315 (C) $430 (D) $445 (E) $475

3. As a graduation gift, a young man was given 100 shares of stock worth $27.50 apiece. Within a year the price of the stock had risen by 8%. How much more were the stocks worth at the end of the first year than when they were given to the young man?

(A) $110 (B) $220 (C) $1220 (D) $2750 (E) $2970

8. SETS

A **set** is any collection of well-defined objects called elements.

A set which contains only a finite number of elements is called a **finite set**; a set which contains an infinite number of elements is called an **infinite set**. Often the sets are designated by listing their elements. For example: $\{a, b, c, d\}$ is the set which contains elements a, b, c, and d. The set of positive integers is $\{1, 2, 3, 4, \ldots\}$.

Venn diagrams can represent sets. These diagrams are circles which help to visualize the relationship between members or objects of a set.

PROBLEM

In a certain Broadway show audition, it was asked of 30 performers if they knew how to either sing or dance, or both. If 20 auditioners said they could dance and 14 said they could sing, how many could sing and dance?

SOLUTION

Divide the 30 people into 3 sets: those who dance, those who sing and those who dance and sing. S is the number of people who both sing and dance. So $20 - S$ represents the number of people who dance and $14 - S$ represents the number of people who sing.

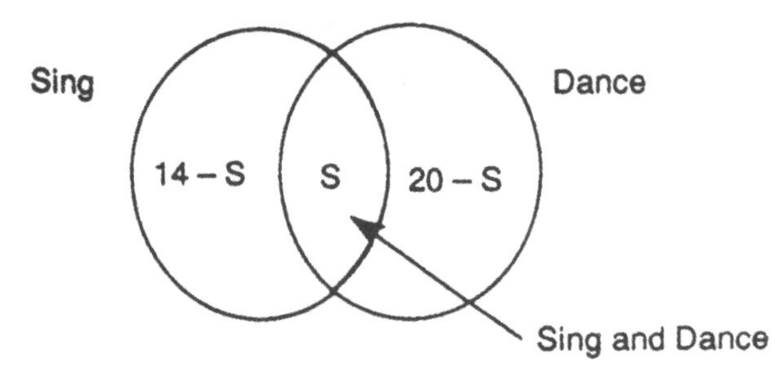

The equation for this problem is as follows:

$$(20 - S) + S + (14 - S) = 30.$$
$$20 + 14 - 30 = S$$
$$34 - 30 = S$$
$$4 = S$$

So, 4 people in the audition both sing and dance.

DRILL: SETS

1. In a small school there are 147 sophomores. Of this number, 96 take both Biology and Technology I. Eighty-three take both Chemistry and Technology I. How many students are taking Technology I?

(A) 32 (B) 51 (C) 64 (D) 83 (E) 96

2. In a survey of 100 people, 73 owned only stocks. Six of the people invested in both stocks and bonds. How many people owned bonds only?

(A) 6 (B) 21 (C) 73 (D) 94 (E) 100

3. On a field trip, the teachers counted the orders for a snack and sent the information in with a few people. The orders were for 77 colas only and 39 fries only. If there were 133 orders, how many were for colas and fries?

(A) 17 (B) 56 (C) 77 (D) 95 (E) 150

9. GEOMETRY

PROBLEM

A boy knows that his height is 6 ft. and his shadow is 4 ft. long. At the same time of day, a tree's shadow is 24 ft. long. How high is the tree? (See following figure.)

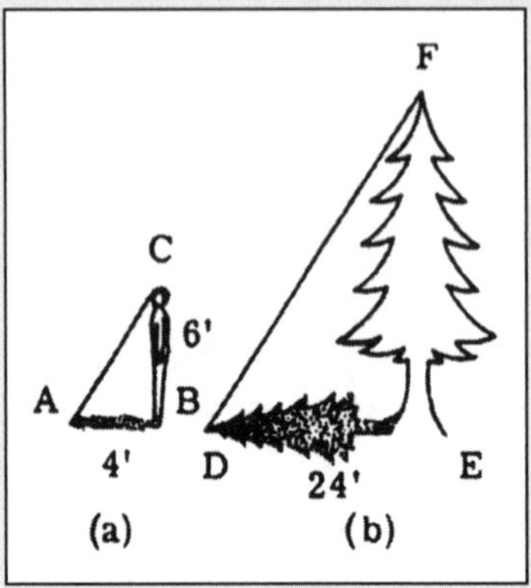

SOLUTION

Show that $\Delta ABC \approx \Delta DEF$, and then set up a proportion between the known sides AB and DE, and the sides BC and EF.

First, assume that both the boy and the tree are $\perp$ to the earth. Then, $\overline{BC} \perp \overline{BA}$ and $\overline{EF} \perp \overline{ED}$. Hence,

$$\angle ABC \cong \angle DEF.$$

Since it is the same time of day, the rays of light from the sun are incident on both the tree and the boy at the same angle, relative to the earth's surface. Therefore,

$$\angle BAC \cong \angle EDF.$$

We have shown, so far, that 2 pairs of corresponding angles are congruent. Since the sum of the angles of any triangle is 180°, the third pair of corresponding angles is congruent (i.e. $\angle ACB \cong \angle DFE$). By the Angle Angle Angle (AAA) Theorem

$$\angle ABC \approx \angle DEF.$$

By definition of similarity,

$$\frac{FE}{CB} = \frac{ED}{BA}.$$

$CB = 6'$, $ED = 24'$, and $BA = 4'$. Therefore,

$$FE = (6')\,(24'/4') = 36'.$$

DRILL: GEOMETRY

1. Δ *PQR* is a scalene triangle. The measure of $\angle P$ is 8 more than twice the measure of $\angle R$. The measure of $\angle Q$ is two less than three times the measure of $\angle R$. Determine the measure of $\angle Q$.

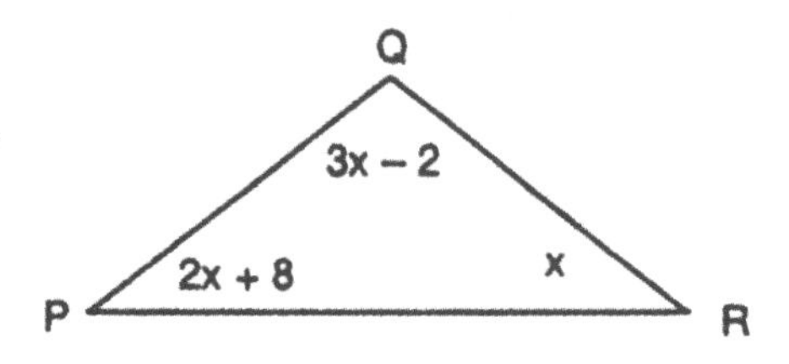

(A) 29 (B) 53 (C) 60 (D) 85 (E) 174

2. Angle *A* and angle *B* are supplementary. The measure of angle *B* is 5 more than four times the measure of angle *A*. Find the measure of angle *B*.

(A) 35 (B) 125 (C) 140 (D) 145 (E) 155

3. Triangle *RUS* is isosceles with base $\overline{SU}$. Each leg is 3 less than 5 times the length of the base. If the perimeter of the triangle is 60 cm, find the length of a leg.

(A) 6 (B) 12 (C) 27 (D) 30 (E) 33

10. MEASUREMENT

When measurement problems are presented in either metric or English units which involve conversion of units, the appropriate data will be given in the problem.

PROBLEM

The Eiffel Tower is 984 feet high. Express this height in meters, in kilometers, in centimeters, and in millimeters.

SOLUTION

A meter is equivalent to 39.370 inches. In this problem, the height of the tower in feet must be converted to inches and then the inches can be converted to meters. There are 12 inches in 1 foot. Therefore, feet can be converted to inches by using the factor 12 inches/1 foot.

$$984 \text{ feet} \times 12 \text{ inches/1 foot} = 11808$$

Once the height is found in inches, this can be converted to meters by the factor 1 meter/39.370 inches.

$$11808 \text{ inches} \times 1 \text{ meter/39.370 inches} = 300 \text{ m.}$$

Therefore, the height in meters is 300 m.

There are 1,000 meters in one kilometer. Meters can be converted to kilometers by using the factor 1 km/1000 m.

$$300 \text{ m} \times 1 \text{ km/1000 m} = .300 \text{ km.}$$

As such, there are .300 kilometers in 300 m.

There are 100 centimeters in 1 meter, thus meters can be converted to centimeters by multiplying by the factor 100 cm/1 m.

$$300 \text{ m} \times 100 \text{ cm/1 m} = 300 \times 10^2 \text{ cm}.$$

There are 30,000 centimeters in 300 m.

There are 1,000 millimeters in 1 meter; therefore, meters can be converted to millimeters by the factor 1000 mm/1 m.

$$300 \text{ m} \times 1{,}000 \text{ mm/1 m} = 300 \times 10^3 \text{ mm}.$$

There are 300,000 millimeters in 300 meters.

PROBLEM

The unaided eye can perceive objects which have a diameter of 0.1 mm. What is the diameter in inches?

SOLUTION

From a standard table of conversion factors, one can find that 1 inch = 2.54 cm. Thus, cm can be converted to inches by multiplying by 1 inch/2.54 cm. Here, one is given the diameter in mm, which is .1 mm. Millimeters are converted to cm by multiplying the number of mm by .1 cm/1 mm. Solving for cm, you obtain:

$$0.1 \text{ mm} \times .1 \text{ cm/1 mm} = .01 \text{ cm}.$$

Solving for inches:

$$0.01 \text{ cm} \times {}^{1 \text{ inch}}/_{2.54 \text{ cm}} = 3.94 \times 10^{-3} \text{ inches}.$$

Measurements are numbers that describe the amount or quantity of some attribute. These attributes can be time, distance, speed, weight, area, volume, or temperature. Further, these attributes can be measured in different units. Distance can be expressed in various units of measurement including inches, feet, yards, and miles. Distance is a one-dimensional measurement. Weight is a common measurement that can be expressed in grams, ounces, pounds, and tons. Area is an important measurement that can be expressed in terms of square inches, square feet, and square yards. In addition, area can be measured in square miles and acres. Area is a two-dimensional measurement. Volume is a three-dimensional measurement. For solids, volume can be measured in terms of cubic inches, cubic feet, or cubic yards. For liquids, volume can be measured in ounces, cups, pints, quarts, and gallons. Temperature can be measured in degrees Fahrenheit or degrees Centigrade.

EXAMPLE

If there are 5,280 feet in one mile, how many yards are there in 4 miles?

First, divide the number of feet in a mile by 3 in order to find the number of yards in a mile: $5{,}280 \div 3 = 1{,}760$ yards. Second, multiply 1,760 by 4 in order to find the number of yards in 4 miles: $1{,}760 \times 4 = 7{,}040$ yards.

EXAMPLE

How many pounds are there in 25 tons?

There are 2,000 pounds in one ton, so we multiply 2,000 by 25 to find the number of pounds in 25 tons: $2{,}000 \times 25 = 50{,}000$ pounds.

EXAMPLE

Candle-Wick is a company that manufactures candles. Their standard candle weighs 5 ounces. In order to calculate shipping charges for a case of candles, a worker needs to know the weight of 24 candles.

24	candles	20	ounces
$\times 5$	ounces per candle	$\div 16$	ounces per pound
120	ounces	7.5	pounds, which equals 7 pounds and 8 ounces

EXAMPLE

Calculate the number of square yards in an area that measures 162 square feet.

162	square feet
$\div 9$	square feet per square yard
18	square yards

EXAMPLE

How many ounces are there in 2 and 3/4 cups?

Express 2 and 3/4 in decimal form: 2.75. Because there are 8 ounces in a cup, there are 22 ounces in 2.75 cups: $2.75 \times 8 = 22$.

EXAMPLE

How many gallons are there in 246 quarts?

There are 4 quarts in one gallon, so 246 quarts is the same amount as 61.5 gallons: $246 \div 4 = 61.5$ gallons.

EXAMPLE

One inch is approximately equal to 2.54 centimeters. Calculate the approximate number of centimeters in 5 inches.

2.54 centimeters $\times 5 = 12.7$ centimeters

EXAMPLE

One cubic inch is approximately equal to 16.38 cubic centimeters. Calculate the approximate number of cubic centimeters in 5 cubic inches.

16.38 cubic centimeters × 5 = 81.9 cubic centimeters

EXAMPLE

There are two commonly used temperature scales. The Fahrenheit temperature scale is widely used in the United States, but the Celsius temperature scale (also called the Centigrade scale) is equally useful.

Water boils at 212 degrees Fahrenheit and 100 degrees Celsius, and freezes at 32 degrees Fahrenheit and 0 degrees Celsius.

At –40 degrees, the Fahrenheit and Celsius temperature scales meet.

A Celsius temperature can be converted in a Fahrenheit temperature using the following formula:

$$F = \frac{9}{5}C + 32$$

Conversely, Fahrenheit temperature can be converted in a Celsius temperature using the formula:

$$C = \frac{5}{9}(F - 32)$$

EXAMPLE

A traveler from the United States is visiting a country that uses the Celsius temperature scale. The temperature on the thermometer reads 25 degrees Celsius. Calculate the temperature in degrees Fahrenheit.

$$F = \frac{9}{5}(25) + 32$$

$$F = \frac{225}{5} + 32$$

$$45 + 32 = 77 \text{ degrees Fahrenheit}$$

EXAMPLE

A traveler visiting the United States comes from a foreign country that uses the Celsius temperature scale. The temperature on the thermometer reads 86 degrees Fahrenheit. Calculate the temperature in degrees Celsius.

$$C = \frac{5}{9}(86 - 32)$$

$$C = \frac{5}{9}(54)$$

$$\frac{270}{9} = 30 \text{ degrees Celsius}$$

PROBLEM

How many ounces are there in 4 and 3/4 pounds?

SOLUTION

Convert 4 and 3/4 into its decimal form: 4.75. Because there are 16 ounces in a pound, we can find the number of ounces in 4.75 pounds by multiplying: $4.75 \times 16 = 76$.

PROBLEM

A parent is planning a birthday party for a child. The parent has 3 and 1/4 gallons of fruit punch. How many cups is this?

SOLUTION

There are 4 quarts in a gallon and 4 cups in a quart. Therefore, there are 16 cups in a gallon. $3.25 \times 16 = 52$ cups.

PROBLEM

A thermometer reads 20 degrees Celsius. How many degrees Fahrenheit is this?

SOLUTION

Use the following formula to solve this problem:

$$F = \frac{9}{5}C + 32$$

$$F = \frac{9}{5}(20) + 32$$

$$F = \frac{180}{5} + 32$$

$$36 + 32 = 68 \text{ degrees Fahrenheit}$$

PROBLEM

The temperature on a Fahrenheit thermometer reads 59 degrees. How many degrees Celsius is this?

SOLUTION

Use the following formula to solve this problem:

$$C = \frac{5}{9}(F - 32)$$

$$C = \frac{5}{9}(59 - 32)$$

$$C = \frac{5}{9}(27)$$

$$\frac{135}{9} = 15 \text{ degrees Celsius}$$

DRILL: MEASUREMENT

1. A brick walkway measuring 3 feet by 11 feet is to be built. The bricks measure 4 inches by 6 inches. How many bricks will it take to complete the walkway?

(A) 132 (B) 198 (C) 330 (D) 1927 (E) 4752

2. A wall to be papered is three times as long as it is wide. The total area to be covered is 192 ft^2. Wallpaper comes in rolls that are 2 feet wide by 8 feet long. How many rolls will it take to cover the wall?

(A) 8 (B) 12 (C) 16 (D) 24 (E) 32

3. A bottle of medicine containing 2 kg is to be poured into smaller containers that hold 8 grams each. How many of these smaller containers can be filled from the 2 kg bottle?

(A) 0.5 (B) 1 (C) 5 (D) 50 (E) 250

11. DATA INTERPRETATION

Data interpretation questions test the ability to read graphs and tables and apply information given in order to come to a decision. The questions are based upon graphs or tables. Most of the graphs presented are grid graphs, pie charts, r graphs.

le of Weight Distribution of a 70,000–Gram Man
(Weights of some organs given)

	Weight in Grams	Organ	Weight in Grams
	10,000	Muscles	30,000
	5,000	Intestinal tract	2,000
	1,700	Lungs	1,000
	1,500		

t of the weight of the blood is made up of cells, what percent (to enth) of the total body weight is made up of blood cells?

(B) 3.6 (C) 1.4

(E) 9.9

ion represents the total body weight if the weight of the skeleton l by S grams?

(B) $70000S$ (C) $60S$

(E) Cannot be determined

;

40% of the weight of blood is made up of cells, then the weight nes 5000 grams or 2000 grams. So, to find the percent of the is made up of blood cells, form the following ratio and change

.028 = 2.8%

he solution one needs to set up a proportion. Thus, let x de- ight in grams and S denote the weight of the skeleton. Then, on can be formed.

$$\frac{\text{ton}}{\text{ht}} = \frac{10{,}000 \text{ grams}}{70{,}000 \text{ grams}} = \frac{S}{x}$$

TEGIES

Try no o much time understanding the data. Become more famil- you read and answer the questions.

- It is more efficient to estimate rather than to compute the average.
- When a question seems too involved, it is helpful to break it down into parts.
- Avoid figuring large computations by estimating products and quotients.
- Become familiar with the different types of charts discussed above.

PROBLEM

In which year was the least number of bushels of wheat produced? (See figure below.)

SOLUTION

By inspection of the graph, we find that the shortest bar representing wheat production is the one representing the wheat production for 1993. Thus, the least number of bushels of wheat was produced in 1993.

Number of bushels (to the nearest 5 bushels) of wheat and corn produced by farm RQS from 1992 - 2002.

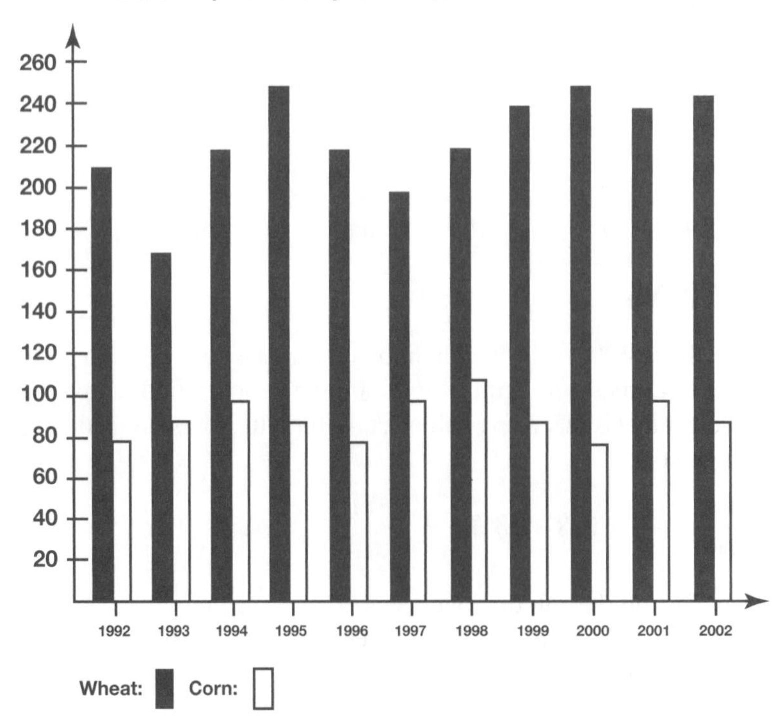

PROBLEM

What was the ratio of wheat production in 2002 to that of 1992?

SOLUTION

From the graph representing wheat production, the number of bushels of wheat produced in 1992 is equal to 210 bushels. This number can be found by locating the bar on the graph representing wheat production in 1992 and then drawing a horizontal line from the top of that bar to the vertical axis. The point where this horizontal line meets the vertical axis represents the number of bushels of wheat produced in 1992. This number on the vertical axis is 210. Similarly, the graph indicates that the number of bushels of wheat produced in 2002 is equal to 245 bushels.

Thus, the ratio of wheat production in 2002 to that of 1992 is 245 to 210, which can be written as $^{245}/_{210}$. Simplifying this ratio to its simplest form yields

$$\frac{245}{210} = \frac{5 \times 7 \times 7}{2 \times 3 \times 5 \times 7} = \frac{7}{2 \times 3} = \frac{7}{6} \text{ or } 7{:}6$$

DRILL: DATA INTERPRETATION

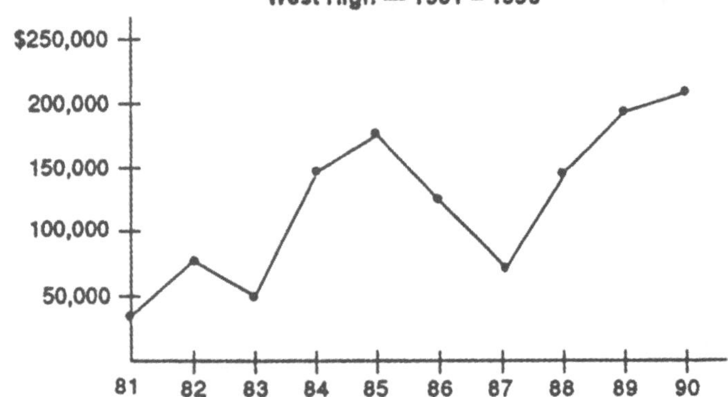

1. What was the approximate amount of scholarship money awarded in 1985?

   (A) $150,000 (B) $155,000 (C) $165,000
   (D) $175,000 (E) $190,000

2. By how much did the scholarship money increase between 1987 and 1988?

   (A) $25,000 (B) $30,000 (C) $50,000
   (D) $55,000 (E) $75,000

3. By how much did the mileage increase for Car 2 when the new product was used? (See figure on following page.)

(A) 5 mpg (B) 6 mpg (C) 7 mpg (D) 10 mpg (E) 12 mpg

4. Which car's mileage increased the most in this test?

   (A) Car 1 (B) Car 2 (C) Car 3
   (D) Cars 1 and 2 (E) Cars 2 and 3

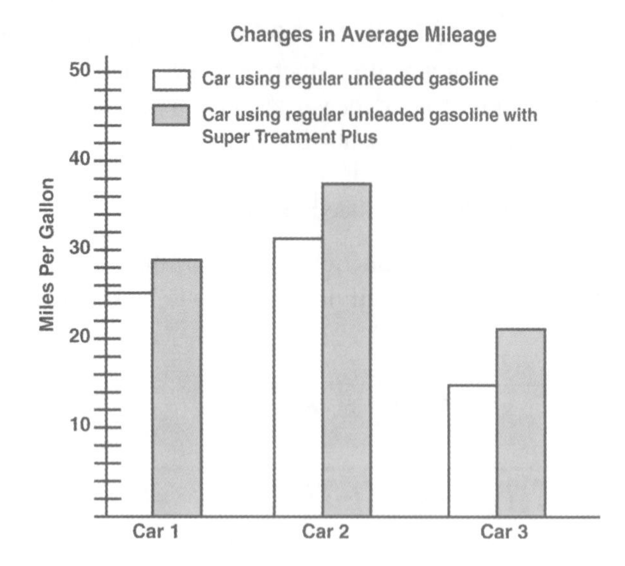

5. According to the bar graph, if your car averages 25 mph, what mileage might you expect with the new product?

(A) 21 mpg (B) 29 mpg (C) 31 mpg (D) 35 mpg (E) 37 mpg

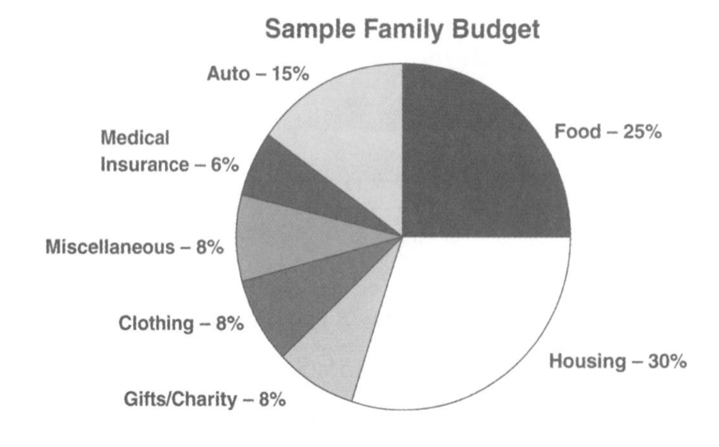

6. Using the budget shown, a family with an income of $1500 a month would plan to spend what amount on housing?

(A) $300 (B) $375 (C) $450 (D) $490 (E) $520

7. In this sample family budget, how does the amount spent on an automobile compare to the amount spent on housing?

(A) $^1/_3$ (B) $^1/_2$ (C) $^2/_3$ (D) $1^1/_2$ (E) 2

8. A family with a monthly income of $1240 spends $125 a month on clothing. By what amount do they exceed the sample budget?

(A) $1.00 (B) $5.20 (C) $10.00 (D) $25.80 (E) $31.75

CALORIE CHART — BREADS

Bread	Amount	Calories
French Bread	2 oz	140
Bran Bread	1 oz	95
Whole Wheat	1 oz	115
Oatmeal Bread	0.5 oz	55
Raisin Bread	1 oz	125

9. One dieter eats two ounces of french bread. A second dieter eats two ounces of bran bread. The second dieter has consumed how many more calories than the first dieter?

(A) 40 (B) 45 (C) 50 (D) 55 (E) 65

10. One ounce of whole wheat bread has how many more calories than an ounce of oatmeal bread?

(A) 5 (B) 15 (C) 60 (D) 75 (E) 125

12. SCALED DRAWINGS

A scaled drawing is a drawing of an object that is proportional to the size of the actual object. Scaled drawings are based on a certain scale of conversion. If you know this scale, you can determine the measurements of the actual objects.

EXAMPLE

The scaled drawing of a volleyball court shown below is drawn using the scale 1 centimeter is equal to 2 meters.

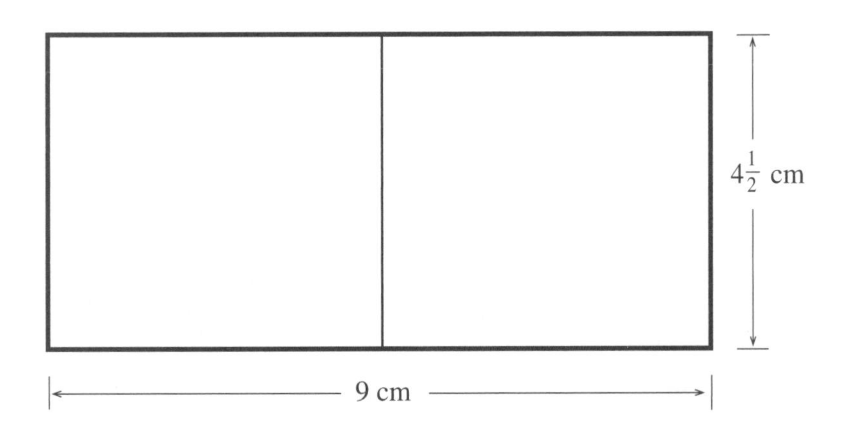

Calculate the area of this volleyball court in square meters.

In the scaled drawing, length is equal to 9 centimeters and width is equal to 4.5 centimeters. The scale is "1 centimeter is equal to 2 meters." So, the length is 18 meters and the width is 9 meters. To find area, multiply length times width: $18 \times 9 = 162$. The volleyball court is 162 square meters.

EXAMPLE

1 inch equals 24 feet in the scaled drawing of a tennis court (shown below).

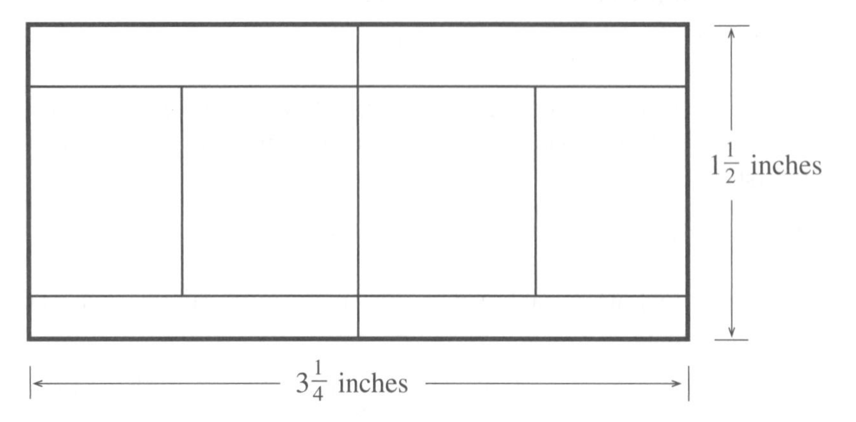

Calculate the length of this tennis court in feet.

In the scaled drawing, length is equal to $3^1/_4$ inches (in decimal form, 3.25 inches.) The scale is "1 inch equals 24 feet." In order to find the length in feet, multiply the number of inches times 24: $3.25 \times 24 = 78$ feet.

PROBLEM

In the scaled drawing of a soccer field below, 1 centimeter (cm) = 10 meters.

What is the area of the Penalty Box in square meters (not including the Goal Box)?

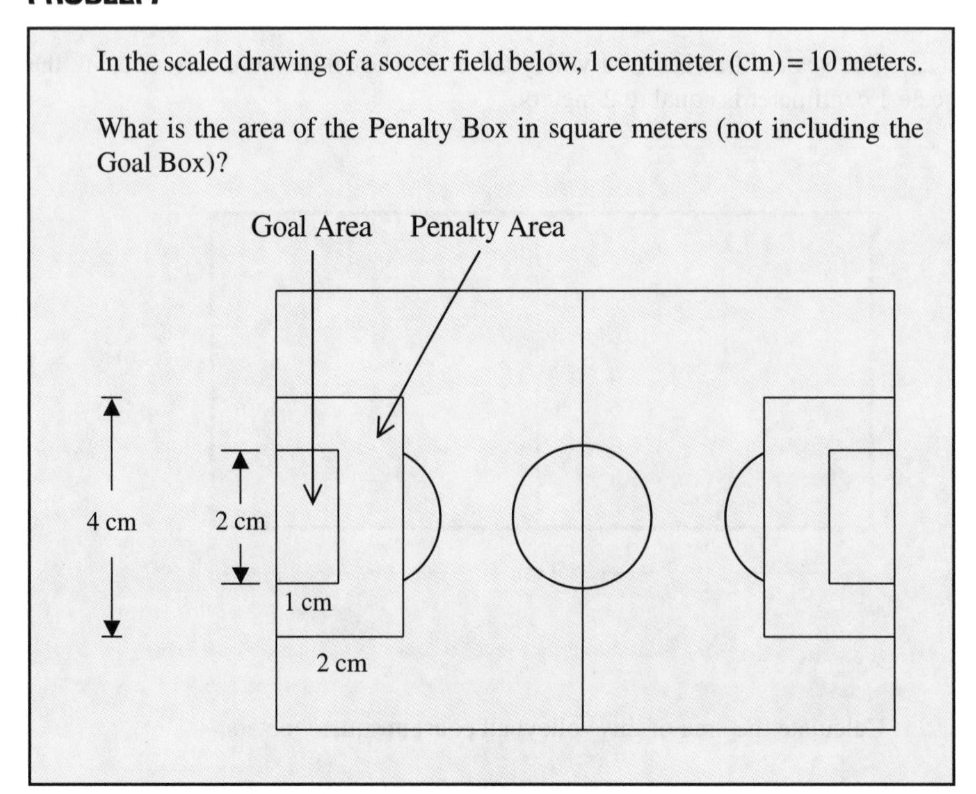

SOLUTION

First, calculate the large area that includes both the Penalty Box and the Goal Box. 4 cm represents 40 meters, and 2 cm represents 20 meters. $40 \times 20 = 800$ square meters.

Second, calculate the area of the Goal Box. 2 centimeters represents 20 meters, and 1 cm represents 10 meters. $20 \times 10 = 200$ square meters.

Finally, to find the area of Penalty Box alone, subtract the smaller area from the larger area. $800 - 200 = 600$ square meters.

---

13. PROBLEMS WITH MISSING OR MISLEADING INFORMATION

PROBLEM

A truck made 100 different deliveries in one week, using a total of 40 gallons of gasoline. What was the average number of miles per gallon?

What other information is needed in order to solve this problem?

SOLUTION

In order to calculate the average number of miles per gallon, you need to know two different variables: miles and gallons. This question gives you the number of gallons, but *not* the number of miles traveled. The fact that the truck made 100 deliveries is irrelevant information.

PROBLEM

Delivery Company	Customer Complaints	Months in Business
SPEEDY	100	25
ON TIME	10	2

On Time used the data in the table above to support their claim, "We have one-tenth the number of customer complaints that Speedy has." Why is this claim misleading?

SOLUTION

The number of complaints is not as important as the average number of complaints. This is because an average takes into consideration how long a company has been in business. Speedy has 100/25, or an average of 4 complaints per month. However, On Time has 10/2, or an average of 5 complaints per month. Therefore, on the average, On Time has more complaints per month. It is misleading for a company to say that they have less complaints than another company when they have not been in business as long.

14. MAKING INFERENCES

In a survey, a small sample of people are asked some questions. The results of this survey are used to infer how a large population of people are feeling. Surveys are done to infer who will win presidential elections. They are also used to infer how large audiences feel about television shows and movies.

EXAMPLE

A survey asked a group of people to name their favorite pet. The results from this survey are shown in the chart.

FAVORITE PET

Pet	Number of People
Bird	8
Cat	24
Fish	16
Tortoise	5
Dog	20
Hamster	7

If a large survey of 200 people were conducted, how many people are expected to say that their favorite pet is a fish?

First, find the proportion of people who have a fish by dividing the number of people who have a fish by the total number of people surveyed: $16 \div 80 = 0.2$. Next, multiply the number of people in the large survey by this proportion: $200 \times 0.2 = 40$ people.

EXAMPLE

The graph below shows the number of oranges produced by a grower in California for the years 1997, 1999, and 2001.

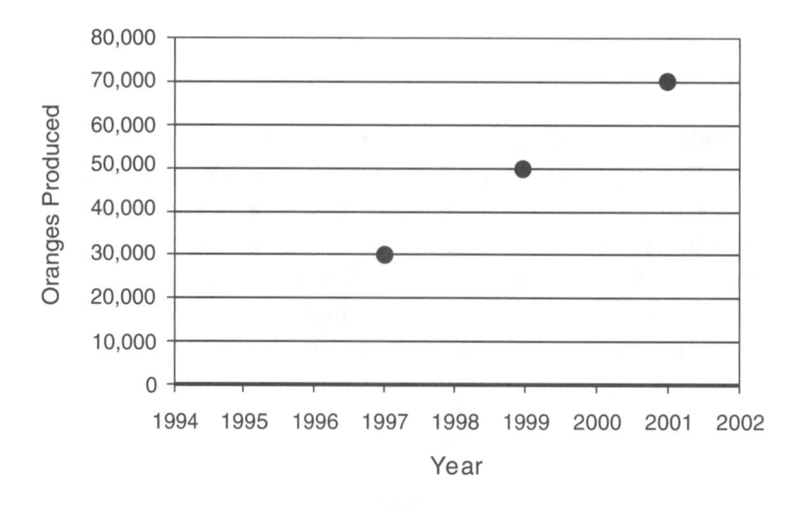

From this graph, find the probable number of oranges produced by this grower in 1995.

Draw a line through the points given but also extend this line downward. Next, find the year 1995 on the x-axis and then go up until you hit the line that you have drawn. Then, read the number of oranges produced on the y-axis. The answer is 10,000 oranges.

PROBLEM

The chart shows the results of a survey that asked a group of eighth-grade students to indicate who has their vote for class president.

Candidate	Number of Students
Rachael	18
Stephanie	21
Daniel	12
Bobby	9

Based on the survey results, how many of the 400 eighth-grade students in the whole school would vote for Stephanie?

SOLUTION

The first step is to figure out what proportion of the survey would vote for Stephanie. This is done by dividing the number of students that voted for Stephanie in the survey by the total number of students in the survey: $21 \div 60 = 0.35$. Now, multiply this proportion by the total number of students in the school: $0.35 \times 400 = 140$ students.

WORD PROBLEM DRILLS

ANSWER KEY

DRILL—ALGEBRAIC
1. (D) 2. (A) 3. (B)

DRILL—RATE
1. (C) 2. (E) 3. (A)

DRILL—WORK
1. (B) 2. (D) 3. (D)

DRILL—MIXTURE
1. (C) 2. (D) 3. (B)

DRILL—INTEREST
1. (E) 2. (C) 3. (B)

DRILL—DISCOUNT
1. (D) 2. (E) 3. (B)

DRILL—PROFIT
1. (B) 2. (A) 3. (B)

DRILL—SETS
1. (A) 2. (B) 3. (A)

DRILL—GEOMETRY
1. (D) 2. (D) 3. (C)

DRILL—MEASUREMENT
1. (B) 2. (B) 3. (E)

DRILL—DATA INTERPRETATION
1. (D) 4. (E) 7. (B) 10. (A)
2. (E) 5. (B) 8. (D)
3. (B) 6. (C) 9. (C)

QUANTITATIVE ABILITY

Quantitative Ability questions include quantitative comparison, discrete quantitative, and data interpretation skills. Quantitative Ability measures basic mathematical ability, quantitative reasoning, problem solving, and understanding of mathematical concepts. These questions require a review of basic arithmetic, algebra, geometry, and familiarity with word problems.

The following material introduces the types of questions you may be asked in the section.

QUANTITATIVE COMPARISON

The quantitative comparison questions challenge ability to determine accurately and quickly the relative sizes of two quantities. It also measures the ability to see whether enough information is given in order to pick a correct choice.

The directions to be observed for this topic are as follows:

Numbers: All numbers are real numbers.

Figures: Position of points, angles, regions, etc. are assumed to be in the order shown and angle measures are assumed to be positive.

Lines: Assume that lines shown as straight are indeed straight.

Directions: Each of the following given set of quantities is placed into either column A or B. Compare the two quantities to decide whether:

A. the quantity in Column A is greater

B. the quantity in Column B is greater

C. the two quantities are equal

D. the relationship cannot be determined from the information given

Note: Do not choose (E) since there are only four choices.

Common Information: Information which relates to one or both given quantities is centered above the two columns. A symbol which appears in both columns will indicate the same item in Column A and Column B.

EXAMPLES

	Column A	Column B
	The length of a ruler is "L". This value is increased by 10%, and then decreased by 10%.	
1.	L	final length

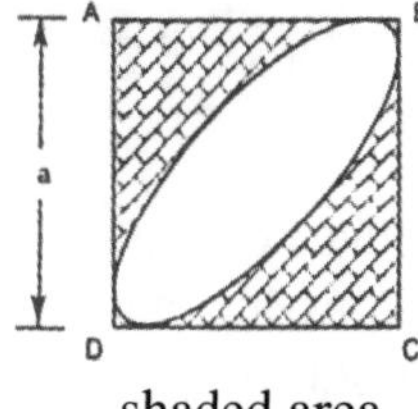

ABCD is a square. The unshaded area represents the intersection of two quadrants.

2. shaded area | unshaded area

SUGGESTED STRATEGIES:

- Memorize and understand the directions since they present abundant information. You will be able to save more time to work on the problems instead of spending time reading the directions.
- It is better to simplify both given quantities as little as possible to come to a conclusion rather than extensively work out the problem. This will save time and still will allow you to arrive at the correct answer.
- The answer can be reduced to three choices [(A), (B), or (C)] if both quantities being compared present no variables. If this is the case, the answer can never be (D), which states the relationship **cannot be determined from the information given.**
- If it is established that (A) is greater when considering certain numbers but (B) is greater when considering other certain numbers, then choose (D) immediately.
- If it is established that (A) is greater when considering certain numbers, then you can rule out choices (B) and (C) immediately.
- If it is established that (B) is greater when considering certain numbers, then you can rule out choices (A) and (C) immediately.

SOLUTIONS

1. **(A)** The original length = "L". When the ruler increases by 10%, the length will be

$$L + 10\% \text{ of } L = L + (10/100)L = 1.1\ L$$

Then the ruler is decreased by 10%:

$$1.1L - 10\% \text{ of } 1.1L = 1.1L - (10/100)1.1L = .99L$$

Note: In the second part, the 10% refers to $1.1L$. Therefore, the final length is $.99L$ and the answer is (A) because $L > .99L$.

2. **(B)** First, evaluate the shaded area with the following procedure:

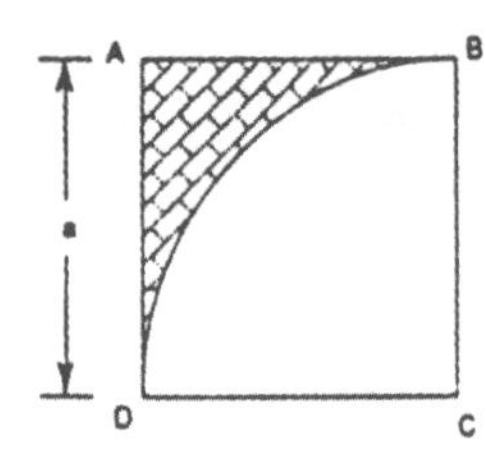

The half-shaded area can be expressed by

half-shaded area = (Sq. area – Quadrant area)

half-shaded area = $a^2 - \pi a^2/4 = a^2(1 - \pi/4)$

because the quadrant represents one-fourth of the area of a circle with the radius a. Therefore, the shaded area in the problem will be

shaded area = $2a^2 (1 - \pi/4) \approx .43a^2$

The unshaded area can be expressed by

unshaded area = (Sq. area – Shaded area)

unshaded area = $a^2 - 2a^2(1 - \pi/4)$

unshaded area = $a^2(\pi/2 - 1) \approx .57a^2$

Therefore, the unshaded area > the shaded area. The answer is (B).

DISCRETE QUANTITATIVE

Discrete Quantity questions test the ability to apply given information in an abstract situation in order to solve a problem.

Directions: For the following questions, select the best answer choice to the given question.

EXAMPLES

1. Which of the following has the smallest value?

   (A) $^{1}/_{0.2}$ (B) $^{0.1}/_{2}$ (C) $^{0.2}/_{1}$

   (D) $^{0.2}/_{0.1}$ (E) $^{2}/_{0.1}$

2. A square is inscribed in a circle of area 18π. What is the length of a side of the square?

   (A) 6 (B) 3 (C) $3\sqrt{2}$

   (D) $6\sqrt{2}$ (E) Cannot be determined

SUGGESTED STRATEGIES:

- First determine what form the answer should be put in by noting the forms used in the answer choices.

- If the question requires an approximation, get an idea of the degree of approximation by scanning the multiple choices.

SOLUTIONS

1. **(B)** Note that

$\frac{.1}{2} = \frac{.1 \times 10}{2 \times 10} = \frac{1}{20}$ for Response (B). For choice (A),

$\frac{1}{.2} = \frac{1 \times 10}{.2 \times 10} = \frac{10}{2} = 5$ which is larger than $\frac{1}{20}$. For choice (C),

$\frac{.2}{1} = \frac{.2 \times 10}{1 \times 10} = \frac{2}{10} = \frac{1}{5}$ which is larger than $\frac{1}{20}$. For choice (D),

$\frac{.2}{.1} = \frac{.2 \times 10}{.1 \times 10} = \frac{2}{1} = 2$ which is larger than $\frac{1}{20}$. For choice (E),

$\frac{2}{.1} = \frac{2 \times 10}{.1 \times 10} = \frac{20}{1} = 20$ which is larger than $\frac{1}{20}$.

2. **(A)** The formula for the area of a circle is $A = \pi r^2$. Since the area of the square is 18π, then it is true that

$$\pi r^2 = 18\pi \quad \text{or} \quad r^2 = 18 \quad \text{or} \quad r = \sqrt{18} = 3\sqrt{2}.$$

Then, the diameter of the circle is

$$2r = 2(3\sqrt{2}) = 6\sqrt{2}.$$

Using the Pythagorean Theorem:

$$x^2 + x^2 = (6\sqrt{2})^2$$

$$2x^2 = 36(2); x^2 = 36$$

$x = 6$ is the length of the side of the square.

SPATIAL SENSE/ RELATIONSHIPS

In representing a three-dimensional object we make use of a method that is universally understood. This method is a standardized technique, often referred to as orthographic projection, which represents the object of a series of views projected on planes.

Fig. 1 below shows relationships between the three principal planes which define/represent the object. The method shown allows three-dimensional objects to be represented by two-dimensional views.

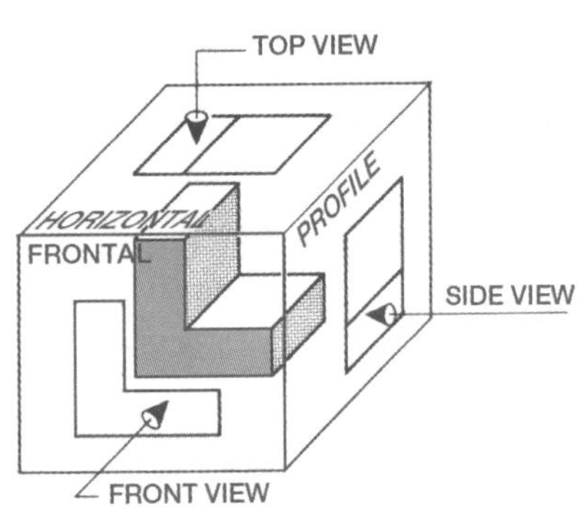

Fig. 1 — The Principal Projection Planes.

PROBLEM

Given Fig. 2, draw the front, top, and right side views of the object.

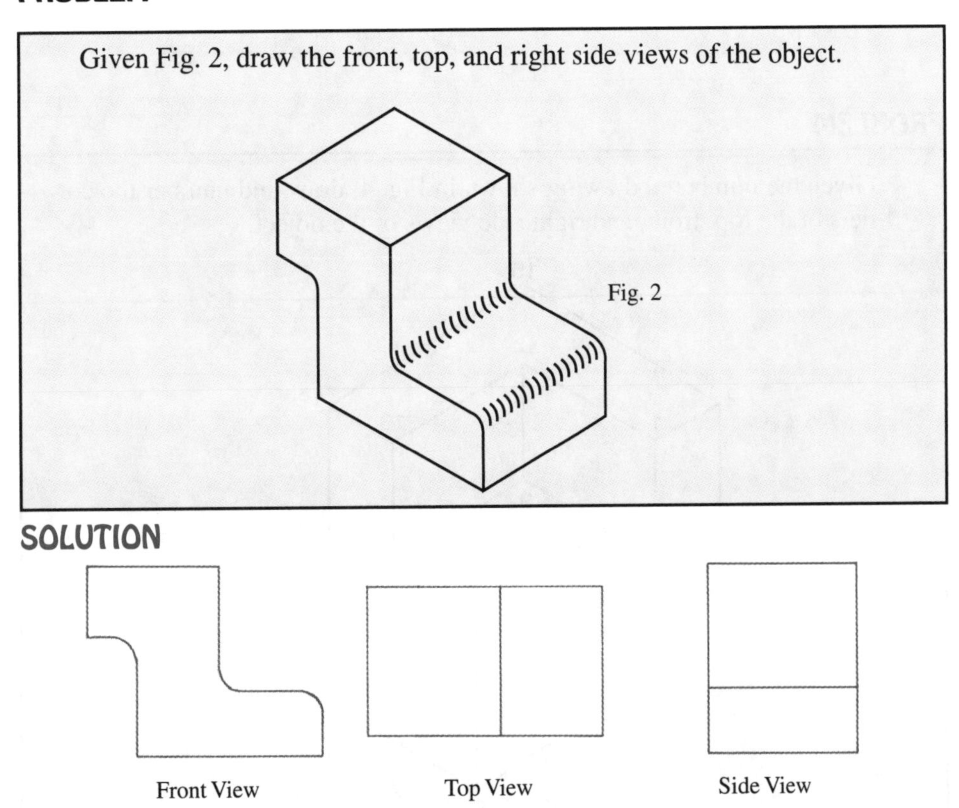

PROBLEM

From the drawing given in Fig. 3, draw the front, top, and right side views.

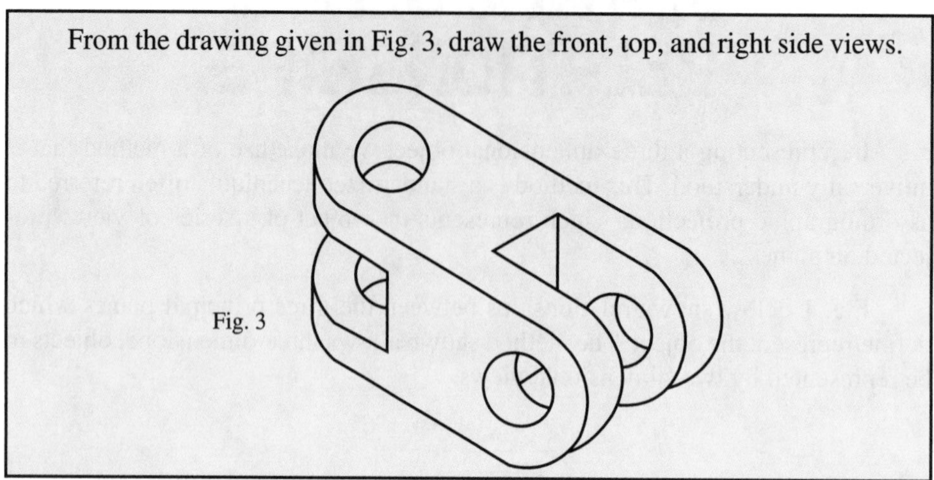

SOLUTION

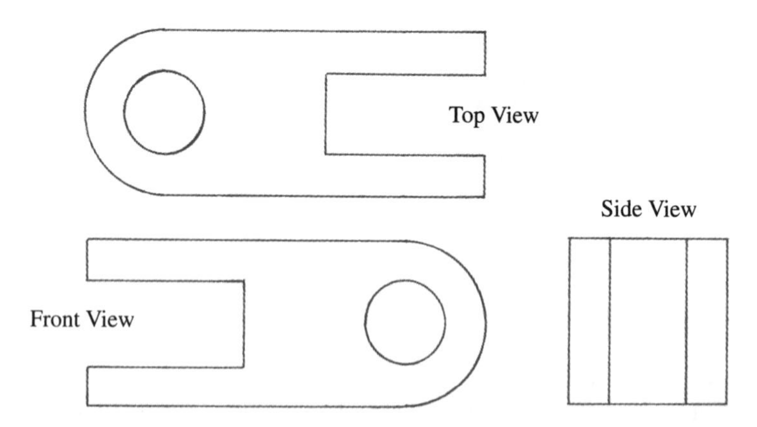

PROBLEM

Given the numbered drawing shown in Fig. 4, draw and number the corners of the top, front, and right side views of the object.

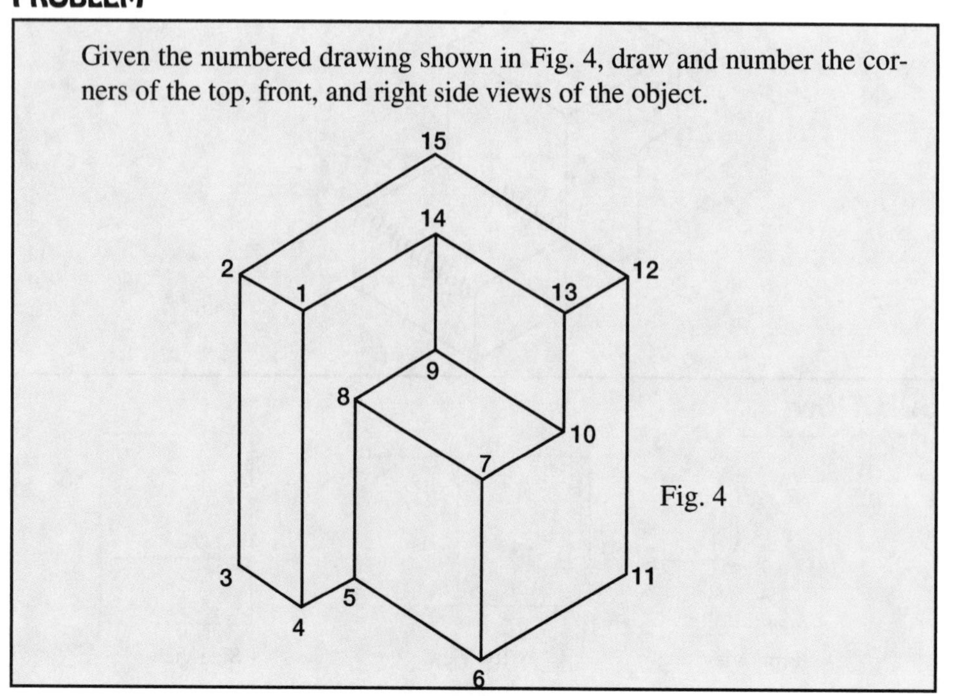

SOLUTION

In order to draw the different views asked for in the problem, first determine which planes are horizontal and which planes are vertical (refer to Fig. 4). For example, planes 1-2-15-12-13-14 and 7-8-9-10 are horizontal planes while planes 6-7-10-13-12-11 and 4-1-14-9-8-5 are vertical. Keep in mind that a vertical plane will appear as a line in the top view while a horizontal plane will appear as a line in the front and side views. To draw the top view first draw the horizontal plane 1-2-15-12-13-14. The horizontal plane 7-8-9-10 lies directly below it. This means that lines 8-9 and 9-10 of the lower plane appear to be on the same lines as lines 1-14 and 14-13 as seen from the top. Therefore, plane 7-8-9-10 can be drawn by using lines 1-14 and 14-13 as references. The same type of analysis can be made for both the front and right side views. Notice that this problem would be made more complicated had inclined planes been included in the given drawing. Had this been the case, the lines representing the inclined planes would have to be located by using horizontal and vertical planes for reference.

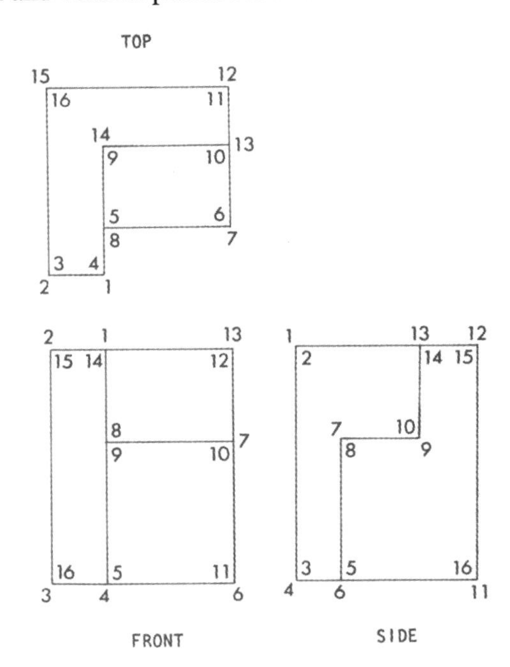

PROBLEM

Draw the front, top, and right views of the object represented in Fig. 5. Use the "Glass Box" method.

Fig. 5

SOLUTION

The "Glass Box" method of drawing the projections of an object makes use of an imaginary box which surrounds the given object. The box is placed so that the faces of the "glass box" are parallel to the principal sides of the given object. When the "glass box" has been properly placed, each of the faces of the object is projected perpendicularly to the corresponding face of the "glass box." Figure 6 shows what the "glass box" looks like after all the sides of the object have been projected. Figure 7 shows the three required views.

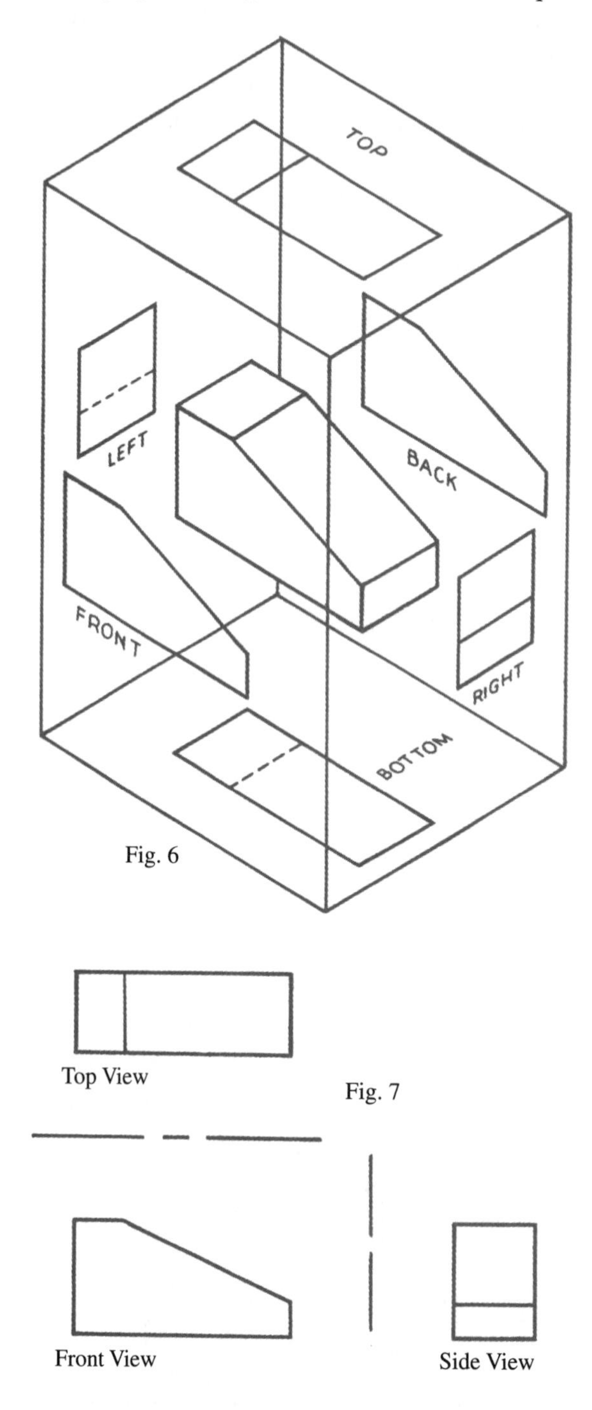

Fig. 6

Fig. 7

DRILL: SPATIAL SENSE/RELATIONSHIPS

1. Fig. 8 is a top view of a combination of stacked cubes. Numbers in the squares correspond to the number of cubes in the stacks below the top plane. Which one of the configurations—A, B, C, or D—corresponds to the top view?

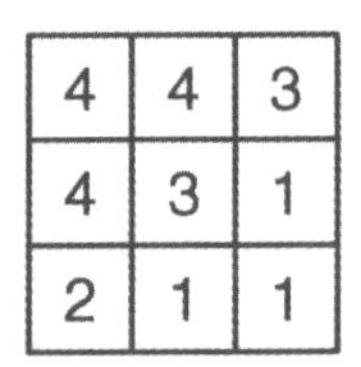

Fig. 8

A

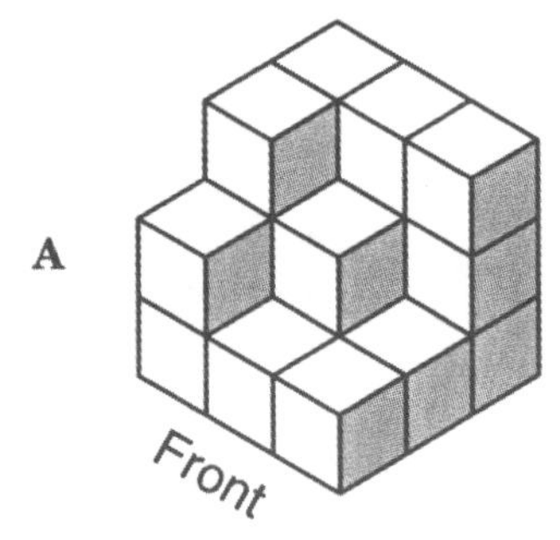

C

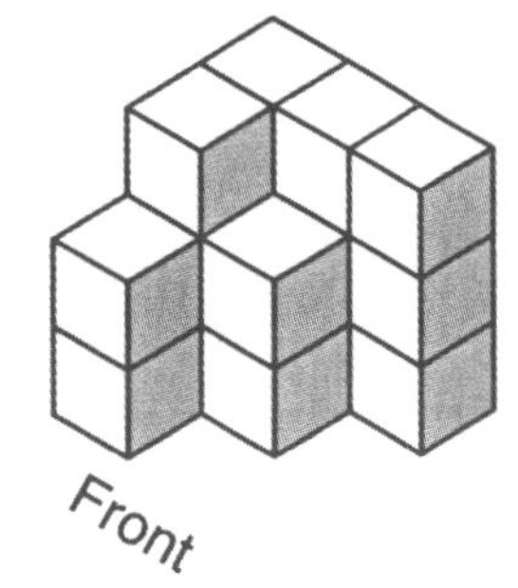

B

D

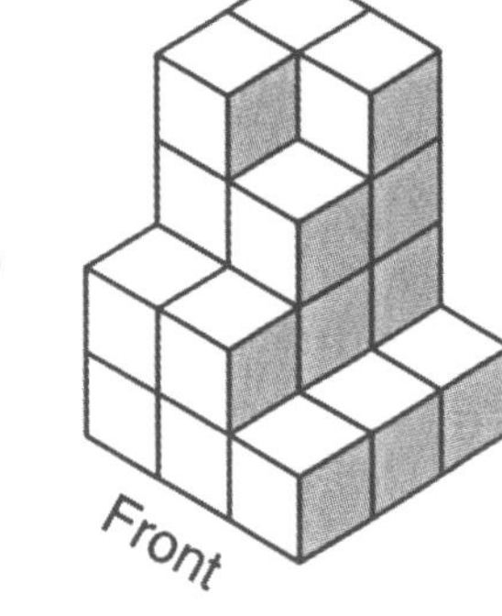

2. The triangle in Fig. 9 is reflected across the y-axis. Which one of the configurations, A to E, is the result?

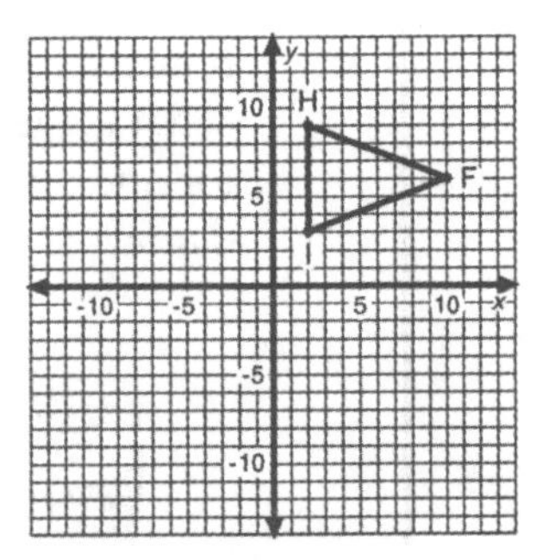

Fig. 9

A.

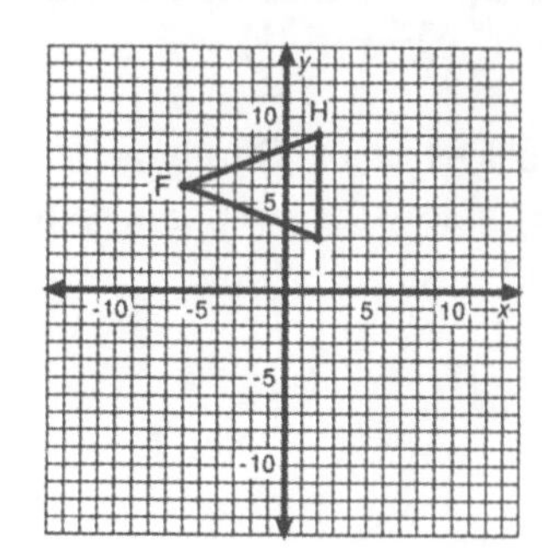
y
10
H
F
5
I
-10
-5
5
10
x
-5
-10

D.

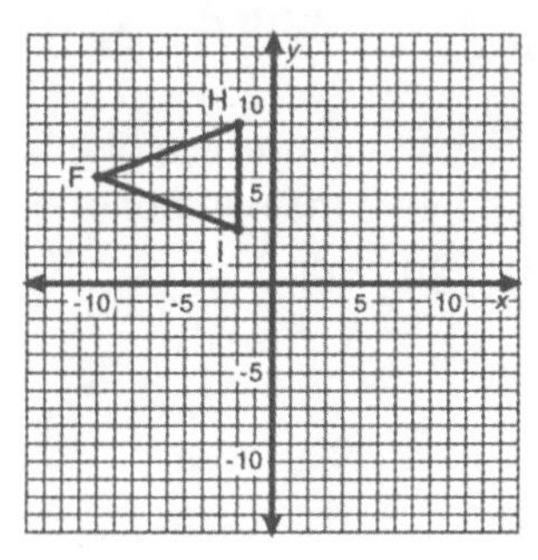
y
H
10
F
5
I
-10
-5
5
10
x
-5
-10

B.

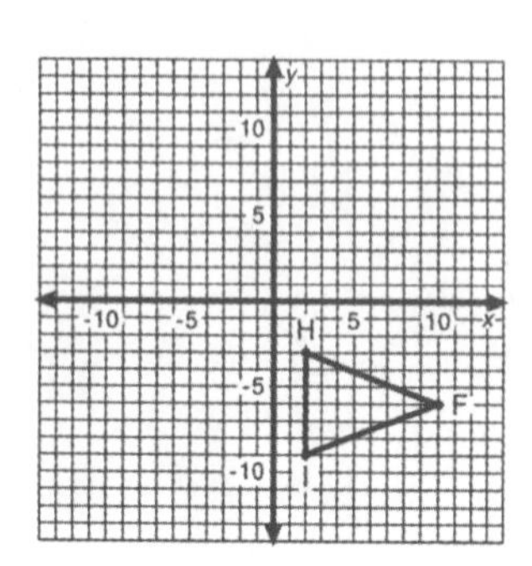
y
10
5
-10
-5
5
10
x
H
-5
F
-10
I

E.

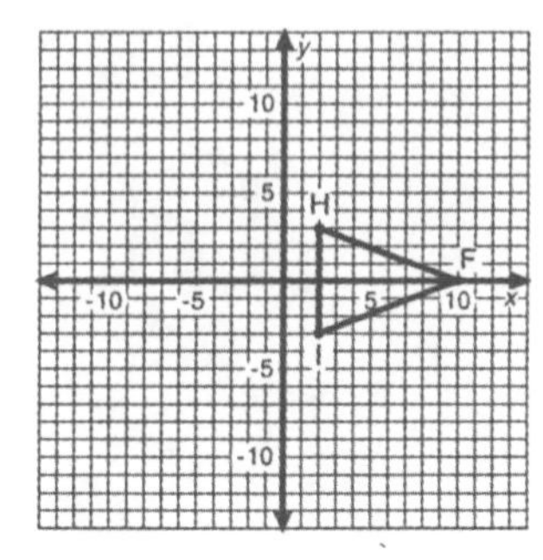
y
10
5
H
F
-10
-5
5
10
x
I
-5
-10

C.

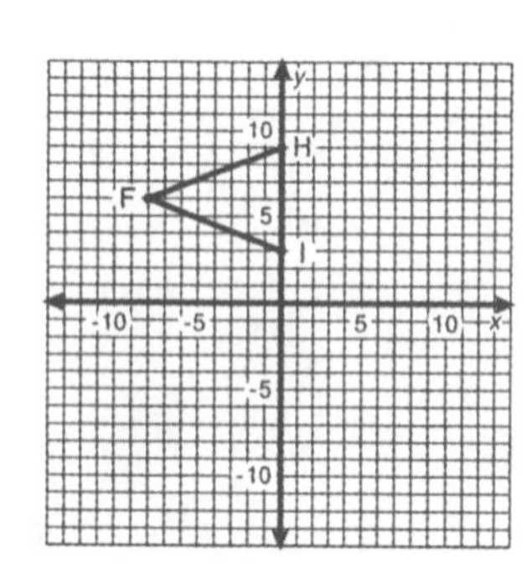
y
10
H
F
5
I
-10
-5
5
10
x
-5
-10

3. In Fig. 10, which transformation will map ΔABC onto ΔDEF?

A. Reflect ΔABC over the y-axis and shift up 6 spaces.

B. Reflect ΔABC over the x-axis and shift up 6 spaces.

C. Reflect ΔABC over the y-axis and shift down 6 spaces.

D. Reflect ΔABC over the y-axis, reflect over the x-axis, and shift down 4 spaces.

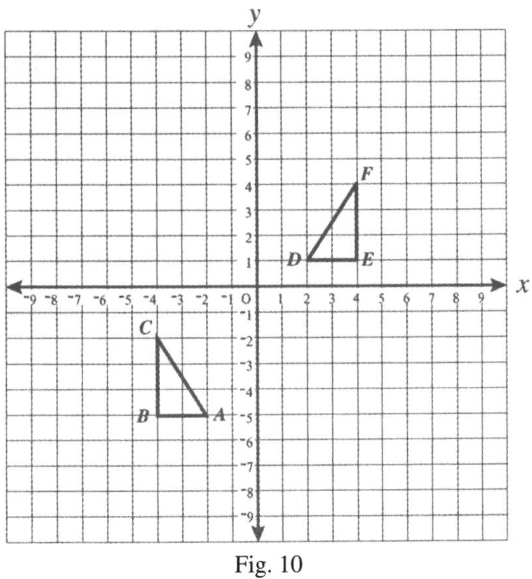

Fig. 10

---

4. Fig. 11 is constructed from cube blocks. Which drawing—A, B, C, or D—represents the top view of the structure?

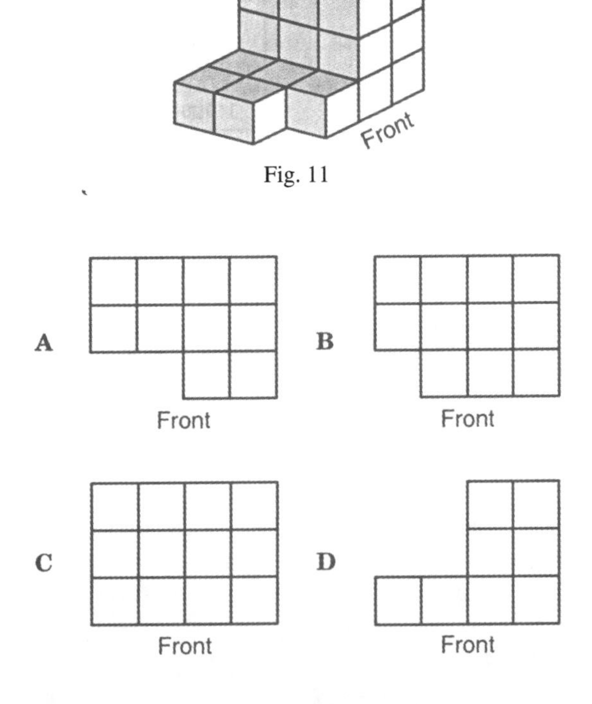

Fig. 11

SPATIAL SENSE/ RELATIONSHIPS DRILL

1. (B) 2. (D) 3. (A) 4. (B)

STATISTICS

WHAT IS STATISTICS?

Statistics is the science of assembling, organizing, and analyzing data, as well as drawing conclusions about what the data means. Often these conclusions come in the form of predictions, and this makes statistics particularly important.

The first task of a statistician is to define the population and to choose the sample. In most cases it is impractical or impossible to examine the entire group, which is called the **population** or universe. This being the case, one must instead examine a part of the population called a **sample**. The chosen sample has to reflect as closely as possible the characteristics of the population. For example, statistical methods were used in testing the Salk vaccine, which protects children against polio. The sample consisted of 400,000 children. Half of the children, chosen at random, received the Salk vaccine; the other half received an inactive solution. After the study was completed it turned out that 50 cases of polio appeared in the group which received the vaccine and 150 cases appeared in the group which did not receive the vaccine. Based on that study of a sample of 400,000 children it was decided that the Salk vaccine was effective for the entire population.

A population can be finite or infinite.

EXAMPLE

The population of all television sets in the USA is finite.

The population consisting of all possible outcomes in successive tosses of a coin is infinite.

Descriptive or deductive statistics collects data concerning a given group and analyzes it without drawing conclusions.

From an analysis of a sample, inductive statistics or statistical inference draws conclusions about the population.

VARIABLES

Variables are usually denoted by symbols, such as x, y, z, t, a, M, etc. A variable can assume any of its prescribed values. A set of all the possible values of a variable is called its domain.

EXAMPLE

For a toss of a die the domain is $\{1, 2, 3, 4, 5, 6\}$.

The variable can be discrete or continuous.

EXAMPLE

The income of an individual is a discrete variable. It can be \$1,000 or \$1,000.01 but it cannot be between these two numbers.

EXAMPLE

The height of a person can be 70 inches or 70.1 inches; it can also assume any value between these two numbers. Hence, height is a continuous variable.

Similarly, we are dealing with discrete data or continuous data. Usually, countings and enumerations yield discrete data, while measurements yield continuous data.

FUNCTIONS

Y is a function of X if for each value of X there corresponds one and only one value of Y. We write

$$Y = f(X)$$

to indicate that Y is a function of X. The variable X is called the independent variable and the variable Y is called the dependent variable.

EXAMPLE

The distance s traveled by a car moving with a constant speed is a function of time t.

$$s = f(t)$$

The functional dependence can be depicted in the form of a table or an equation.

EXAMPLE

Mr. Brown can present income from his real estate investment in the form of a table.

Year	Income
1995	18,000
1996	17,550
1997	17,900
1998	18,200
1999	18,600
2000	19,400
2001	23,500
2002	28,000

DATA DESCRIPTION: GRAPHS

Graphs illustrate the dependence between variables. We shall discuss linear graphs, bar graphs, and pie graphs.

Repeated measurements yield data, which must be organized according to some principle. The data should be arranged in such a way that each observation can fall into one and only one category. A simple graphical method of presenting data is the pie chart, which is a circle divided into parts which represent categories.

EXAMPLE

2002 Budget

38% came from individual income taxes
28% from social insurance receipts
13% from corporate income taxes
12% from borrowing
5% from excise taxes
4% other

Excise Tax 5%
Other 4%
Borrowing 12%
Individual Income Tax 38%
Corporate Income Tax 13%
28% Social Insurance Receipts

This data can also be presented in the form of a bar chart or bar graph.

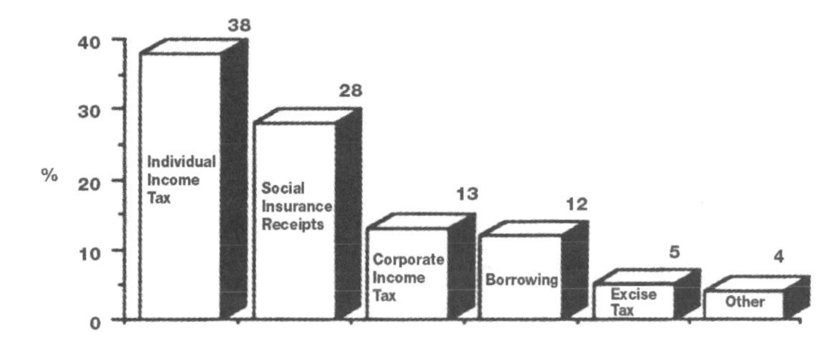

EXAMPLE

The population of the United States for the years 1860, 1870, ..., 1950, 1960 is shown in the table below,

Year	1860	1870	1880	1890	1900	1910
Population in millions	31.4	39.8	50.2	62.9	76.0	92.0
Year	1920	1930	1940	1950	1960	
Population in millions	105.7	122.8	131.7	151.1	179.3	

and in this graph,

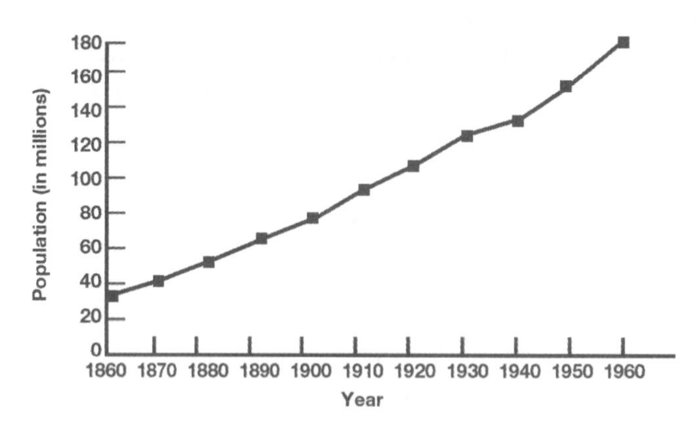

and in this bar chart.

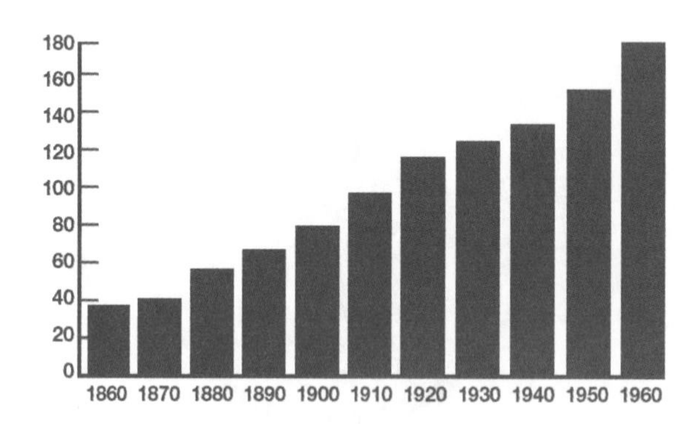

EXAMPLE

A quadratic function is given by

$$y = x^2 + x - 2$$

We compute the values of y corresponding to various values of x.

x	–3	–2	–1	0	1	2	3
y	4	0	–2	–2	0	4	10

From this table the points of the graph are obtained:

$$(-3, 4)\ (-2, 0)\ (-1, -2)\ (0, -2)\ (1, 0)\ (2, 4)\ (3, 10)$$

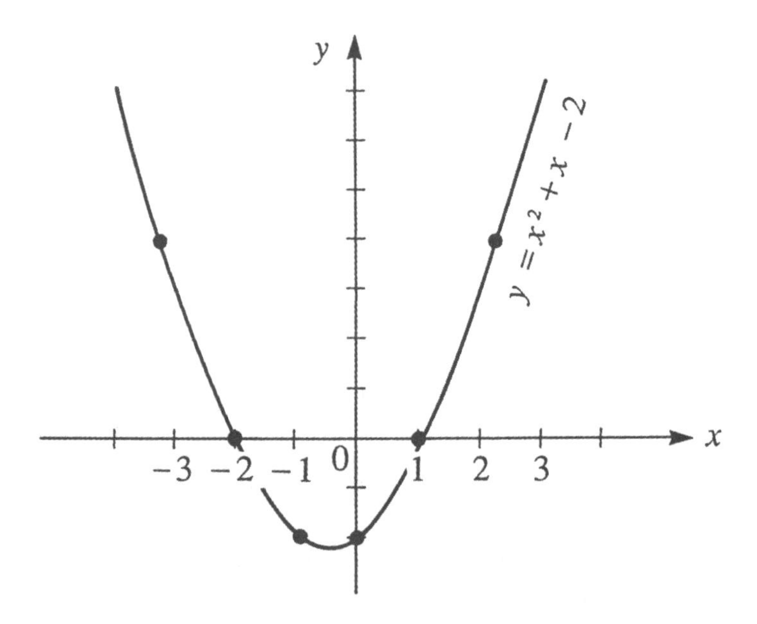

The curve shown is called a parabola. The general equation for a parabola is

$$y = ax^2 + bx + c, a \neq 0$$

where a, b, and c are constants.

FREQUENCY DISTRIBUTIONS

CLASS INTERVALS AND CLASS LIMITS

A set of measurements which has not been organized numerically is called raw data. An arrangement of raw numerical data in descending or ascending order of magnitude is called an array. The difference between the largest and smallest numbers in a set of data is called the range of the data.

EXAMPLE

One hundred families were chosen at random and their yearly income was recorded.

Table 1
Income of 100 Families

Income in Thousands	Number of Families
10 – 14	3
15 – 19	12
20 – 24	19
25 – 29	20
30 – 34	23
35 – 39	18
40 – 44	5

Total 100

The range of the measurements is divided by the number of class intervals desired. In this case we have seven classes. The number of individuals belonging to each class is called the class frequency.

For example, the class frequency of the class 35 – 39 is 18. The list of classes together with their class frequencies is called a frequency table or frequency distribution. Table 1 is a frequency distribution of the income of 100 families. In Table 1 the labels 10 – 14 , 15 – 19, …, 40 – 44 are called class intervals. The numbers 14 and 10 are called class limits; 10 is the lower class limit and 14 is the upper class limit.

In some cases, open class intervals are used, for example, "40 thousand and over." In Table 1 income is recorded to the nearest thousand. Hence, the class interval 30 – 34 includes all incomes from 29,500 to 34,499. The exact numbers 29,500 and 34,499 are called true class limits or class boundaries. The smaller is the lower class boundary and the larger is the upper class boundary. Class boundaries are often used to describe the classes. The difference between the lower and upper class boundaries is called the class width or class size. Usually, all class intervals of a frequency distribution have equal widths.

The class mark (or class midpoint) is the midpoint of the class interval.

$$\text{Class Midpoint} = \frac{\text{Lower Class Limit} + \text{Upper Class Limit}}{2}$$

The class mark of the class interval 35 – 39 is

$$\frac{35+39}{2} = 37$$

For purposes of mathematical analysis, all data belonging to a given class interval are assumed to coincide with the class mark.

GUIDELINES FOR CONSTRUCTING CLASS INTERVALS AND FREQUENCY DISTRIBUTIONS

1. Find the range of the measurements, which is the difference between the largest and the smallest measurements.
2. Divide the range of the measurements by the approximate number of class intervals desired. The number of class intervals is usually between 5 and 20, depending on the data.

   Then round the result to a convenient unit, which should be easy to work with. This unit is a common width for the class intervals.
3. The first class interval should contain the smallest measurement and the last class interval should contain the largest measurement. No measurement should fall on a point of division between two intervals.
4. Determine the number of measurements which fall into each class interval; that is, find the class frequencies.

EXAMPLE

The 25 measurements given below represent the sulphur level in the air for a sample of 25 days. The units used are parts per million.

Table 2

27	32	28	32	31
35	28	44	45	36
33	40	41	36	35
39	37	39	37	44
41	41	35	35	33

The lowest reading is 27 and the highest is 45.

Thus, the range is $45 - 27 = 18$. It will be convenient to use 5 class intervals $\frac{18}{5} = 3.6$ We round 3.6 to 4. The width for the class intervals is 4. The class intervals are labeled as follows: 26 – 29, 30 – 33, 34 – 37, 38 – 41, and 42 – 45.

Now, we can construct the frequency table and compute the class frequency for each class. The relative frequency of a class is defined as the frequency of the class divided by the total number of measurements. Table 3 shows the classes, frequencies and relative frequencies of the data in Table 2.

Table 3

Class Interval	Frequency	Relative Frequency
26 – 29	3	0.12
30 – 33	5	0.20
34 – 37	8	0.32
38 – 41	6	0.24
42 – 45	3	0.12
Total	25	1.00

FREQUENCY HISTOGRAMS AND RELATIVE FREQUENCY HISTOGRAMS

Polygons

A frequency histogram (or histogram) is a set of rectangles placed in the coordinate system. The vertical axis is labeled with the frequencies and the horizontal axis is labeled with the class intervals. Over each class interval a rectangle is drawn with a height such that the area of the rectangle is proportional to the class frequency.

Often the height is numerically equal to the class frequency. Based on the results of Table 3, we obtain the histogram shown below.

Table 4

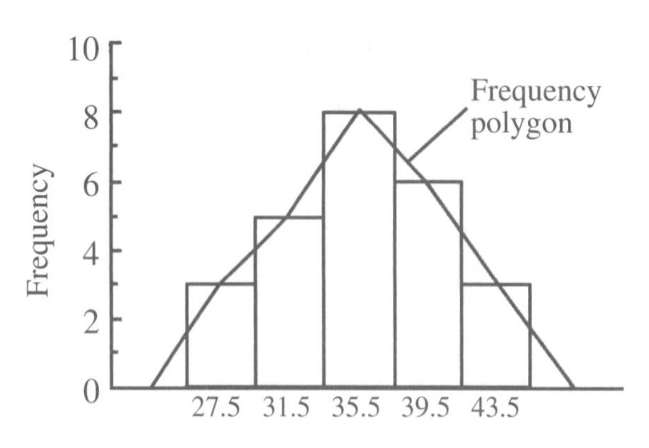

A frequency polygon is obtained by connecting midpoints of the tops of the rectangles of the histogram. The area bounded by the frequency polygon and the x-axis is equal to the sum of the areas of the rectangles in the histogram.

The relative frequency histogram is similar to the frequency histogram. Here, the vertical axis shows relative frequency. A rectangle is constructed over each class interval with a height equal to the class relative frequency. Based on Table 3, we obtain the relative frequency histogram shown in Table 5.

Table 5

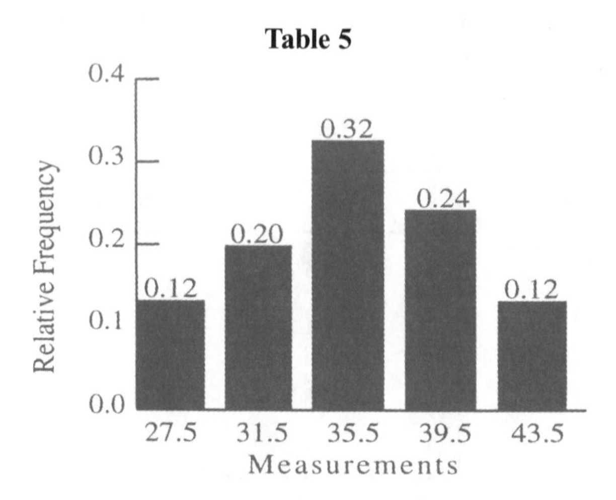

CUMULATIVE FREQUENCY DISTRIBUTIONS

Definition of Cumulative Frequency

The total frequency of all values less than the upper class boundary of a given interval is called the cumulative frequency up to and including that interval. When we know the class intervals and their corresponding frequencies we can compute the cumulative frequency distribution.

Consider the class intervals and frequencies contained in Table 3, from which we compute the cumulative frequency distribution. See Table 6.

Table 6

Parts of Sulphur	Number of Days
Less than 26	0
Less than 30	3
Less than 34	8
Less than 38	16
Less than 42	22
Less than 46	25

Using the coordinate system we can present the cumulative frequency distribution graphically. Such a graph is called a cumulative frequency polygon or ogive. The ogive obtained from Table 6 is shown in Table 7.

Table 7

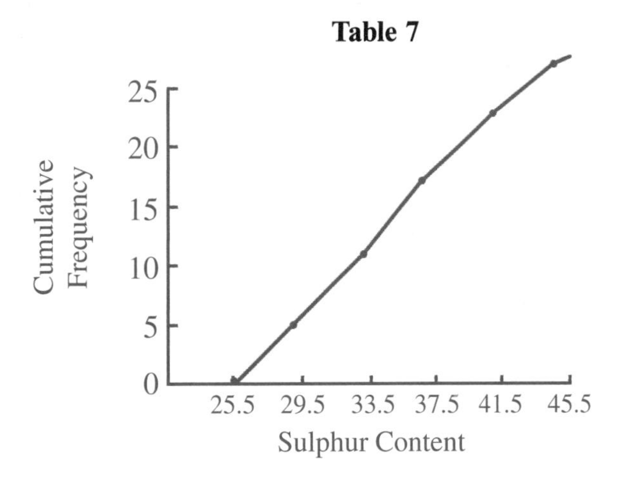

For a very large number of observations it is possible to choose very small class intervals and still have a significant number of observations falling within each class.

The sides of the frequency polygon or relative frequency polygon get smaller as class intervals get smaller. Such polygons closely approximate curves. Such curves are called frequency curves or relative frequency curves. Usually frequency curves can be obtained by increasing the number of class intervals, which requires a larger sample.

TYPES OF FREQUENCY CURVES

In applications we find that most of the frequency curves fall within the popular bell-shaped or symmetrical frequency curve (see figure on next page).

Note that observations equally distant from the maximum have the same frequency. The normal curve has a symmetrical frequency curve.

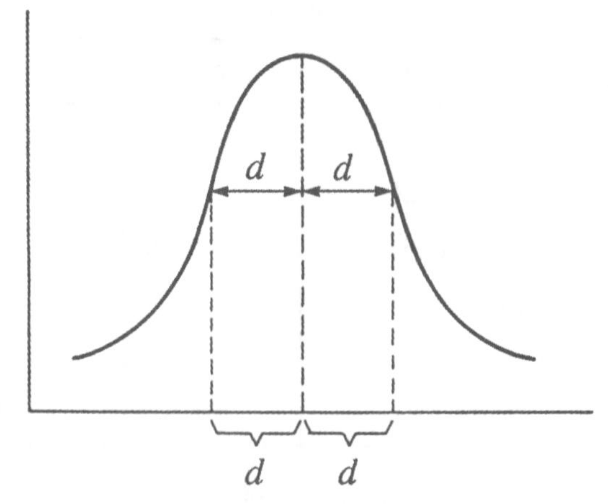

Bell-shaped or Symmetrical

PROBLEM

Discuss and distinguish between discrete and continuous values.

SOLUTION

The kinds of numbers that can take on any fractional or integer value between specified limits are categorized as continuous, whereas values that are usually restricted to whole-number values are called discrete. Thus, if we identify the number of people who use each of several brands of toothpaste, the data generated must be discrete. If we determine the heights and weights of a group of college men, the data generated is continuous.

However, in certain situations, fractional values are also integers. For example, stock prices are generally quoted to the one-eighth of a dollar. Since other fractional values between, say, 24.5 and 24.37 cannot occur, these values can be considered discrete. However, the discrete values that we consider are usually integers.

PROBLEM

(a) Suppose a manufacturer conducts a study to determine the average retail price being charged for his product in a particular market area. Is such a variable discrete or continuous? (b) In conjunction with the previous study, the manufacturer also wants to determine the number of units sold in the area during the week in which an advertising campaign was conducted. Is this variable discrete or continuous?

SOLUTION

(a) Since an average may take any fractional value, the average retail price is continuous.

(b) We have a count. This variable must be discrete. The number of units sold is a discrete variable.

PROBLEM

Twenty students are enrolled in the foreign language department, and their major fields are as follows: Spanish, Spanish, French, Italian, French, Spanish, German, German, Russian, Russian, French, German, German, German, Spanish, Russian, German, Italian, German, Spanish.

(a) Make a frequency distribution table.

(b) Make a frequency histogram.

SOLUTION

The frequency distribution table is constructed by writing down the major field and next to it the number of students.

Major Field	Number of Students
German	7
Russian	3
Spanish	5
French	3
Italian	2
Total	20

A histogram follows:

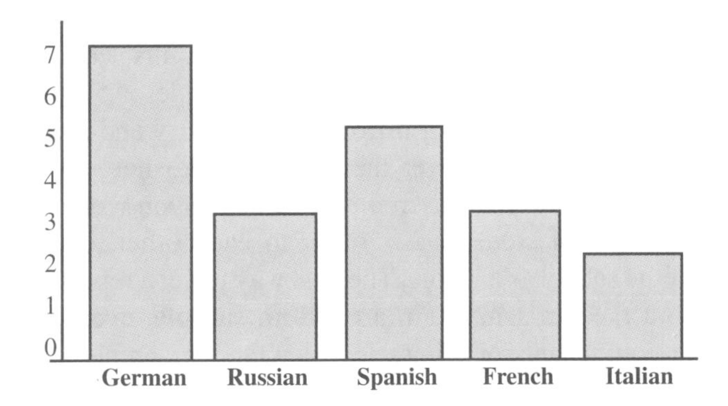

In the histogram, the fields are listed and spaced evenly along the horizontal axis. Each specific field is represented by a rectangle, and all have the same width. The height of each, identified by a number on the vertical axis, corresponds to the frequency of that field.

DEDUCTIVE AND INDUCTIVE REASONING

Deductive and inductive reasoning are opposite ends of the same argument. The argument is "how to relate a part to a whole." Deductive reasoning involves knowing all the possible outcomes of an experiment and then making a judgment. For example, let's say that you dump a large jar of mixed jellybeans out on a table and count and classify each according to flavor. Then, you put the beans back into the jar and mix well. You can now make such statements as, "The probability of reaching into that jar and randomly pulling out a blueberry jellybean is equal to the total number of blueberry jellybeans divided by the total number of jellybeans. In deductive reasoning, you argue from the general to the specific.

In inductive reasoning, you argue from the specific to the general. Sometimes, you don't have the time or energy to count all those jellybeans. So, you reach in, grab a handful, and count and classify it. From that handful, you guess at what is in the jar.

SAMPLING

A **population** is a well-defined group of people or things. A **sample** is a small part of a population. Often times, a scientist is interested in estimating values of a population from the values gained from a sample. This is a valuable tool as long as the sample is **representative** of the population. This means that the sample must be like the population in important ways.

For example, let's say you want to do a survey to predict the outcome of an upcoming school election. You could survey 50 students and ask them which candidate they plan to vote for. Then, based on the results, you could predict a winner. The question is, "How do you select that sample of 50 people from the whole school?" You could hang out in front of the library and simply survey the first 50 students that walk by. However, these students may not be representative of the population. If you selected a classroom of students, you would have the same problem. A classroom of students is similar to each other, but not necessary representative of the school as a whole. The best way to get a representative sample would be to use a random sample. In a **random sample**, every member of the population has an equal chance of being selected. A random sample could be obtained by writing the names of all the students at the school on individual slips of paper. Put these names in a hat, mix them up, and randomly select 50 names. These 50 students should be surveyed. Last, the results from this survey could be generalized to represent the population of the whole school.

STATISTICAL PLOTS AND GRAPHS

A **box-and-whiskers** plot is a graph that displays five statistics. A minimum score, a maximum score, and three percentiles. A percentile value for a score tells you the percentage of scores lower than it. The beginning of the box is the score at the 25th percentile. The end of the box represents the 75th percentile. The score inside of the box is the median, or the score at the 50th percentile. Attached to the box you will find two whiskers. The score at the end of the left whisker is the minimum score. The score at the end of the right whisker is the maximum score.

EXAMPLE

In the box-and-whiskers plot below, the minimum score is 70, the score at the 25th percentile is 78, the median score is 82, the score at the 75th percentile is 90, and the maximum score is 94.

Scores on a Test

70 80 90 100

PROBLEM

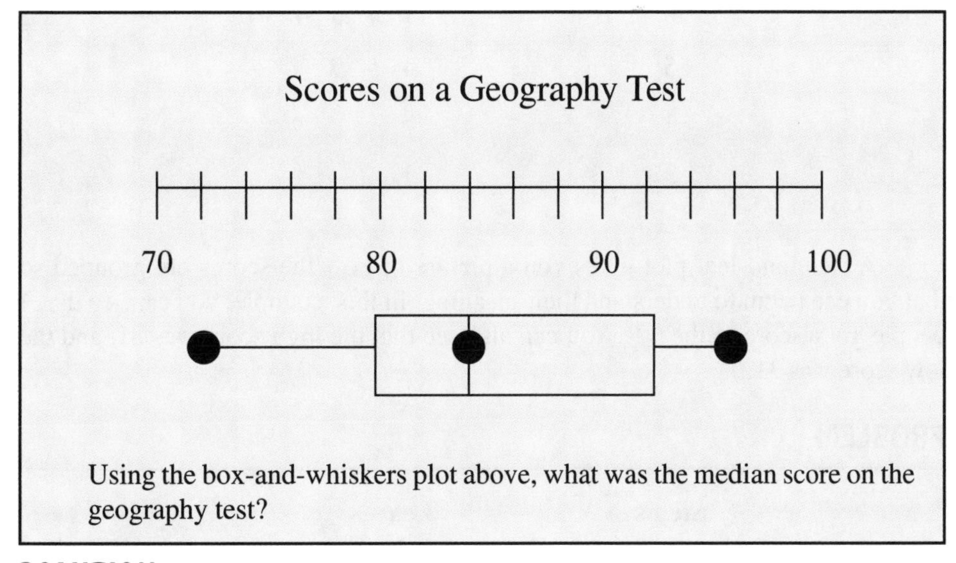

Using the box-and-whiskers plot above, what was the median score on the geography test?

SOLUTION

The median score is 84. On a box-and-whiskers plot, the median score is the score on the inside of the box.

EXAMPLE

A **stem-and-leaf plot** is a way of displaying scores in groups. A stem-and-leaf plot gives you a picture of the scores, as well as the actual numbers themselves in a compact form. In this type of plot, a score is broken into a stem and a leaf. The leaf consists of the smallest digit and the stem consists of the remaining larger digits.

Task: Create a stem and leaf plot using the following scores.

Scores: 64, 48, 61, 81, 63, 59, 70, 54, 76, 61, 55, 31

Solution: The first step is to take these scores and create a set of "ranked ordered scores," ordering the scores from smallest to largest. Notice that the minimum score is 31, while the maximum score is 81.

Ranked Ordered Scores: 31, 48, 54, 55, 59, 61, 61, 63, 64, 70, 76, 81

The second step is to list the range of scores for the stems in a column. The stem of our smallest score (31) is 3, and that of our largest score (81) is 8. List all of the whole numbers between 3 and 8.

The third step is to put the leaves on the stems. Take each score, one at a time, and put the last digit in a column next to its stem. For example, the last digit of 31 is 1, so put a 1 next to its stem of 3. The last digit of 48 is 8, so put an 8 next to its stem of 4. Do this for the remaining scores.

Stems	Leaves
8	1
7	0 6
6	1 1 3 4
5	4 5 9
4	8
3	1

A stem-and-leaf plot gives you a picture of how the scores are grouped so that you can begin to understand their meaning. In this example, you can see that 4 people got a score in the 60s. You can also see that the high score was 81, and the low score was 31.

PROBLEM

Stems	Leaves
6	5
5	3 6
4	0 1 7
3	2 9
2	4

The stem-and-leaf plot above was created using what set of scores?

SOLUTION

In this stem-and-leaf plot, the stems are the first digit, and the leaves are the remaining digits. Starting from the bottom, the first score is 24, the second score is 32, then comes 39, 40, 41, 47, 53, 56, and 65.

EXAMPLE

A **scatter-plot** is a graph that shows the relationship between two variables. A scatter-plot is a set of (x, y) coordinates. Each coordinate is a point on the graph. x represents a value of one variable, while y represents the value of another variable. Remember, a variable is just a measurement that can take on more than one value. A scatter-plot is useful because in one picture you can see if there is a relationship between two variables. It has been said that, "A picture is worth a thousand words." Likewise, "A graph is worth a thousand numbers."

Given: Variable x represents Grade Level. Variable y represents Hours of Homework each week.

Task: Using the data below, construct a scatter-plot.

Question: What is the relationship between these two variables?

x	**Grade Level**	1	2	3	4	5	6	7	8	9	10	11	12
y	**Hours of Homework**	2	3	3	6	4	10	7	10	12	9	14	15

Answer: You can think of these two variables as one set of (x, y) coordinates on a graph. Remember, a coordinate is just a point. Graph the following points: (1, 2), (2, 3), (3, 3), (4, 6), (5, 4), (6, 10), (7, 7), (8, 10), (9, 12), (10, 9), (11, 14), and (12, 15). This graph shows that as grade level increases, the number of homework hours per week tends to increase.

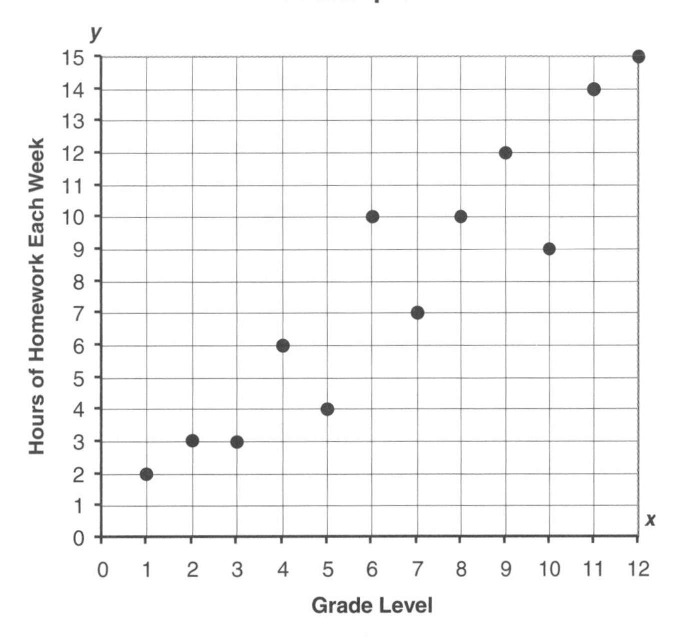

A statistic called **correlation** tells you if two measurements go together along a straight line. A scatter-plot is one way of looking at correlation. There are three different types of correlation.

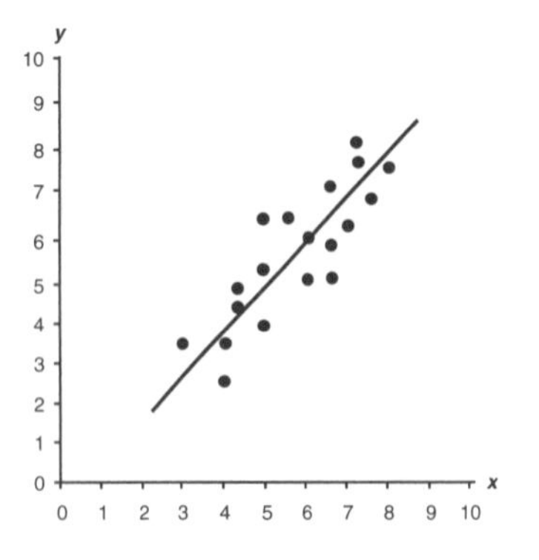

POSITIVE CORRELATION

As one measurement increases, the other measurement also increases.

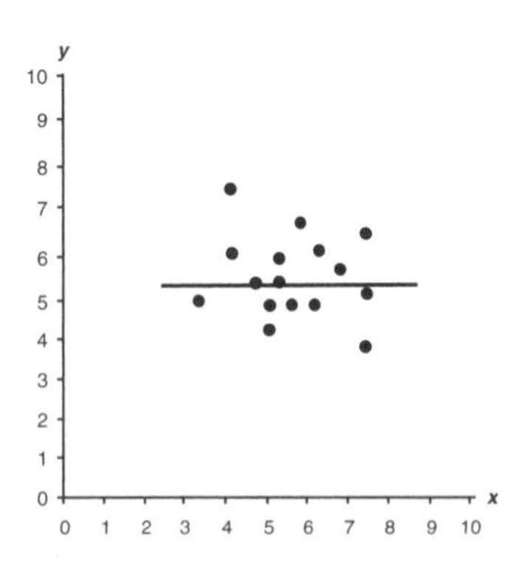

ZERO CORRELATION

The two measurements are not related to each other along a straight line.

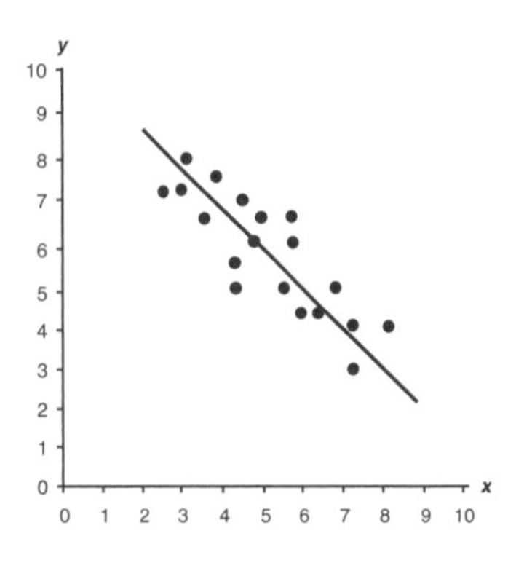

NEGATIVE CORRELATION

As one measurement increases, the other measurement decreases.

PROBLEM

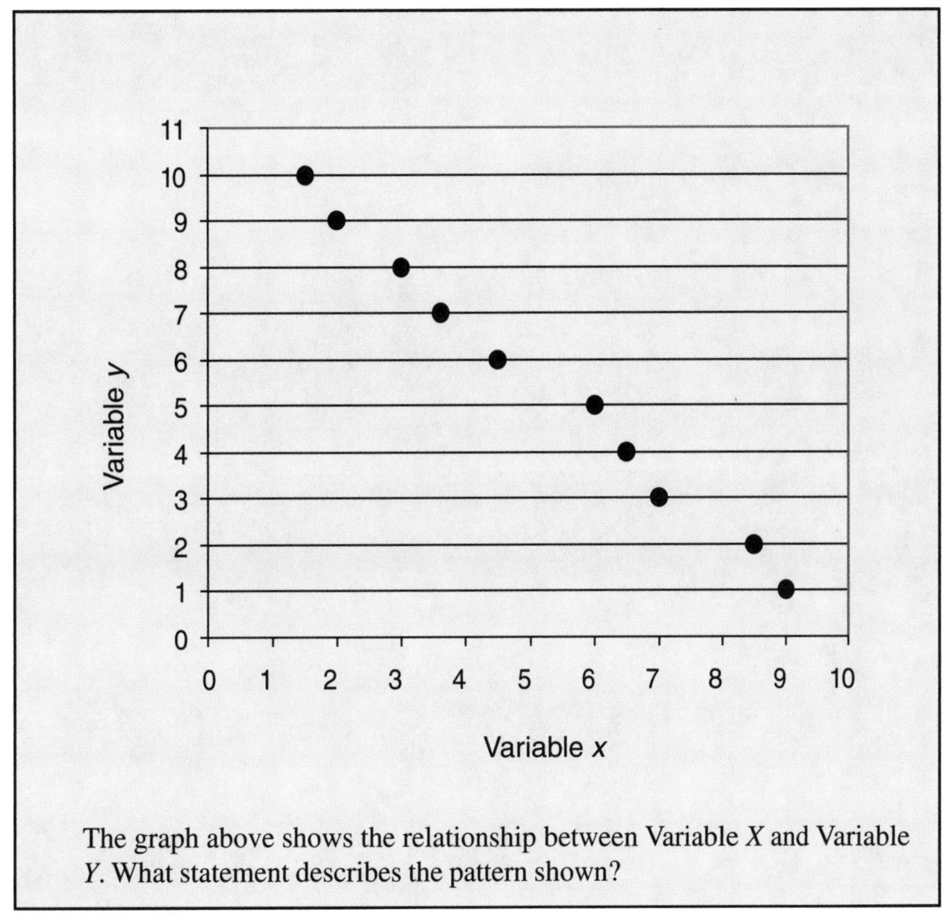

The graph above shows the relationship between Variable *X* and Variable *Y*. What statement describes the pattern shown?

SOLUTION

As Variable *X* increases, Variable *Y* decreases. This represents a negative correlation.

YOUR ANSWER SHEET FOR THE

ARITHMETIC DIAGNOSTIC TEST

FOR CLASS AND HOMEWORK ASSIGNMENTS

RECORD YOUR ANSWERS BY FILLING IN THE APPROPRIATE OVALS.

1. Ⓐ Ⓑ Ⓒ Ⓓ Ⓔ	26. Ⓐ Ⓑ Ⓒ Ⓓ Ⓔ
2. Ⓐ Ⓑ Ⓒ Ⓓ Ⓔ	27. Ⓐ Ⓑ Ⓒ Ⓓ Ⓔ
3. Ⓐ Ⓑ Ⓒ Ⓓ Ⓔ	28. Ⓐ Ⓑ Ⓒ Ⓓ Ⓔ
4. Ⓐ Ⓑ Ⓒ Ⓓ Ⓔ	29. Ⓐ Ⓑ Ⓒ Ⓓ Ⓔ
5. Ⓐ Ⓑ Ⓒ Ⓓ Ⓔ	30. Ⓐ Ⓑ Ⓒ Ⓓ Ⓔ
6. Ⓐ Ⓑ Ⓒ Ⓓ Ⓔ	31. Ⓐ Ⓑ Ⓒ Ⓓ Ⓔ
7. Ⓐ Ⓑ Ⓒ Ⓓ Ⓔ	32. Ⓐ Ⓑ Ⓒ Ⓓ Ⓔ
8. Ⓐ Ⓑ Ⓒ Ⓓ Ⓔ	33. Ⓐ Ⓑ Ⓒ Ⓓ Ⓔ
9. Ⓐ Ⓑ Ⓒ Ⓓ Ⓔ	34. Ⓐ Ⓑ Ⓒ Ⓓ Ⓔ
10. Ⓐ Ⓑ Ⓒ Ⓓ Ⓔ	35. Ⓐ Ⓑ Ⓒ Ⓓ Ⓔ
11. Ⓐ Ⓑ Ⓒ Ⓓ Ⓔ	36. Ⓐ Ⓑ Ⓒ Ⓓ Ⓔ
12. Ⓐ Ⓑ Ⓒ Ⓓ Ⓔ	37. Ⓐ Ⓑ Ⓒ Ⓓ Ⓔ
13. Ⓐ Ⓑ Ⓒ Ⓓ Ⓔ	38. Ⓐ Ⓑ Ⓒ Ⓓ Ⓔ
14. Ⓐ Ⓑ Ⓒ Ⓓ Ⓔ	39. Ⓐ Ⓑ Ⓒ Ⓓ Ⓔ
15. Ⓐ Ⓑ Ⓒ Ⓓ Ⓔ	40. Ⓐ Ⓑ Ⓒ Ⓓ Ⓔ
16. Ⓐ Ⓑ Ⓒ Ⓓ Ⓔ	41. Ⓐ Ⓑ Ⓒ Ⓓ Ⓔ
17. Ⓐ Ⓑ Ⓒ Ⓓ Ⓔ	42. Ⓐ Ⓑ Ⓒ Ⓓ Ⓔ
18. Ⓐ Ⓑ Ⓒ Ⓓ Ⓔ	43. Ⓐ Ⓑ Ⓒ Ⓓ Ⓔ
19. Ⓐ Ⓑ Ⓒ Ⓓ Ⓔ	44. Ⓐ Ⓑ Ⓒ Ⓓ Ⓔ
20. Ⓐ Ⓑ Ⓒ Ⓓ Ⓔ	45. Ⓐ Ⓑ Ⓒ Ⓓ Ⓔ
21. Ⓐ Ⓑ Ⓒ Ⓓ Ⓔ	46. Ⓐ Ⓑ Ⓒ Ⓓ Ⓔ
22. Ⓐ Ⓑ Ⓒ Ⓓ Ⓔ	47. Ⓐ Ⓑ Ⓒ Ⓓ Ⓔ
23. Ⓐ Ⓑ Ⓒ Ⓓ Ⓔ	48. Ⓐ Ⓑ Ⓒ Ⓓ Ⓔ
24. Ⓐ Ⓑ Ⓒ Ⓓ Ⓔ	49. Ⓐ Ⓑ Ⓒ Ⓓ Ⓔ
25. Ⓐ Ⓑ Ⓒ Ⓓ Ⓔ	50. Ⓐ Ⓑ Ⓒ Ⓓ Ⓔ

CLASS AND HOMEWORK ASSIGNMENTS

ARITHMETIC DIAGNOSTIC TEST

This diagnostic test is designed to help you determine your strengths and your weaknesses in arithmetic. Follow the directions for each part and check your answers. Record your answers on the preceding page designated "Your Answer Sheet."

50 Questions

DIRECTIONS: Choose the correct answer for each of the following problems. Fill in each answer on the answer sheet.

1. What part of three fourths is one tenth?

   (A) $^1/_8$ (B) $^{15}/_2$ (C) $^2/_{15}$
   (D) $^3/_{40}$ (E) None of the above

2. One number is 2 more than 3 times another. Their sum is 22. Find the numbers.

   (A) 8, 14 (B) 2,20 (C) 5, 17
   (D) 4, 18 (E) 10, 12

3. What is the median of the following group of scores?

   27, 27, 26, 26, 26, 26, 18, 13, 36, 36, 30, 30, 30, 27, 29

   (A) 30 (B) 26 (C) 25.4
   (D) 27 (E) 36

4. What percent of 260 is 13?

   (A) .05% (B) 5% (C) 50%
   (D) .5% (E) 20%

5. Subtract: $4\ ^1/_3 - 1\ ^5/_6$

   (A) $3\ ^2/_3$ (B) $2^1/_2$ (C) $3^1/_2$
   (D) $2\ ^1/_6$ (E) None of the above

6. What is the product of $(\sqrt{3}+6)$ and $(\sqrt{3}-2)$?

   (A) $9+4\sqrt{3}$ (B) -9 (C) $-9+4\sqrt{3}$
   (D) $-9+2\sqrt{3}$ (E) 9

7. The number missing in the series, 2, 6, 12, 20, x, 42, 56 is:

   (A) 36 (B) 24 (C) 30

   (D) 38 (E) 40

8. What is the value of the following expression: $\dfrac{1}{1+\dfrac{1}{1+\dfrac{1}{4}}}$

   (A) $^9/_5$ (B) $^5/_9$ (C) $^1/_2$

   (D) 2 (E) 4

9. Which of the following has the smallest value?

   (A) $^1/_{0.2}$ (B) $^{0.1}/_2$ (C) $^{0.2}/_1$

   (D) $^{0.2}/_{0.1}$ (E) $^2/_{0.1}$

10. Which is the smallest number?

    (A) $5 \times 10^{-3} / 3 \times 10^{-3}$ (B) $.3 / .2$

    (C) $.3 / 3 \times 10^{-3}$ (D) $5 \times 10^{-2} / .1$

    (E) $.3 / 3 \times 10^{-1}$

11. $10^3 + 10^5 =$

    (A) 10^8 (B) 10^{15} (C) 20^8

    (D) 2^{15} (E) 101,000

12. How many digits are in the standard numeral for $2^{31} \times 5^{27}$?

    (A) 31 (B) 29 (C) 28

    (D) 26 (E) 25

13. $475{,}826 \times 521{,}653 + 524{,}174 \times 521{,}653 =$

    (A) 621,592,047,600 (B) 519,697,450,000

    (C) 495,652,831,520 (D) 521,653,000,000

    (E) 524,174,000,000

14. How many ways can you make change for a quarter?

    (A) 8 (B) 9 (C) 10

    (D) 12 (E) 14

15. The sixtieth digit in the decimal representation of $^1/_7$ is

    (A) 1 (B) 4 (C) 2

    (D) 5 (E) 7

16. What is the least prime number which is a divisor of $7^9 + 11^{25}$?

(A) 1 (B) 2 (C) 3
(D) 5 (E) $7^9 + 11^{25}$

17. Evaluate $10 - 5[2^3 + 27 \div 3 - 2(8 - 10)]$

(A) -95 (B) 105 (C) 65
(D) -55 (E) -85

18. Fifteen percent of what number is 60?

(A) 9 (B) 51 (C) 69
(D) 200 (E) 400

19. Which is the largest fraction: $^1/_5$, $^2/_9$, $^2/_{11}$, $^4/_{19}$, $^4/_{17}$?

(A) $^1/_5$ (B) $^2/_9$ (C) $^2/_{11}$
(D) $^4/_{19}$ (E) $^4/_{17}$

20. How many of the scores 10, 20, 30, 35, 55 are larger than their arithmetic mean score?

(A) None (B) One (C) Two
(D) Three (E) Four

21. Evaluate $\left(2^{1-\sqrt{3}}\right)^{1+\sqrt{3}}$

(A) 4 (B) -4 (C) 16
(D) $^1/_2$ (E) $^1/_4$

22. $\dfrac{2^{100} + 2^{98}}{2^{100} - 2^{98}} =$

(A) 2^{198} (B) 2^{99} (C) 64
(D) 4 (E) $^5/_3$

23. What is the least natural number which is a multiple of each number from 1 to 10?

(A) 3,628,800 (B) 5040 (C) 840
(D) 1,260 (E) 2,520

24. If in ΔABC, $AB = BC$ and angle A has measure 46°, then angle B has measure

(A) 46° (B) 92° (C) 88°
(D) 56° (E) 23°

25. What is the last digit in the number 3^{2000}?

(A) 0 (B) 1 (C) 3

(D) 7 (E) 9

26. In the set of integers 1000, 1001, 1002, …, 9998, 9999, how many of the numbers do not contain the digit 5?

(A) 6,561 (B) 5,000 (C) 9,000

(D) 4,500 (E) 5,832

27. $15{,}561 \div 25 + 9{,}439 \div 25 =$

(A) 997 (B) 1,000 (C) 1,002

(D) 1,005 (E) 1,005.08

28. What is the units digit for 4^{891}?

(A) 4 (B) 6 (C) 8

(D) 0 (E) 1

29. $\frac{1}{1\times 2}+\frac{1}{2\times 3}+\frac{1}{3\times 4}+\ldots+\frac{1}{99\times 100}=$

(A) $^{49}/_{50}$ (B) $^{74}/_{75}$ (C) $^{98}/_{99}$

(D) $^{99}/_{100}$ (E) $^{101}/_{100}$

30. $1 + 2 + 3 + 4 + \ldots + 99 =$

(A) 4,700 (B) 4,750 (C) 4,850

(D) 4,900 (E) 4,950

31. The decimal $.24\overline{24}$ expressed as a fraction is

(A) $^{8}/_{33}$ (B) $^{6}/_{25}$ (C) $^{1}/_{4}$

(D) $^{303}/_{1250}$ (E) $^{121}/_{500}$

32. $\frac{2^{-4}+2^{-1}}{2^{-3}}=$

(A) $9/2^{7}$ (B) $9/2^{-1}$ (C) 1/2

(D) 2^{-3} (E) 9/2

33. What is the smallest positive number that leaves a remainder of 2 when the number is divided by 3, 4 or 5?

(A) 22 (B) 42 (C) 62

(D) 122 (E) 182

34. What part of three eighths is one tenth?

(A) $^{1}/_{8}$ (B) $^{15}/_{2}$ (C) $^{4}/_{15}$

(D) $^{3}/_{40}$ (E) None of the above

35. $(^{2}/_{3}) + (^{5}/_{9}) =$

(A) $^{7}/_{12}$ (B) $^{11}/_{9}$ (C) $^{7}/_{3}$

(D) $^{7}/_{9}$ (E) $^{11}/_{3}$

36. Add $^{3}/_{6} + ^{2}/_{6}$

(A) $^{1}/_{12}$ (B) $^{5}/_{6}$ (C) $^{5}/_{12}$

(D) $^{8}/_{9}$ (E) $^{9}/_{8}$

37. Change 125.937% to a decimal

(A) 1.25937 (B) 12.5937 (C) 125.937

(D) 1259.37 (E) 12593.7

38. What is the ratio of 8 feet to 28 inches?

(A) $^{1}/_{7}$ (B) $^{7}/_{1}$ (C) $^{24}/_{7}$

(D) $^{6}/_{7}$ (E) $^{7}/_{2}$

39. Using order of operations, solve: $3 \times 6 - 12/2 =$

(A) – 9 (B) 3 (C) 6

(D) 12 (E) 18

40. The most economical price among the following prices is

(A) 10 oz. for 16¢ (B) 2 oz. for 3¢

(C) 4 oz. for 7¢ (D) 20 oz. for 34¢

(E) 8 oz. for 13¢

41. Change $4^{5}/_{6}$ to an improper fraction.

(A) $^{5}/_{24}$ (B) $^{9}/_{6}$ (C) $^{29}/_{6}$

(D) $^{30}/_{4}$ (E) $^{120}/_{6}$

42. If the sum of four consecutive integers is 226, then the smallest of these numbers is

(A) 55 (B) 56 (C) 57

(D) 58 (E) 59

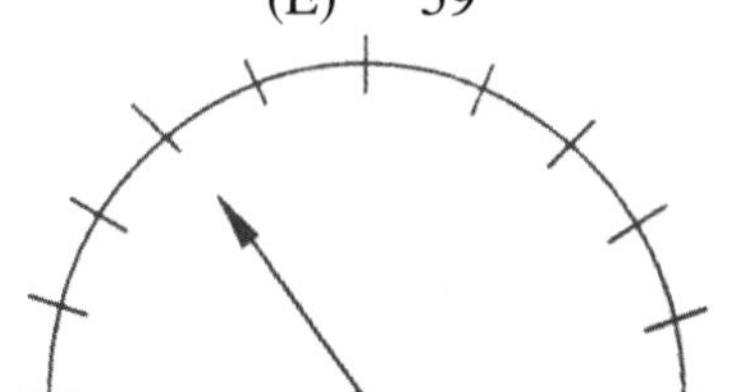

43. How much time is left on the parking meter shown on the previous page?

(A) 8 minutes (B) 9 minutes (C) 10 minutes

(D) 12 minutes (E) 15 minutes

44. $15{,}561 \div 25 - 9{,}561 \div 25 =$

(A) 997 (B) 240 (C) 1,002

(D) 1,005 (E) 1,005.08

45. $4\% \times 4\% =$

(A) 0.0016% (B) 0.16% (C) 1.6%

(D) 16% (E) 160%

46. Which of the following numbers is not between $.\overline{85}$ and $.\overline{86}$?

(A) $.\overline{851}$ (B) $.\overline{859}$ (C) .859

(D) $.\overline{861}$ (E) .861

47. Change the fraction $^7/_8$ to a decimal.

(A) .666 (B) .75 (C) .777

(D) .875 (E) 1.142

48. $\sqrt{75} - 3\sqrt{48} + \sqrt{147} =$

(A) $3\sqrt{3}$ (B) $7\sqrt{3}$ (C) 0

(D) 3 (E) $\sqrt{3}$

49. The following ratio: 40 seconds : $1^1/_2$ minutes : $^1/_6$ hour, can be expressed in lowest terms as

(A) 4 : 9 : 60 (B) 4 : 9 : 6 (C) 40 : 90 : 60

(D) $^2/_3 : 1^1/_2 : 10$ (E) 60 : 9 : 4

50. Simplify $6\sqrt{7} + 4\sqrt{7} - \sqrt{5} + 5\sqrt{7}$

(A) $10\sqrt{7}$ (B) $15\sqrt{7} - \sqrt{5}$ (C) $15\sqrt{21} - \sqrt{5}$

(D) $15\sqrt{16}$ (E) 60

YOUR ANSWER SHEET FOR THE

ALGEBRA DIAGNOSTIC TEST

FOR CLASS AND HOMEWORK ASSIGNMENTS

RECORD YOUR ANSWERS BY FILLING IN THE APPROPRIATE OVALS.

1.	Ⓐ	Ⓑ	Ⓒ	Ⓓ	Ⓔ	26.	Ⓐ	Ⓑ	Ⓒ	Ⓓ	Ⓔ
2.	Ⓐ	Ⓑ	Ⓒ	Ⓓ	Ⓔ	27.	Ⓐ	Ⓑ	Ⓒ	Ⓓ	Ⓔ
3.	Ⓐ	Ⓑ	Ⓒ	Ⓓ	Ⓔ	28.	Ⓐ	Ⓑ	Ⓒ	Ⓓ	Ⓔ
4.	Ⓐ	Ⓑ	Ⓒ	Ⓓ	Ⓔ	29.	Ⓐ	Ⓑ	Ⓒ	Ⓓ	Ⓔ
5.	Ⓐ	Ⓑ	Ⓒ	Ⓓ	Ⓔ	30.	Ⓐ	Ⓑ	Ⓒ	Ⓓ	Ⓔ
6.	Ⓐ	Ⓑ	Ⓒ	Ⓓ	Ⓔ	31.	Ⓐ	Ⓑ	Ⓒ	Ⓓ	Ⓔ
7.	Ⓐ	Ⓑ	Ⓒ	Ⓓ	Ⓔ	32.	Ⓐ	Ⓑ	Ⓒ	Ⓓ	Ⓔ
8.	Ⓐ	Ⓑ	Ⓒ	Ⓓ	Ⓔ	33.	Ⓐ	Ⓑ	Ⓒ	Ⓓ	Ⓔ
9.	Ⓐ	Ⓑ	Ⓒ	Ⓓ	Ⓔ	34.	Ⓐ	Ⓑ	Ⓒ	Ⓓ	Ⓔ
10.	Ⓐ	Ⓑ	Ⓒ	Ⓓ	Ⓔ	35.	Ⓐ	Ⓑ	Ⓒ	Ⓓ	Ⓔ
11.	Ⓐ	Ⓑ	Ⓒ	Ⓓ	Ⓔ	36.	Ⓐ	Ⓑ	Ⓒ	Ⓓ	Ⓔ
12.	Ⓐ	Ⓑ	Ⓒ	Ⓓ	Ⓔ	37.	Ⓐ	Ⓑ	Ⓒ	Ⓓ	Ⓔ
13.	Ⓐ	Ⓑ	Ⓒ	Ⓓ	Ⓔ	38.	Ⓐ	Ⓑ	Ⓒ	Ⓓ	Ⓔ
14.	Ⓐ	Ⓑ	Ⓒ	Ⓓ	Ⓔ	39.	Ⓐ	Ⓑ	Ⓒ	Ⓓ	Ⓔ
15.	Ⓐ	Ⓑ	Ⓒ	Ⓓ	Ⓔ	40.	Ⓐ	Ⓑ	Ⓒ	Ⓓ	Ⓔ
16.	Ⓐ	Ⓑ	Ⓒ	Ⓓ	Ⓔ	41.	Ⓐ	Ⓑ	Ⓒ	Ⓓ	Ⓔ
17.	Ⓐ	Ⓑ	Ⓒ	Ⓓ	Ⓔ	42.	Ⓐ	Ⓑ	Ⓒ	Ⓓ	Ⓔ
18.	Ⓐ	Ⓑ	Ⓒ	Ⓓ	Ⓔ	43.	Ⓐ	Ⓑ	Ⓒ	Ⓓ	Ⓔ
19.	Ⓐ	Ⓑ	Ⓒ	Ⓓ	Ⓔ	44.	Ⓐ	Ⓑ	Ⓒ	Ⓓ	Ⓔ
20.	Ⓐ	Ⓑ	Ⓒ	Ⓓ	Ⓔ	45.	Ⓐ	Ⓑ	Ⓒ	Ⓓ	Ⓔ
21.	Ⓐ	Ⓑ	Ⓒ	Ⓓ	Ⓔ	46.	Ⓐ	Ⓑ	Ⓒ	Ⓓ	Ⓔ
22.	Ⓐ	Ⓑ	Ⓒ	Ⓓ	Ⓔ	47.	Ⓐ	Ⓑ	Ⓒ	Ⓓ	Ⓔ
23.	Ⓐ	Ⓑ	Ⓒ	Ⓓ	Ⓔ	48.	Ⓐ	Ⓑ	Ⓒ	Ⓓ	Ⓔ
24.	Ⓐ	Ⓑ	Ⓒ	Ⓓ	Ⓔ	49.	Ⓐ	Ⓑ	Ⓒ	Ⓓ	Ⓔ
25.	Ⓐ	Ⓑ	Ⓒ	Ⓓ	Ⓔ	50.	Ⓐ	Ⓑ	Ⓒ	Ⓓ	Ⓔ

CLASS AND HOMEWORK ASSIGNMENTS

ALGEBRA DIAGNOSTIC TEST

This diagnostic test is designed to help you determine your strengths and your weaknesses in algebra. Follow the directions for each part and check your answers. Record your answers on the preceding page designated "Your Answer Sheet."

50 Questions

DIRECTIONS: Choose the correct answer for each of the following problems. Fill in each answer on the answer sheet.

1. The value of B in the equation $A = (h/2)(B + b)$ is:

   (A) $(2A - b)h$ (B) $2h/A - b$ (C) $2A - b$

   (D) $2A/h - b$ (E) None of the above

2. Which of the following integers is the square of an integer for every integer x?

   (A) $x^2 + x$ (B) $x^2 + 1$ (C) $x^2 + 2x$

   (D) $x^2 + 2x - 4$ (E) $x^2 + 2x + 1$

3. Each of the integers h, m and n is divisible by 3. Which of the following integers is ALWAYS divisible by 9?

   I. hm

   II. $h + m$

   III. $h + m + n$

   (A) I only (B) II only (C) III only

   (D) II and III only (E) I, II, and III

4. What is the factorization of $x^2 + ax - 2x - 2a$?

   (A) $(x + 2)(x - a)$ (B) $(x - 2)(x + a)$ (C) $(x + 2)(x + a)$

   (D) $(x - 2)(x - a)$ (E) None of the above

5. What is the value of x in the equation

   $\sqrt{5x - 4} - 5 = -1$?

   (A) 2 (B) 5 (C) No value

   (D) 4 (E) -4

6. The number missing in the series, 2, 6, 12, 20, 30, 42, x is:

(A) 36 (B) 24 (C) 30

(D) 38 (E) 56

7. If $T = 2\pi\sqrt{\frac{L}{g}}$, then L is equal to

(A) $\frac{T^2}{2\pi g}$ (B) $\frac{T^2 g}{2\pi}$ (C) $\frac{T^2 g}{4\pi^2}$

(D) $\frac{T^2 g}{4\pi}$ (E) $\frac{T^2}{4\pi^2 g}$

8. $1 + \frac{y}{x-2y} - \frac{y}{x+2y} =$

(A) 0 (B) 1

(C) $\frac{1}{(x-2y)(x+2y)}$ (D) $\frac{2x-y}{(x-2y)(x+2y)}$

(E) $\frac{x^2}{(x-2y)(x+2y)}$

9. If $0 < a < 1$ and $b > 1$, which is the largest value?

(A) $^a/_b$ (B) $^b/_a$ (C) $(^a/_b)^2$

(D) $(^b/_a)^2$ (E) Cannot be determined

10. Given $\frac{(\alpha + x) + y}{x + y} = \frac{\beta + y}{y}$, $\frac{x}{y} = ?$

(A) α/β (B) β/α (C) $\beta/\alpha - 1$

(D) $\alpha/\beta - 1$ (E) 1

11. If n is an integer, which of the following represents an odd number?

(A) $2n + 3$ (B) $2n$ (C) $2n + 2$

(D) $3n$ (E) $n + 1$

12. Which of the following statements are true, if

$x + y + z = 10$

$y \geq 5$

$4 \geq z \geq 3$

I. $x < z$

II. $x > y$

III. $x + z \leq y$

(A) I only (B) II only (C) III only

(D) I and III (E) I, II, and III

13. $\sqrt{X\sqrt{X\sqrt{X}}} = ?$

(A) $X^{7/8}$ (B) $X^{7/4}$ (C) $X^{15/16}$

(D) $X^{3/4}$ (E) $X^{15/8}$

14. If $v = \pi b^2 (r - {}^{b}/_{3})$, then r is equal to

(A) $\frac{v}{\pi b^2} + \frac{b}{3}$ (B) $\frac{v}{\pi b^2} + \frac{b}{3\pi}$ (C) $\frac{v}{\pi b^2} + 3b$

(D) $v + \frac{b}{3}$ (E) $v + \frac{\pi b}{3}$

15. If ${}^{a}/_{x} - {}^{b}/_{y} = c$ and $xy = {}^{1}/_{c}$, then $bx = ?$

(A) $1 - ay$ (B) ay (C) $ay + 1$

(D) $ay - 1$ (E) $2ay$

16. If $z = x^a$, $y = x^b$ then $z^b y^a = ?$

(A) $x^{(ab)^2}$ (B) x^{ab} (C) x^0

(D) x^{2ab} (E) x

17. The mean (average) of the numbers 50, 60, 65, 75, x and y is 65. What is the mean of x and y?

(A) 67 (B) 70 (C) 71

(D) 73 (E) 75

18. If x and 10 are relatively prime natural numbers, then x could be a multiple of

(A) 9 (B) 18 (C) 4

(D) 25 (E) 14

19. A first square has a side of length x while the length of a side of a second square is two units greater than the length of a side of the first square. What is an expression for the sum of the areas of the two squares?

(A) $2x^2 + 4x + 4$ (B) $x^2 + 2$ (C) $x^2 + 4$

(D) $2x^2 + 2x + 2$ (E) $2x^2 + 3x + 4$

20. If a and b each represent a nonzero real number and if

$$x = \frac{a}{|a|} + \frac{b}{|b|} + \frac{ab}{|ab|}$$

the set of all possible values for x is

(A) $\{-3, -2, -1, 1, 2, 3\}$ (B) $\{3, -1, -2\}$

(C) $\{3, -1, -3\}$ (D) $\{3, -1\}$

(E) $\{3, 1, -1\}$

21. If $x - y = 9$ then $3x - 3y - 1 =$

(A) 23 (B) 24 (C) 25

(D) 26 (E) 28

22. $4^{x-3} = \left(\sqrt{2}\right)^x$ The value of x is

(A) 0 (B) 5 (C) 4

(D) $^1/_2$ (E) 3

23. Find the first term of the arithmetic progression whose third term a_3 is 7 and whose eighth term a_8 is 17.

(A) 0 (B) 2 (C) 3

(D) 1 (E) 4

24. If $x = -2y$ and $2x - 6y = 5$, then $\frac{1}{x} + \frac{1}{y} =$

(A) $^3/_2$ (B) -3 (C) -1

(D) $-^3/_2$ (E) 3

25. If $f(x) = 2x - 5$ then $f(x + h) =$

(A) $2x + h - 5$ (B) $2h - 5$ (C) $2x + 2h - 5$

(D) $2x - 2h + 5$ (E) $2x - 5$

26. If $a + b = 3$ and $2b + c = 2$, then $2a - c =$

(A) -4 (B) -1 (C) 1

(D) 4 (E) 5

27. If $x > {}^1/_5$, then

(A) x is greater than 1. (B) x is greater than 5.

(C) $^1/_x$ is greater than 5. (D) $^1/_x$ is less than 5.

(E) None of the above statements is true.

28. If $f(x) = x^2 + 3x + 2$, then $[f(x + a) - f(x)]/a =$

(A) $2x + a + 3$ (B) $(x + a)^2 - x^2$ (C) $a^2 + 2ax + 3a$

(D) $2x + a$ (E) $2x + 3$

29. If $x + 2y > 5$ and $x < 3$, then $y > 1$ is true

(A) never. (B) only if $x = 0$. (C) only if $x > 0$.

(D) only if $x < 0$. (E) always.

30. If $x + y = 8$ and $xy = 6$, then $1/x + 1/y =$

(A) $1/8$ (B) $1/6$ (C) $1/4$

(D) $4/3$ (E) 8

31. If $x^{64} = 64$ then $x^{32} =$

(A) 8 or – 8 (B) 12 or – 12 (C) 16

(D) 32 or – 32 (E) 48

32. If $\sqrt{x-1} = 2$ then $(x - 1)^2 =$

(A) 4 (B) 6 (C) 8

(D) 10 (E) 16

33. If $2^x = \dfrac{16^2 \times 8^3}{2^{19}}$ then $x =$

(A) – 3 (B) – 2 (C) 1

(D) 2 (E) 3

34. If $2^{(6x-8)} = 16$ then $x =$

(A) 2 (B) 4 (C) 10

(D) 1 (E) 6

35. $\sqrt{X\sqrt{X\sqrt{X^2}}} = ?$

(A) X (B) $X^{7/4}$ (C) $X^{15/16}$

(D) $X^{3/4}$ (E) $X^{15/8}$

36. The quotient of $(x^2 - 5x + 3)/(x + 2)$ is:

(A) $x - 7 + 17/(x + 2)$ (B) $x - 3 + 9/(x + 2)$

(C) $x - 7 - 11/(x + 2)$ (D) $x - 3 - 3/(x + 2)$

(E) $x + 3 - 3(x + 2)$

37. If x and y are two different real numbers and $xz = yz$, then what is the value of z?

(A) $x - y$ (B) 1 (C) x / y

(D) y / x (E) 0

38. If $2a + 2b = 1$, and $6a - 2b = 5$, which of the following statements is true?

(A) $3a - b = 5$ (B) $a + b > 3a - b$ (C) $a + b = -2$

(D) $a + b < 3a - b$ (E) $a + b = -1$

39. Which of the following equations can be used to find a number n, such that if you multiply it by 3 and take 2 away, the result is 5 times as great as if you divide the number by 3 and add 2?

(A) $3n - 2 = 5 + (n/3 + 2)$ (B) $3n - 2 = 5\,(n/3 + 2)$

(C) $3n - 2 = 5n/3 + 2$ (D) $5(3n - 2) = n/3 + 2$

(E) $5n - 2 = n/3 + 2$

40. If $3/2x = 5$, then $2/3 + x =$

(A) $10/3$ (B) 4 (C) $15/2$

(D) 8 (E) 12

41. If $x + y = 12$ and $x^2 + y^2 = 126$, then $xy =$

(A) 9 (B) 10 (C) 11

(D) 13 (E) 16

42. If $\frac{7a - 5b}{b} = 7$, then $\frac{4a + 6b}{2a}$ equals

(A) $15/4$ (B) 4 (C) $17/4$

(D) 5 (E) 6

43. The fraction

$$\frac{7x - 11}{x^2 - 2x - 15}$$

was obtained by adding the two fractions

$$\frac{A}{x - 5} + \frac{B}{x + 3}.$$

The values of A and B are:

(A) $A = 7x, B = 11$ (B) $A = -11, B = 7x$

(C) $A = 3, B = 4$ (D) $A = 5, B = -3$

(E) $A = -5, B = 3$

44. What number must be added to 28 and 36 to give an average of 29?

(A) 23 (B) 32 (C) 21

(D) 4 (E) 5

45. Solve for x:

$$\frac{5}{x} = \frac{2}{x - 1} + \frac{1}{x(x - 1)}.$$

(A) -1 (B) 0 (C) 1

(D) 2 (E) 3

46. If $2X + Y = 2$ and $X + 3Y > 6$, then

(A) $Y \geq 2$ (B) $Y > 2$ (C) $Y < 2$

(D) $Y \leq 2$ (E) $Y = 2$

47. The expression $(x + y)^2 + (x - y)^2$ is equivalent to

(A) $2x^2$ (B) $4x^2$ (C) $2(x^2 + y^2)$

(D) $2x^2 + y^2$ (E) $x^2 + 2y^2$

48. If $x + y = {}^1/_k$ and $x - y = k$, what is the value of $x^2 - y^2$?

(A) 4 (B) 1 (C) 0

(D) k^2 (E) $\frac{1}{k^2}$

49. If $3^{a-b} = {}^1/_9$ and $3^{a+b} = 9$, then $a =$

(A) -2 (B) 0 (C) 1

(D) 2 (E) 3

50. If $\frac{3}{X-1} = \frac{2}{X+1}$, then $X =$

(A) -5 (B) -1 (C) 0

(D) 1 (E) 5

YOUR ANSWER SHEET FOR THE

GEOMETRY DIAGNOSTIC TEST

FOR CLASS AND HOMEWORK ASSIGNMENTS

RECORD YOUR ANSWERS BY FILLING IN THE APPROPRIATE OVALS.

1.	Ⓐ Ⓑ Ⓒ Ⓓ Ⓔ	26.	Ⓐ Ⓑ Ⓒ Ⓓ Ⓔ
2.	Ⓐ Ⓑ Ⓒ Ⓓ Ⓔ	27.	Ⓐ Ⓑ Ⓒ Ⓓ Ⓔ
3.	Ⓐ Ⓑ Ⓒ Ⓓ Ⓔ	28.	Ⓐ Ⓑ Ⓒ Ⓓ Ⓔ
4.	Ⓐ Ⓑ Ⓒ Ⓓ Ⓔ	29.	Ⓐ Ⓑ Ⓒ Ⓓ Ⓔ
5.	Ⓐ Ⓑ Ⓒ Ⓓ Ⓔ	30.	Ⓐ Ⓑ Ⓒ Ⓓ Ⓔ
6.	Ⓐ Ⓑ Ⓒ Ⓓ Ⓔ	31.	Ⓐ Ⓑ Ⓒ Ⓓ Ⓔ
7.	Ⓐ Ⓑ Ⓒ Ⓓ Ⓔ	32.	Ⓐ Ⓑ Ⓒ Ⓓ Ⓔ
8.	Ⓐ Ⓑ Ⓒ Ⓓ Ⓔ	33.	Ⓐ Ⓑ Ⓒ Ⓓ Ⓔ
9.	Ⓐ Ⓑ Ⓒ Ⓓ Ⓔ	34.	Ⓐ Ⓑ Ⓒ Ⓓ Ⓔ
10.	Ⓐ Ⓑ Ⓒ Ⓓ Ⓔ	35.	Ⓐ Ⓑ Ⓒ Ⓓ Ⓔ
11.	Ⓐ Ⓑ Ⓒ Ⓓ Ⓔ	36.	Ⓐ Ⓑ Ⓒ Ⓓ Ⓔ
12.	Ⓐ Ⓑ Ⓒ Ⓓ Ⓔ	37.	Ⓐ Ⓑ Ⓒ Ⓓ Ⓔ
13.	Ⓐ Ⓑ Ⓒ Ⓓ Ⓔ	38.	Ⓐ Ⓑ Ⓒ Ⓓ Ⓔ
14.	Ⓐ Ⓑ Ⓒ Ⓓ Ⓔ	39.	Ⓐ Ⓑ Ⓒ Ⓓ Ⓔ
15.	Ⓐ Ⓑ Ⓒ Ⓓ Ⓔ	40.	Ⓐ Ⓑ Ⓒ Ⓓ Ⓔ
16.	Ⓐ Ⓑ Ⓒ Ⓓ Ⓔ	41.	Ⓐ Ⓑ Ⓒ Ⓓ Ⓔ
17.	Ⓐ Ⓑ Ⓒ Ⓓ Ⓔ	42.	Ⓐ Ⓑ Ⓒ Ⓓ Ⓔ
18.	Ⓐ Ⓑ Ⓒ Ⓓ Ⓔ	43.	Ⓐ Ⓑ Ⓒ Ⓓ Ⓔ
19.	Ⓐ Ⓑ Ⓒ Ⓓ Ⓔ	44.	Ⓐ Ⓑ Ⓒ Ⓓ Ⓔ
20.	Ⓐ Ⓑ Ⓒ Ⓓ Ⓔ	45.	Ⓐ Ⓑ Ⓒ Ⓓ Ⓔ
21.	Ⓐ Ⓑ Ⓒ Ⓓ Ⓔ	46.	Ⓐ Ⓑ Ⓒ Ⓓ Ⓔ
22.	Ⓐ Ⓑ Ⓒ Ⓓ Ⓔ	47.	Ⓐ Ⓑ Ⓒ Ⓓ Ⓔ
23.	Ⓐ Ⓑ Ⓒ Ⓓ Ⓔ	48.	Ⓐ Ⓑ Ⓒ Ⓓ Ⓔ
24.	Ⓐ Ⓑ Ⓒ Ⓓ Ⓔ	49.	Ⓐ Ⓑ Ⓒ Ⓓ Ⓔ
25.	Ⓐ Ⓑ Ⓒ Ⓓ Ⓔ	50.	Ⓐ Ⓑ Ⓒ Ⓓ Ⓔ

CLASS AND HOMEWORK ASSIGNMENTS

GEOMETRY DIAGNOSTIC TEST

This diagnostic test is designed to help you determine your strengths and your weaknesses in geometry. Follow the directions for each part and check your answers. Record your answers on the preceding page designated "Your Answer Sheet."

50 Questions

DIRECTIONS: Choose the correct answer for each of the following problems. Fill in each answer on the answer sheet.

1. An old picture has dimensions 33 inches by 24 inches. What one length must be cut from each dimension so that the ratio of the shorter side to the longer side is $^2/_3$?

   (A) $4^1/_2$ inches (B) 9 inches (C) 6 inches

   (D) $10^1/_2$ inches (E) 3 inches

2. The greatest area that a rectangle whose perimeter is 52 m can have is

   (A) 12 m^2 (B) 169 m^2 (C) 172 m^2

   (D) 168 m^2 (E) 52 m^2

3. If the triangle *ABC* has angle $A = 35°$ and angle $B = 85°$, then the measure of the angle x in degrees is:

   (A) 85

   (B) 90

   (C) 100

   (D) 120

   (E) 180

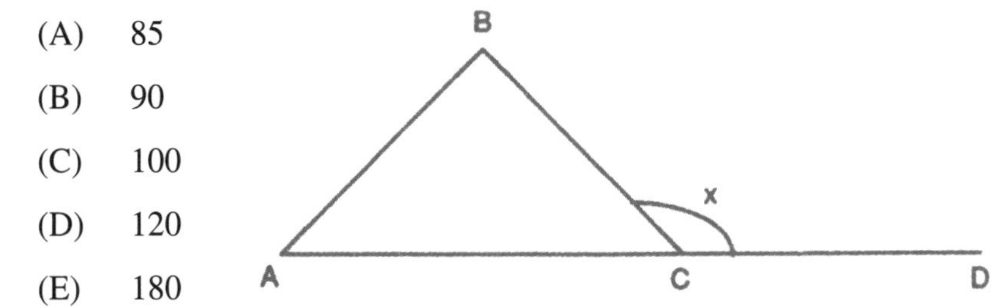

4. In the following figure, 0 is the center of the circle. If arc *ABC* has length 2π, what is the area of the circle?

   (A) 3π

   (B) 6π

   (C) 9π

   (D) 12π

   (E) 15π

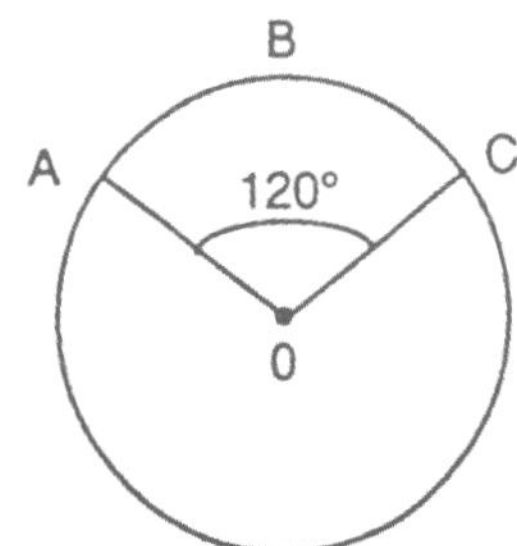

5. If the area of a rectangle is 120 and the perimeter is 44, then the length is

(A) 30 (B) 20 (C) 15

(D 12 (E) 10

6. What is the measure of the angle made by the minute and hour hand of a clock at 3:30?

(A) 60° (B) 75° (C) 90°

(D) 115° (E) 120°

7. A rectangular piece of metal has an area of 35m^2 and a perimeter of 24 m. Which of the following are possible dimensions of the piece?

(A) $^{35}/_2$ m × 2 m (B) 5 m × 7 m (C) 35 m × 1 m

(D) 6 m × 6 m (E) 8 m × 4 m

8. The area of ΔADE is 12 square units. If B is the midpoint of $\overline{AD}$ and C is the midpoint of $\overline{AE}$, what is the area of ΔABC?

(A) 2 square units

(B) 3 square units

(C) $3^1/_2$ square units

(D) 4 square units

(E) 6 square units

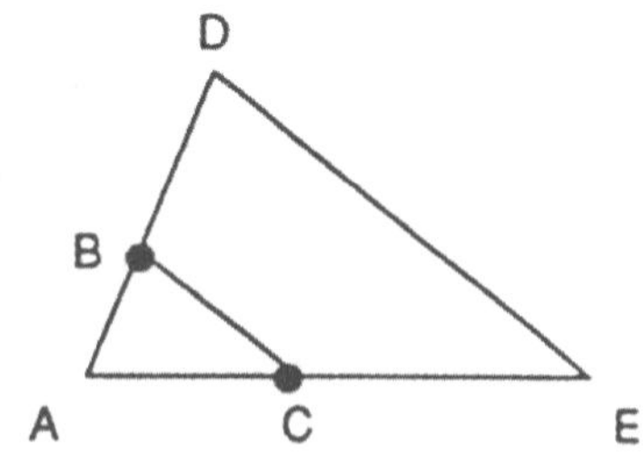

9. If the quadrilateral $ABCD$ has angle $A = 35°$, angle $B = 85°$, and angle $C = 120°$, then the measure of the angle D in degrees is:

(A) 85 (B) 90 (C) 100

(D) 120 (E) 180

10. In the figure shown, two chords of the circle intersect, making the angles shown. What is the value of $x + y$?

(A) 40°

(B) 50°

(C) 80°

(D) 160°

(E) 320°

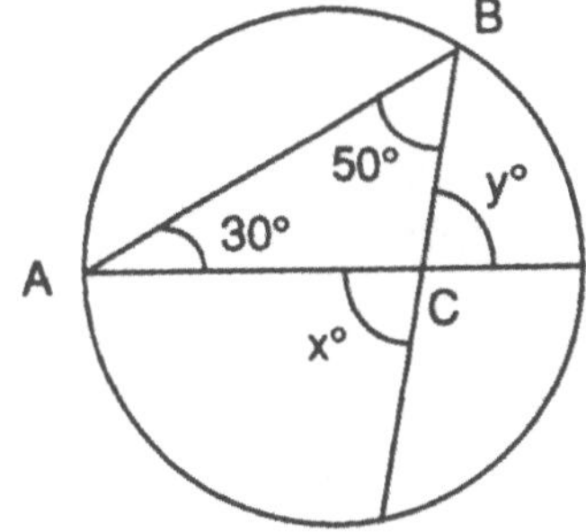

11. In the figure shown, three chords of the circle intersect making the angles shown. What is the value of θ?

(A) 35°

(B) 45°

(C) 60°

(D) 75°

(E) 80°

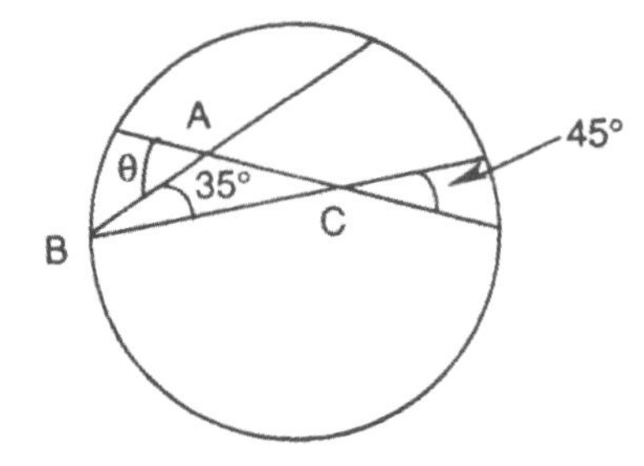

12. If a triangle of base 6 units has the same area as a circle of radius 6 units, what is the altitude of the triangle?

(A) π (B) 3π (C) 6π

(D) 12π (E) 36π

13. A cube consists of 96 square feet. What is the volume of the cube in cubic feet?

(A) 16 (B) 36 (C) 64

(D) 96 (E) 216

14. If the angles of a triangle *ABC* are in the ratio of 3 : 5 : 7, then the triangle is:

(A) acute (B) right (C) isosceles

(D) obtuse (E) equilateral

15. If the measures of the three angles of a triangle are $(3x + 15)°$, $(5x – 15)°$, and $(2x + 30)°$, what is the measure of each angle?

(A) 75° (B) 60° (C) 45°

(D) 25° (E) 15°

16. In the figure shown below, line *l* is parallel to line *m*. If the area of triangle *ABC* is 40 cm^2, what is the area of triangle *ABD*?

(A) Less than 40 cm^2

(B) More than 40 cm^2

(C) The length of segment $\overline{AD}$ times 40 cm^2

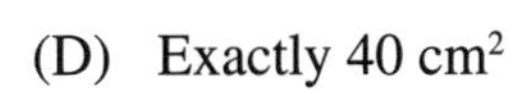

(D) Exactly 40 cm^2

(E) Cannot be determined from the information given

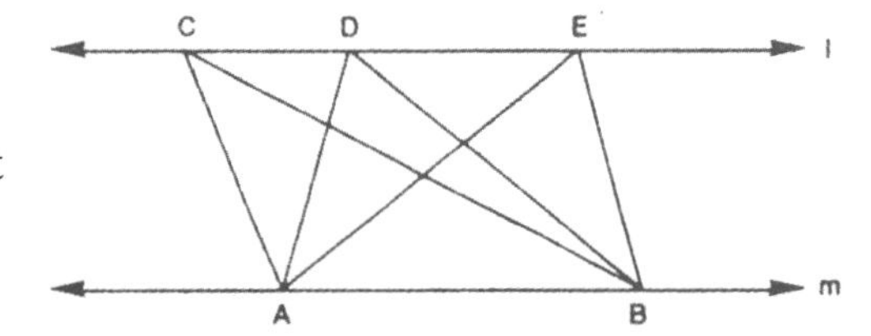

17. In the figure shown, if the length of segment $\overline{EB}$, base of triangle *EBC*, is equal to $^1/_4$ the length of segment $\overline{AB}$ ($\overline{AB}$ is the length of rectangle *ABCD*), and the area of triangle *EBC* is 12 square units, find the area of the hatchmarked region.

(A) 24 square units

(B) 96 square units

(C) 84 square units

(D) 72 square units

(E) 120 square units

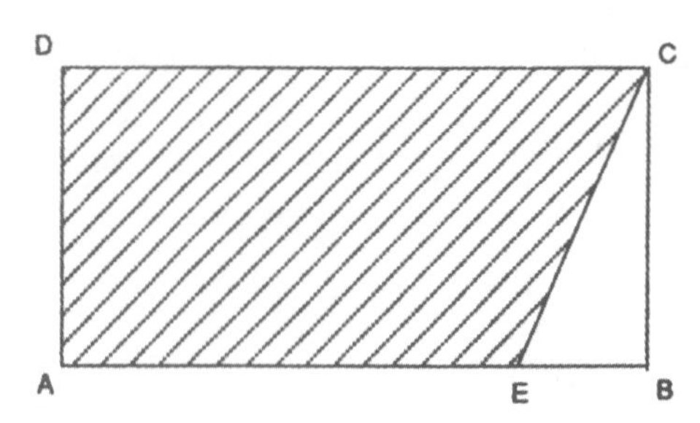

18. What is the perimeter of triangle ABC?

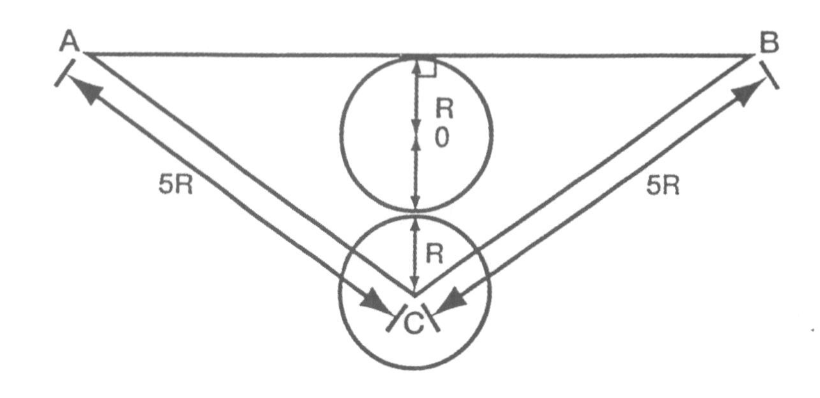

(A) $12R$ (B) $18R^2$ (C) $12R^2$

(D) $18R$ (E) $16R$

19. Which of the following alternatives is correct?

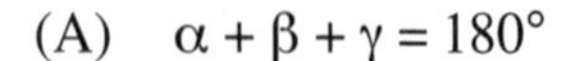

(A) $\alpha + \beta + \gamma = 180°$

(B) $\gamma - \alpha + 180° = \beta$

(C) $\alpha = \beta + \gamma$

(D) $\gamma = \alpha + \beta$

(E) $\alpha = 180° - \beta - \alpha$

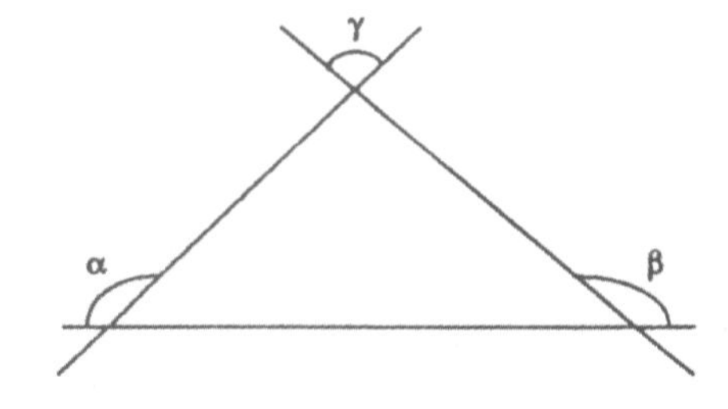

20. In the figure shown, all segments meet at right angles. Find the figure's perimeter in terms of r and s.

(A) $r + s$

(B) $2r + s$

(C) $2s + r$

(D) $r^2 + s^2$

(E) $2r + 2s$

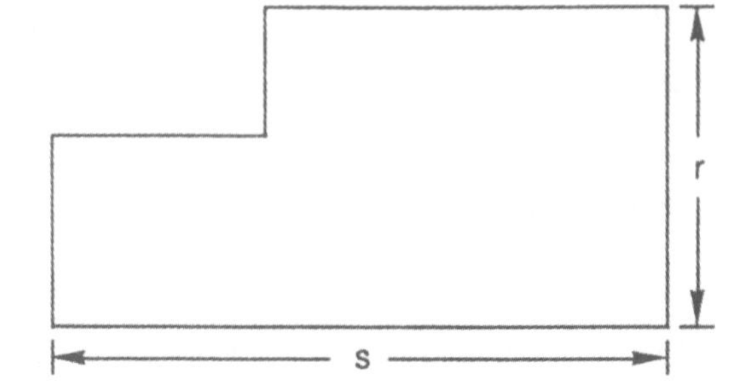

21. If lines *l*, *m*, and *n* intersect at point *P*, express $x + y$ in terms of a.

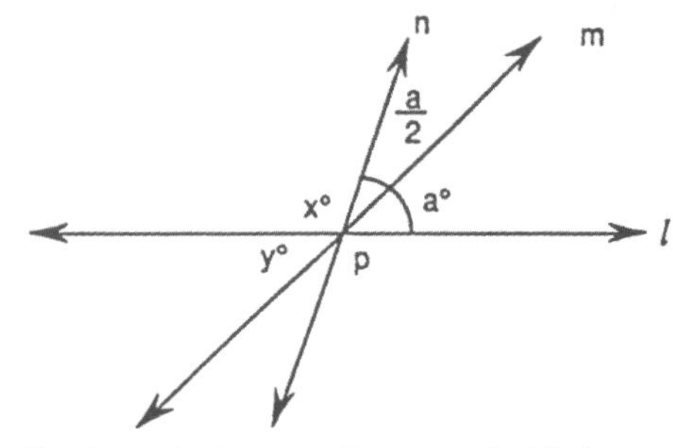

(A) $180 - a/2$

(B) $a/2 - 180$

(C) $90 - a/2$

(D) $a - 180$

(E) $180 - a$

22. The measure of an inscribed angle is equal to one-half the measure of its inscribed arc. In the figure shown, triangle *ABC* is inscribed in circle *O*, and line $\overline{BD}$ is tangent to the circle at point *B*. If the measure of angle *CBD* is 70°, what is the measure of angle *BAC*?

(A) 110°

(B) 70°

(C) 140°

(D) 35°

(E) 40°

23. What is the value of *x*?

(A) 20°

(B) 40°

(C) 60°

(D) 90°

(E) 30°

$L_1 // L_2$

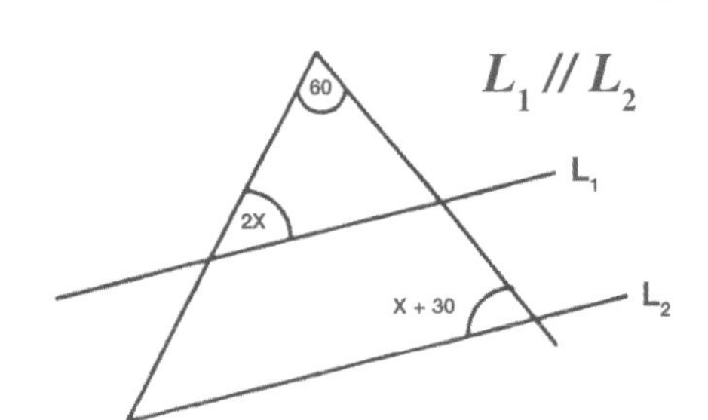

24. A rectangle is divided into three squares, as shown in the diagram. If the long side of the rectangle is equal to 12 cm, what is the area of one of the squares?

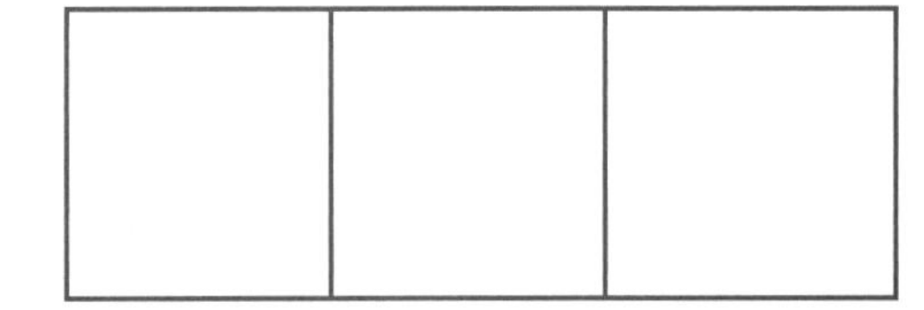

(A) 8 cm^2

(B) 16 cm^2

(C) 32 cm^2

(D) 64 cm^2

(E) 54cm^2

25. In the figure shown, line *r* is parallel to line *l*. Find the measure of angle *RBC*.

(A) 30° (B) 80° (C) 90°

(D) 100° (E) 110°

26. In the five-pointed star shown, what is the sum of the measures of angles *A, B, C, D,* and *E*?

(A) 108°

(B) 72°

(C) 36°

(D) 150°

(E) 180°

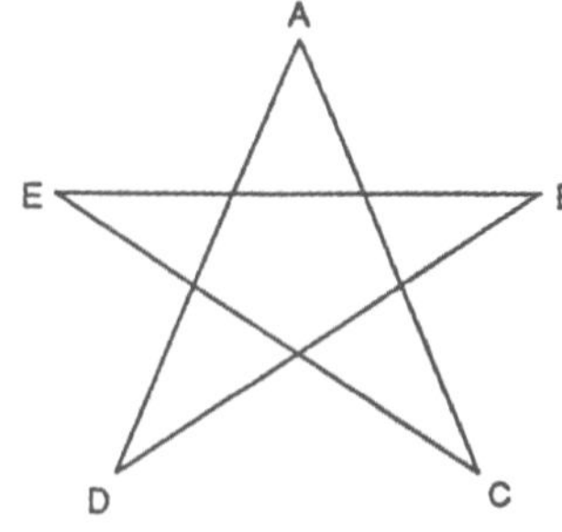

27. A room measures 13 feet by 26 feet. A rug which measures 12 feet by 18 feet is placed on the floor. What is the area of the uncovered portion of the floor?

(A) 554 sq. ft. (B) 216 sq. ft. (C) 100 sq. ft.

(D) 122 sq. ft. (E) 338 sq. ft.

28. Find the area of MNOP

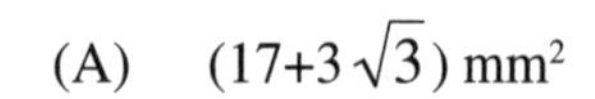

(A) $(17+3\sqrt{3})$ mm^2

(B) 33/2 mm^2

(C) $33\sqrt{3}/2$ mm^2

(D) 33 mm^2

(E) $33\sqrt{3}$ mm^2

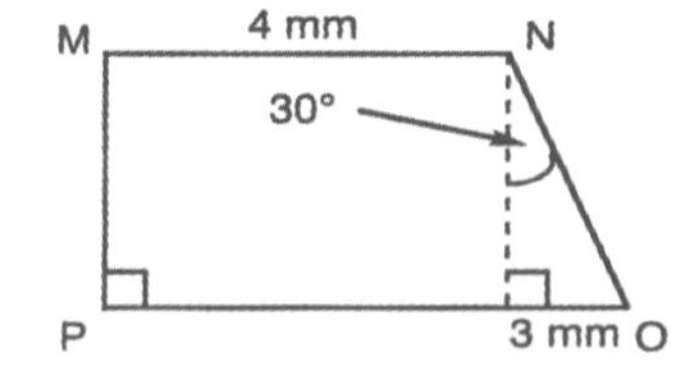

29. In ΔABC, $AB = 6$, $BC = 4$ and $AC = 3$. What kind of a triangle is it?

(A) right and scalene (B) obtuse and scalene

(C) acute and scalene (D) right and isosceles

(E) obtuse and isosceles

30. What is the area of the shaded portion of the rectangle? The heavy dot represents the center of the semicircle.

(A) $200 - 100\pi$ (B) $200 - 25\pi$

(C) $30 - \frac{25\pi}{2}$ (D) $\frac{200 - 25\pi}{2}$

(E) $\frac{400 - 25\pi}{2}$

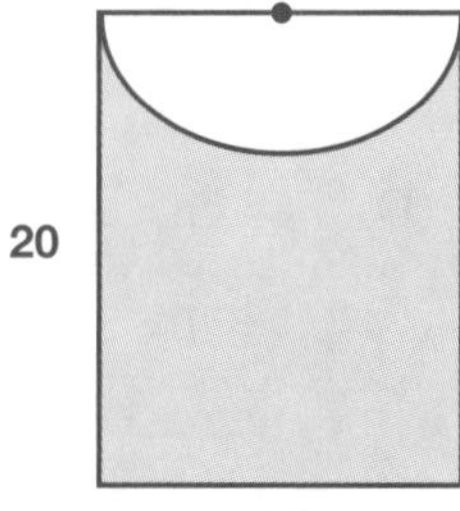

31. Find the area of the isosceles trapezoid.

(A) $250\sqrt{3}$

(B) 150

(C) 250

(D) $125\sqrt{3}$

(E) Area cannot be found.

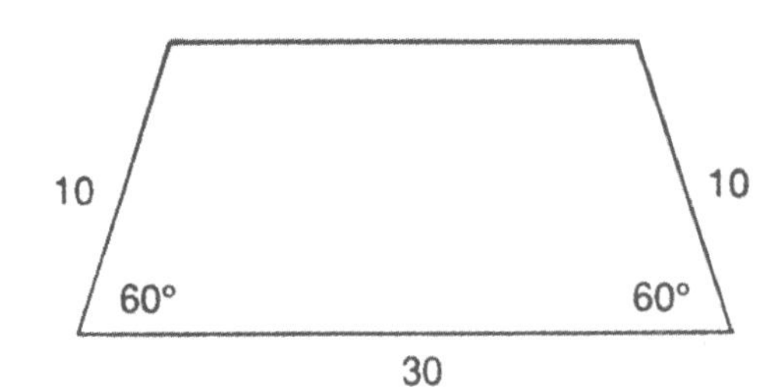

32. If the radius of a sphere is increased by a factor of 3, then the volume of the sphere is increased by a factor of

(A) 3 (B) 6 (C) 9

(D) 18 (E) 27

33. In the diagram shown, ABC is an isosceles triangle. Sides AC and BC are extended through C to E and D to form triangle CDE. The sum of the measures of angles D and E is

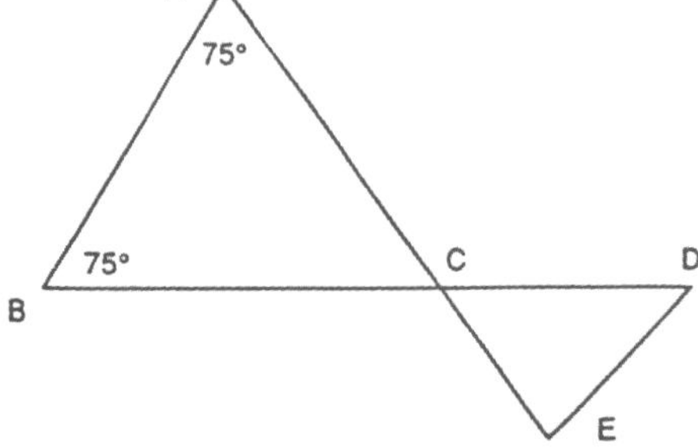

(A) 150°

(B) 105°

(C) 90°

(D) 60°

(E) 30°

34. The box pictured has a square base with side x and a closed top. The surface area of the box is

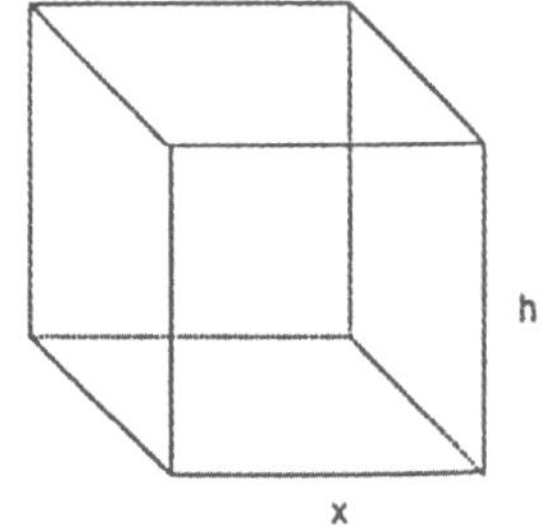

(A) $4x + h$

(B) $4x + 4h$

(C) hx^2

(D) $x^2 + 4xh$

(E) $2x^2 + 4xh$

35. Quadrilaterals $ABCD$ and $AFED$ are squares with sides of length 10 cm. Arc BD and arc DF are quarter circles. What is the area of the shaded region?

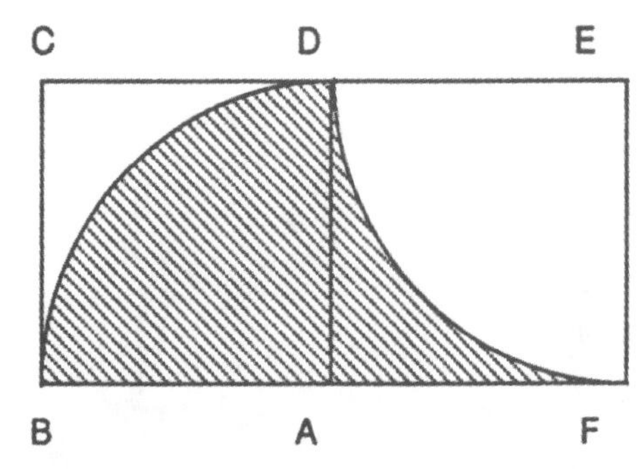

(A) 50 sq. cm

(B) 100 sq. cm

(C) 80 sq. cm

(D) 40 sq. cm

(E) 10 sq. cm

36. If the distance between two adjacent vertical or horizontal dots is 1, what is the perimeter of ΔABC? (See figure following)

    (A) 5

    (B) $\sqrt{3}+\sqrt{10}+\sqrt{11}$

    (C) 8

    (D) 9

    (E) $\sqrt{2}+\sqrt{13}+\sqrt{17}$

37. If the hypotenuse of a right triangle is $x + 1$ and one of the legs is x, then the other leg is

    (A) $\sqrt{2x+1}$ (B) $\sqrt{2x}+1$ (C) $\sqrt{x^2+(x+1)^2}$

    (D) 1 (E) $2x + 1$

38. The measures of the lengths of two sides of an isosceles triangle are x and $2x + 1$. Then, the perimeter of the triangle is

    (A) $4x$ (B) $4x + 1$ (C) $5x + 1$

    (D) $5x + 2$ (E) None of the above

39. Find the length of the diagonal of the rectangular solid shown in the following figure.

    (A) 7

    (B) $2\sqrt{10}$

    (C) $3\sqrt{5}$

    (D) 11

    (E) None of the above.

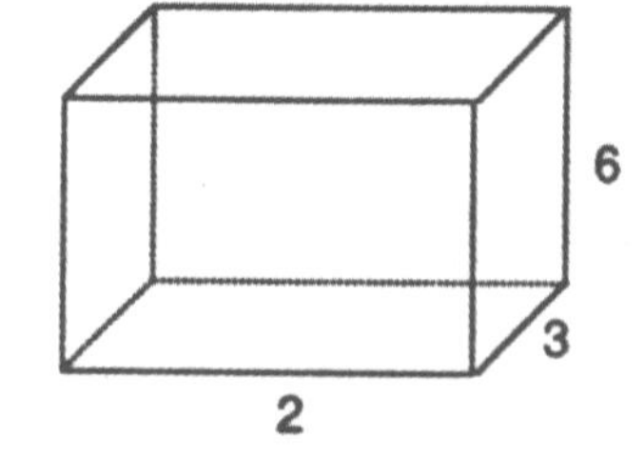

40. Find the area of the shaded portion in the following figure. The heavy dot represents the center of the circle.

    (A) $100\pi - 96$

    (B) $400\pi - 96$

    (C) $400\pi - 192$

    (D) $100\pi - 192$

    (E) $256\pi - 192$

41. $m(\angle A) + m(\angle C) =$

    (A) 160°

    (B) 180°

    (C) 190°

    (D) 195°

    (E) 200°

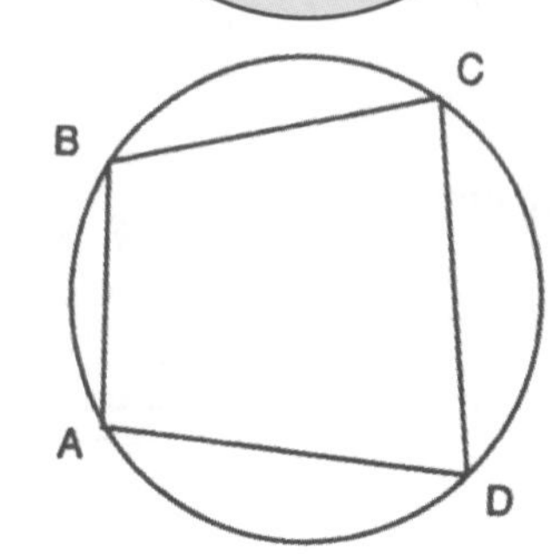

42. The area of the shaded region is:

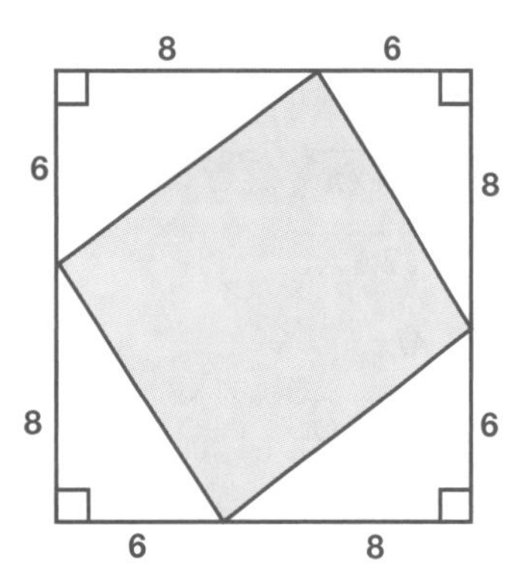

 (A) 25 sq. units

 (B) 36 sq. units

 (C) 49 sq. units

 (D) 100 sq. units

 (E) None of the above

43. In the figure shown, ΔABC is an equilateral triangle. Also, $AC = 3$ and $DB = BE = 1$. Find the perimeter of quadrilateral $ACED$.

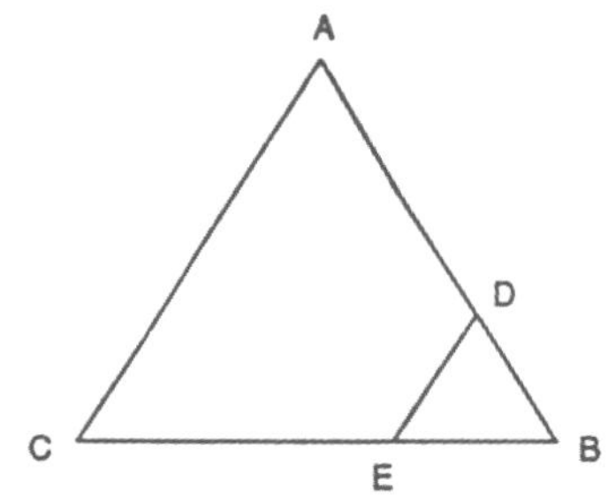

 (A) 6

 (B) $6\frac{1}{2}$

 (C) 7

 (D) $7\frac{1}{2}$

 (E) 8

44. The sum of the exterior angles of the hexagon shown, one angle at each vertex is

 (A) 120°

 (B) 270°

 (C) 360°

 (D) 450°

 (E) None of the above

45. Find the area of the shaded region in the figure, given that $\overline{AB} = \overline{CD} = 4$ and $\overline{BC} = 8$.

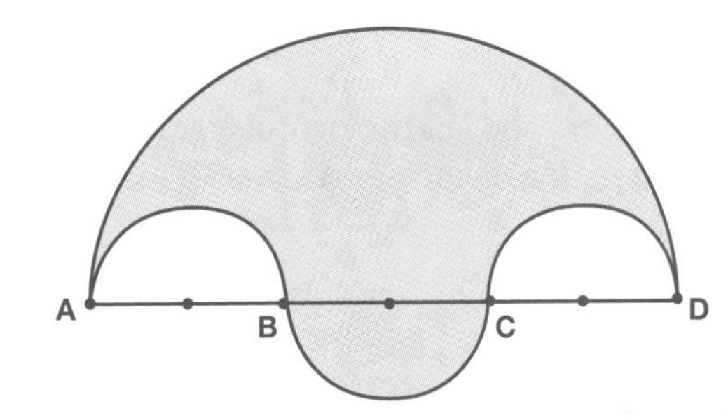

 (A) 40π

 (B) 32π

 (C) 68π

 (D) 76π

 (E) 36π

46. In the figure shown, the area of the inscribed circle is A. What is the length of a side of the square?

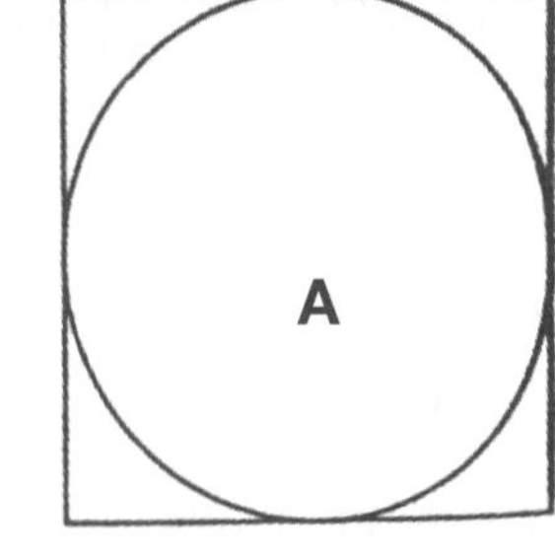

(A) $\sqrt{A/\pi}$

(B) $\sqrt{2A/\pi}$

(C) A/π

(D) $2\sqrt{A/\pi}$

(E) $2A/\pi$

47. In the cube $ABCDEFGH$ with side $AB = 2$, what is the length of diagonal AF?

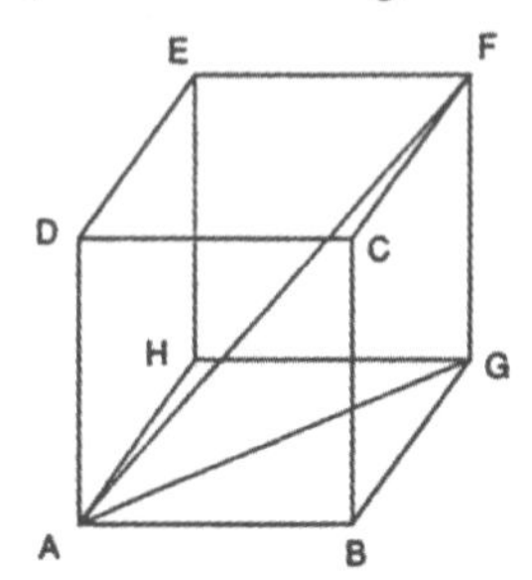

(A) 2

(B) $2\sqrt{2}$

(C) $2\sqrt{3}$

(D) 4

(E) $2\sqrt{5}$

48. Find the area of the shaded region. O is the center of the given circle, whose radius is 6. The distance $\overline{AB} = 6\sqrt{2}$.

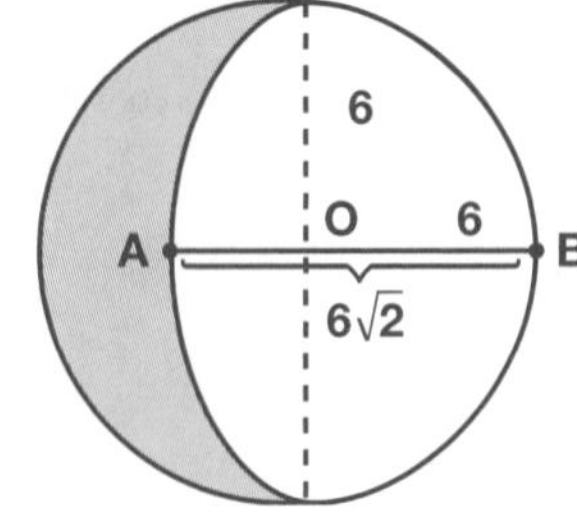

(A) 9π

(B) 72π

(C) 18π

(D) $18\pi - 36$

(E) 36

49. In the given figure, the area of the triangle ABC is

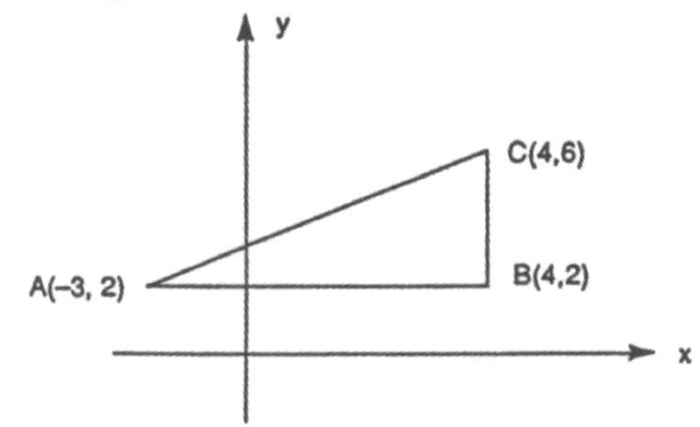

(A) 65

(B) 40

(C) 28

(D) 16

(E) 14

50. In the figure shown, the right-angled figure is a square of length r, and the circular region on top of the square has radius r. The perimeter of the figure is

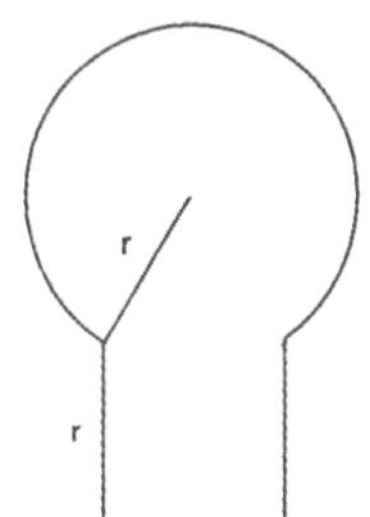

(A) $4r + 2\pi r$

(B) $2r + \pi r/3$

(C) $3r + 2\pi r$

(D) $3r + \pi r/3$

(E) $3r + 5\pi r/3$

YOUR ANSWER SHEET FOR THE

WORD PROBLEMS DIAGNOSTIC TEST

FOR CLASS AND HOMEWORK ASSIGNMENTS

RECORD YOUR ANSWERS BY FILLING IN THE APPROPRIATE OVALS.

1.	Ⓐ Ⓑ Ⓒ Ⓓ Ⓔ	26.	Ⓐ Ⓑ Ⓒ Ⓓ Ⓔ
2.	Ⓐ Ⓑ Ⓒ Ⓓ Ⓔ	27.	Ⓐ Ⓑ Ⓒ Ⓓ Ⓔ
3.	Ⓐ Ⓑ Ⓒ Ⓓ Ⓔ	28.	Ⓐ Ⓑ Ⓒ Ⓓ Ⓔ
4.	Ⓐ Ⓑ Ⓒ Ⓓ Ⓔ	29.	Ⓐ Ⓑ Ⓒ Ⓓ Ⓔ
5.	Ⓐ Ⓑ Ⓒ Ⓓ Ⓔ	30.	Ⓐ Ⓑ Ⓒ Ⓓ Ⓔ
6.	Ⓐ Ⓑ Ⓒ Ⓓ Ⓔ	31.	Ⓐ Ⓑ Ⓒ Ⓓ Ⓔ
7.	Ⓐ Ⓑ Ⓒ Ⓓ Ⓔ	32.	Ⓐ Ⓑ Ⓒ Ⓓ Ⓔ
8.	Ⓐ Ⓑ Ⓒ Ⓓ Ⓔ	33.	Ⓐ Ⓑ Ⓒ Ⓓ Ⓔ
9.	Ⓐ Ⓑ Ⓒ Ⓓ Ⓔ	34.	Ⓐ Ⓑ Ⓒ Ⓓ Ⓔ
10.	Ⓐ Ⓑ Ⓒ Ⓓ Ⓔ	35.	Ⓐ Ⓑ Ⓒ Ⓓ Ⓔ
11.	Ⓐ Ⓑ Ⓒ Ⓓ Ⓔ	36.	Ⓐ Ⓑ Ⓒ Ⓓ Ⓔ
12.	Ⓐ Ⓑ Ⓒ Ⓓ Ⓔ	37.	Ⓐ Ⓑ Ⓒ Ⓓ Ⓔ
13.	Ⓐ Ⓑ Ⓒ Ⓓ Ⓔ	38.	Ⓐ Ⓑ Ⓒ Ⓓ Ⓔ
14.	Ⓐ Ⓑ Ⓒ Ⓓ Ⓔ	39.	Ⓐ Ⓑ Ⓒ Ⓓ Ⓔ
15.	Ⓐ Ⓑ Ⓒ Ⓓ Ⓔ	40.	Ⓐ Ⓑ Ⓒ Ⓓ Ⓔ
16.	Ⓐ Ⓑ Ⓒ Ⓓ Ⓔ	41.	Ⓐ Ⓑ Ⓒ Ⓓ Ⓔ
17.	Ⓐ Ⓑ Ⓒ Ⓓ Ⓔ	42.	Ⓐ Ⓑ Ⓒ Ⓓ Ⓔ
18.	Ⓐ Ⓑ Ⓒ Ⓓ Ⓔ	43.	Ⓐ Ⓑ Ⓒ Ⓓ Ⓔ
19.	Ⓐ Ⓑ Ⓒ Ⓓ Ⓔ	44.	Ⓐ Ⓑ Ⓒ Ⓓ Ⓔ
20.	Ⓐ Ⓑ Ⓒ Ⓓ Ⓔ	45.	Ⓐ Ⓑ Ⓒ Ⓓ Ⓔ
21.	Ⓐ Ⓑ Ⓒ Ⓓ Ⓔ	46.	Ⓐ Ⓑ Ⓒ Ⓓ Ⓔ
22.	Ⓐ Ⓑ Ⓒ Ⓓ Ⓔ	47.	Ⓐ Ⓑ Ⓒ Ⓓ Ⓔ
23.	Ⓐ Ⓑ Ⓒ Ⓓ Ⓔ	48.	Ⓐ Ⓑ Ⓒ Ⓓ Ⓔ
24.	Ⓐ Ⓑ Ⓒ Ⓓ Ⓔ	49.	Ⓐ Ⓑ Ⓒ Ⓓ Ⓔ
25.	Ⓐ Ⓑ Ⓒ Ⓓ Ⓔ	50.	Ⓐ Ⓑ Ⓒ Ⓓ Ⓔ

CLASS AND HOMEWORK ASSIGNMENTS

WORD PROBLEMS DIAGNOSTIC TEST

This diagnostic test is designed to help you determine your strengths and your weaknesses in word problems. Follow the directions for each part and check your answers. Record your answers on the preceding page designated "Your Answer Sheet."

50 Questions

DIRECTIONS: Choose the correct answer for each of the following problems. Fill in each answer on the answer sheet.

1. Two pounds of pears and one pound of peaches cost \$1.40. Three pounds of pears and two pounds of peaches cost \$2.40. How much is the combined cost of one pound of pears and one pound of peaches?

   (A) \$2.00 (B) \$1.50 (C) \$1.60
   (D) \$.80 (E) \$1.00

2. Two dice are thrown, one red and one green. The probability that the number on the red exceeds the number showing on the green by exactly two is

   (A) $\frac{1}{18}$ (B) $\frac{1}{4}$ (C) $\frac{1}{9}$
   (D) $\frac{1}{36}$ (E) $\frac{1}{24}$

3. A man buys a book for \$20 and wishes to sell it. What price should he mark on it if he wishes a 40% discount while making a 50% profit on the cost price?

   (A) \$25 (B) \$30 (C) \$40
   (D) \$50 (E) \$55

4. A man who is 40 years old has three sons, ages 6, 3, and 1. In how many years will the combined age of his three sons equal 80% of his age?

   (A) 5 (B) 10 (C) 15
   (D) 20 (E) 25

5. Peter brought n compact disks for \$$m$. If in a second purchase he paid \$$q$, what was the increment per compact disk? ($q > m$)

   (A) $(m - q)/n$ (B) $(m - q)n$ (C) $(q - m)/n$
   (D) $(q - m)n$ (E) $(n - q)/m$

6. In a three-person, 100-meter race, Amy finishes 10 meters ahead of Brooke and 20 meters ahead of Carol. Assuming constant speeds for each runner, by how many meters did Brooke finish ahead of Carol?

(A) 9 meters (B) 10 meters (C) 11 meters

(D) $11\frac{1}{9}$ meters (E) $11\frac{8}{9}$ meters

7. At a certain restaurant the cost of 3 sandwiches, 7 cups of coffee and 4 pieces of pie is $10.20, while the cost of 4 sandwiches, 8 cups of coffee and 5 pieces of pie is $12.25. What is the cost of a luncheon consisting of one sandwich, one cup of coffee and one piece of pie?

(A) $2.00 (B) $2.05 (C) $2.10

(D) $2.15 (E) $2.25

8. Tilda's car gets 34 miles per gallon of gasoline and Naomi's car gets 8 miles per gallon. When traveling from Washington, D.C. to Philadelphia, they both used a whole number of gallons of gasoline. How far is it from Philadelphia to Washington, D.C.?

(A) 21 miles (B) 32 miles (C) 68 miles

(D) 136 miles (E) 170 miles

9. In a chess match, a win counts 1 point, a draw counts $\frac{1}{2}$ point, and a loss counts 0 points. After 15 games, the winner was 4 points ahead of the loser. How many points did the loser have?

(A) $4\frac{1}{2}$ (B) $5\frac{1}{2}$ (C) 6

(D) 7 (E) 9

10. A sphere with diameter 1 meter has a mass of 120 kilograms. What is the mass, in kilograms, of a sphere of the same kind of material that has a diameter of 2 meters?

(A) 480 (B) 560 (C) 640

(D) 960 (E) 1080

Questions 11–15 refer to the following figure

Portion of Ph.D. Degrees in the Mathematical Sciences Awarded to U.S. Citizens in 1986

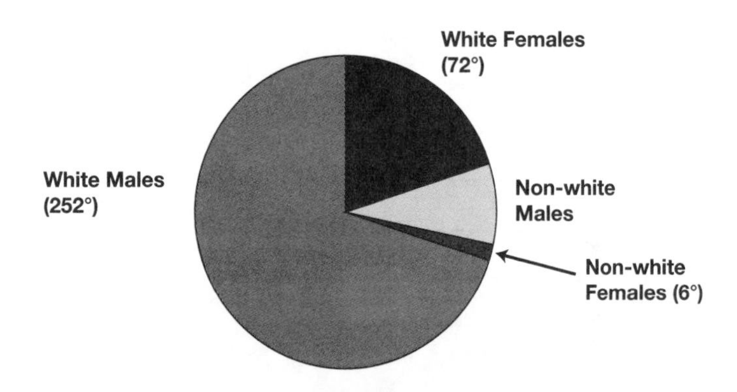

11. What percent of the Ph.D. degrees were awarded in 1986 to non-white males?

(A) 30 (B) $8^1/_3$ (C) $4^1/_6$

(D) 20 (E) None of the above

12. If 4000 Ph.D's were awarded in Mathematical Sciences, how many were awarded to white female U.S. citizens?

(A) 800 (B) 2880 (C) 3200

(D) 1120 (E) None of the above

13. If the 600 white females represent 72° of the figure that depicts the total distribution of Ph.D's awarded in the Mathematical Sciences in the U.S. in 1986, then how many were awarded to white males?

(A) 432 (B) 3000 (C) About 857

(D) 2100 (E) None of the above

14. Given the distribution of Ph.D's awarded in Mathematical Sciences in the U.S. in 1986, what is the ratio of white male's degrees to non-white male's degrees?

(A) 1 to 5 (B) 3.5 to 1 (C) 8.4 to 1

(D) 42 to 1 (E) None of the above

15. If the non-white female category represents 6° of the distribution of a total of 6000 Ph.D's awarded in the Mathematical Sciences, then how many Ph.D's were awarded in this category?

(A) 50 (B) 100 (C) 500

(D) 1000 (E) None of the above

16. A waitress's income consists of her salary and tips. Her salary is $150 a week. During one week that included a holiday her tips were 5/4 of her salary. What fraction of her income for the week came from tips?

(A) 5/8 (B) 5/4 (C) 4/9

(D) 1/2 (E) 5/9

Question 17 refers to the following graph:

Undergraduate Mathematics Enrollments in the U.S., 1965-85
(Thousands of enrollments, fall semester)

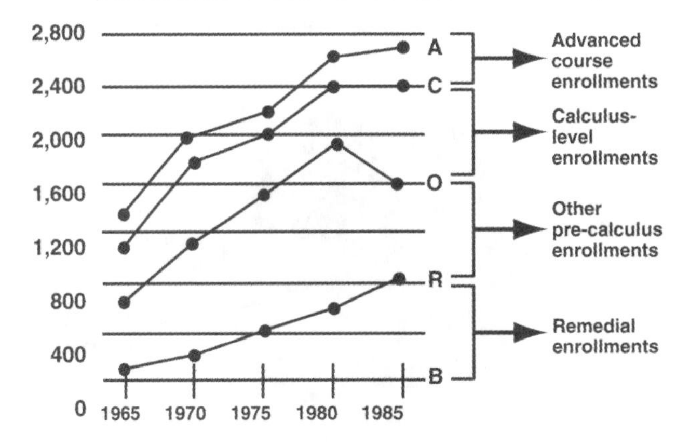

(Note: Area between line segments B and R represent remedial enrollments; between line segments R and O represent other precalculus enrollments; between line segments O and C represent calculus enrollments; and, between line segments C and A represent advanced course enrollments.)

17. What undergraduate enrollments category was fairly constant over the period of the graph?

    (A) Remedial
    (B) Other precalculus
    (C) Calculus level
    (D) Advanced course
    (E) None of the above

18. Jim is twice as old as Susan. If Jim were 4 years younger and Susan were 3 years older, their ages would differ by 12 years. What is the sum of their ages?

    (A) 19
    (B) 42
    (C) 56
    (D) 57
    (E) None of the above

19. Joe and Jim together have 14 marbles. Jim and Tim together have 10 marbles. Joe and Tim together have 12 marbles. What is the maximum number of marbles that any one of these may have?

    (A) 7
    (B) 8
    (C) 9
    (D) 10
    (E) 11

20. Emile receives a flat weekly salary of \$240 plus 12% commission of the total volume of all sales he makes. What must his dollar volume be in a week if he is to make a total weekly salary of \$540?

    (A) \$2880
    (B) \$3600
    (C) \$6480
    (D) \$2500
    (E) \$2000

21. A truck contains 150 small packages, some weighing 1 kg each and some weighing 2 kg each. How many packages weighing 2 kg each are in the truck if the total weight of all the packages is 264 kg?

    (A) 36
    (B) 52
    (C) 88
    (D) 124
    (E) 114

22. I went to 'Las Vegas' Casino and in the first game I lost one third of my money, in the second game I lost half of the rest. If I still have \$1000, how much money did I have when I arrived at the Casino?

    (A) \$1000
    (B) \$2000
    (C) \$3000
    (D) \$6000
    (E) \$12,000

23. A postal truck leaves its station and heads for Chicago, averaging 40 mph. An error in the mailing schedule is spotted and 24 minutes after the truck leaves, a car is sent to overtake the truck. If the car averages 50 mph, how long will it take to catch the postal truck?

(A) 2.6 hours (B) 3 hours (C) 2 hours

(D) 1.5 hours (E) 1.6 hours

24. A gardener wishes to decorate a garden which is 576 square feet in area, and he must decide how to apportion each section of the garden with specific plants. He has decided to set aside 20% of the garden with roses and an additional 32.9 square feet for a tomato patch. How much of the garden is set aside for the roses and tomatoes?

(A) 569.42 square feet (B) 523.1 square feet (C) 148.1 square feet

(D) 52.9 square feet (E) 53.9 square feet

25. A man sold two-thirds of his pencils for 20 cents each. If he has 7 pencils left, how much money did he collect for the pencils he sold?

(A) $2.10 (B) $2.80 (C) $1.40

(D) $2.00 (E) $2.20

26. The cost of digging a basement for a house 30 feet long and 27 feet wide was $975. If the excavating cost was $3.75 per cubic yard, how deep was the basement?

(A) $11^2/_3$ ft (B) $8^2/_3$ ft (C) 11 ft

(D) 9 ft (E) 8 ft

27. Stanley and Mitchell made a bet about Stanley's math test. Mitchell told Stanley that he would give him $8 for each problem he got correct on the test if Stanley would pay him $5 for each problem that he got wrong. There were going to be 26 problems on the test. The next day, after getting his test back, Stanley told Mitchell that neither owed the other any money. How many problems did Stanley get right?

(A) 8 (B) 13 (C) 11

(D) 10 (E) 16

28. The owners of a business divided up the profits as follows: The share of the second was 3 times that of the third and the share of the first was 4 times that of the second. The first received $1,210 more than the third. How much did the first receive?

(A) $1,320 (B) $1,400 (C) $1,550

(D) $2,300 (E) $2,400

29. A company was forced out of business by its competitor. It was able to pay 25 cents on the dollar, but had the company been able to collect a certain debt of $800, it could have paid 30 cents on the dollar. How much did the company owe at the time of its closing?

(A) $1,600 (B) $160,000 (C) $16,000

(D) $1,455 (E) $14,550

30. An investor earns $930 from two accounts in a year. If she has three times as much invested at 8% as she does at 7%, how much does she have invested at 8%?

    (A) $9,000 (B) $3,207 (C) $9,621

    (D) $3,000 (E) $11,500

31. Cyndi invests part of $2,000 in a certificate of deposit that pays simple annual interest of 9% and the remainder in a passbook savings account that pays 5% simple annual interest. If she receives $148 interest in one year how much did she invest in the certificate of deposit?

    (A) $800 (B) $1,000 (C) $1,200

    (D) $1,500 (E) $1,800

32. The average of 12 test scores is 55. When the 2 highest and 2 lowest scores are dropped, the average of the remaining scores is 50. The average of the scores dropped is

    (A) 65 (B) 60 (C) 55

    (D) 52.5 (E) 50

33. A computer is marked up 50 percent and then later marked down 30 percent. If the final price is $3,360, the original price was

    (A) $2,240 (B) $3,200 (C) $4,200

    (D) $4,800 (E) $5,600

34. At $6.75 per hour, the minimum number of hours that Joel needs to work to earn at least $150 is

    (A) 20 (B) 22 (C) 23

    (D) 24 (E) 25

35. Eva can wallpaper a room in 4 hours and Kathy can wallpaper the same room in 6 hours. Kathy works on her own for one hour, and is then joined by Eva. Assuming these rates, how many hours did Kathy and Eva together take to finish the job?

    (A) $2^2/_5$ (B) 2 (C) $1^4/_5$

    (D) $1^1/_5$ (E) 1

36. SPLINT, a "we try harder" telephone company, advertises that it charges 51 cents for the first minute and 34 cents for each additional minute for a long-distance call from New Jersey to New York. If the number of additional minutes after the first minute is x and the cost, in cents, is y, then the equation representing the total cost of the call is $y = 34x + 51$. AC&C, its competitor,

has a cost equation for the same call of the form $y = 31x + 60$. After how many additional minutes will the two companies charge an equal amount?

(A) 9 (B) 3

(C) 2 (D) 4

(E) The two companies will never charge an equal amount because SPLINT is always cheaper than AC&C.

37. At Martin's Market each piece of bubble gum ordinarily sells for 9¢. Mary paid \$1.55 for some of these pieces of bubble gum when they were on sale for \$0.05. How many did she buy?

(A) 21 (B) 28 (C) 31

(D) 35 (E) 37

38. In a four-child family, the four children could all be of one sex (4–0), or there could be three of one sex and one of the other sex (3–1), or there could be two of each sex (2–2). Concerning a four-child family, which of the following is a true sentence?

(A) A family of all of one sex is most likely (4–0).

(B) A family of three of one sex and one of the other sex is most likely (3–1).

(C) A family of two of each sex is most likely (2–2).

(D) The three types of families (4–0), (3-1), and (2-2), are equally likely to occur.

(E) All of the above are false.

39. Mary has $29^1/_2$ yards of material available to make uniforms. Each uniform requires $^3/_4$ yard of material. How many uniforms can she make and how much material will she have left?

(A) 39 uniforms with $^1/_3$ yard left over

(B) 39 uniforms with $^1/_4$ yard left over

(C) 39 uniforms with $^1/_2$ yard left over

(D) 27 uniforms with $^1/_3$ yard left over

(E) 27 uniforms with $^1/_2$ yard left over

40. In a certain city, a taxi ride costs 85¢ for the first $^1/_{10}$ of a mile and 10¢ for every $^1/_{10}$ of a mile after the first. If x is the number of additional $^1/_{10}$-miles after the first $^1/_{10}$ of a mile, write an equation for the cost $f(x)$, in cents of a taxi ride and use it to find the cost of a 2-mile ride.

(A) 175¢ (B) 185¢ (C) 285¢

(D) 200¢ (E) 275¢

41. A pitcher holds 6 times as much water as a paper drinking cup. Three pitchers hold 39 ounces more than 5 drinking cups. How many ounces does a cup hold?

(A) 6 (B) 5 (C) 4
(D) 3 (E) 2

42. At a garage sale, Sarah sold two kitchen gadgets at $2.40 each. Based on the cost, her profit on one was 20% and her loss on the other was 20%. On the sale she:

(A) gained 8¢ (B) lost 8¢ (C) broke even
(D) gained 20¢ (E) lost 20¢

43. A miller took $^1/_{10}$ of the wheat he ground as his fee. How much did he grind if a customer had exactly one bushel left after the fee had been subtracted?

(A) $1^1/_9$ (B) $1^1/_{10}$ (C) $1^1/_3$
(D) $1^1/_2$ (E) $1^1/_5$

44. Of the freshmen at a college, 24% failed remedial mathematics. If 360 students failed remedial mathematics, how many freshmen are enrolled at the college?

(A) 500 (B) 1200 (C) 1500
(D) 18 (E) 81

45. Julie holds 100 shares each of Companies *A, B* and *C*. On January 1 the prices per share, respectively, were $10, $12, and $18. Shares of Company *B* split 2 for 1 and those of Company *C* split 3 for 2. On December 31, the price per share of Companies *A, B* and *C* respectively was $12, $6 and $16. There were no dividends. What was the annual return on Julie's portfolio of stocks?

(A) 16.7% (B) 20% (C) 25%
(D) 30% (E) 33.3%

46. Ten years ago, a house was insured for $10,500, which represented three-fourths of its value. If property values have increased by 150% since then, for how much should the house be insured to represent seven-eighths of its present value?

(A) $17,226.56 (B) $18,375 (C) $30,625
(D) $22,968.75 (E) $35,000

47. The average speed of today's plane is twice that of a plane of 5 years ago and takes $1^1/_4$ hours less time to fly the same distance of 1500 miles. Find the average speed of a plane of 5 years ago.

(A) 450 (B) 1200 (C) 600
(D) 500 (E) 375

48. Tickets for a particular concert cost \$5 each if purchased in advance and \$7 each if bought at the box office on the day of the concert. For this particular concert, 1,200 tickets were sold and the receipts were \$6,700. How many tickets were bought at the box office on the day of the concert?

(A) 500 (B) 700 (C) 600

(D) 350 (E) 200

49. A paint mixture contains 2 parts of green paint and 6 parts of white paint. How many gallons of green paint must be added to 16 gallons of this mixture to obtain a new mixture which has 40% green paint?

(A) 2 (B) 3 (C) 4

(D) 5 (E) 6

50. How many ounces of a metal other than gold must be added to 56 ounces of pure gold to make a composition 70% gold?

(A) 39.2 (B) 16.8 (C) 9.8

(D) 56 (E) 24

Practice Test

Reference Chart

Rectangle

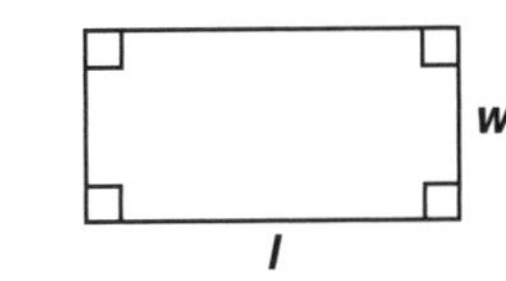

Area = lw

Perimeter = $2\,(l + w)$

Triangle

Area = $\frac{1}{2}bh$

Parallelogram

Area = bh

Pythagorean Formula

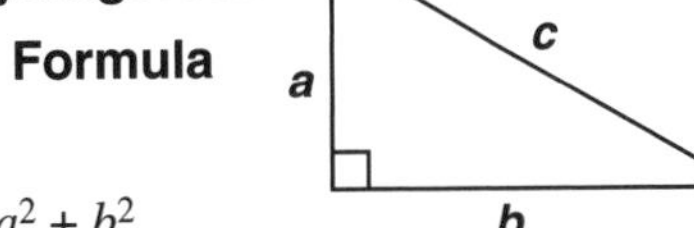

$c^2 = a^2 + b^2$

Circle

Area = πr^2

Circumference = $2\pi r$

Trapezoid

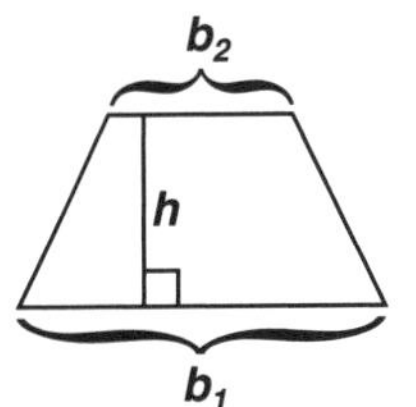

Area = $\frac{1}{2}(b_1 + b_2)h$

Rectangular Prism

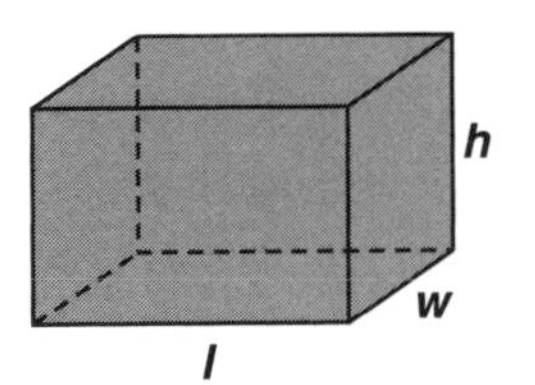

Volume = lwh

Surface Area = $2lw + 2wh + 2lh$

Constant Motion

$d = rt$

Simple Interest

$I = prt$

$\pi \approx 3.14$ or $\frac{22}{7}$

PSSA MATHEMATICS PRACTICE TEST

TIME: 90 Minutes (Note: in the actual test administration, students will be given a brief break after each of the three 30-minute sections.)

Section 1

You may not use a calculator for items 1 – 5.

1. Multiply.

$$\begin{array}{r} 492 \\ \underline{\times\ 847} \end{array}$$

A. 416,524

B. 416,724

C. 416,904

D. 416,924

2. Subtract.

$$17\frac{3}{5} - 8\frac{3}{5} =$$

A. $8\frac{2}{5}$

B. $8\frac{3}{5}$

C. 8

D. 9

3. Divide.

$4 \div 0.16 =$

A. 0.25

B. 2.5

C. 4

D. 25

4. Subtract.

$14 - 4\frac{5}{8} =$

A. $9\frac{3}{8}$

B. $9\frac{5}{8}$

C. $10\frac{3}{8}$

D. $10\frac{5}{8}$

5. A light bulb manufacturing machine produces 72 light bulbs per minute. How much time would it take to make 5,400 light bulbs?

A. 3 hours 20 minutes

B. 1 hour 4 minutes

C. 1 hour 15 minutes

D. 1 hour 45 minutes

You may use a calculator for the rest of the questions on this test.

6. According to the chart below, what percentage of students earned an "A" in this class?

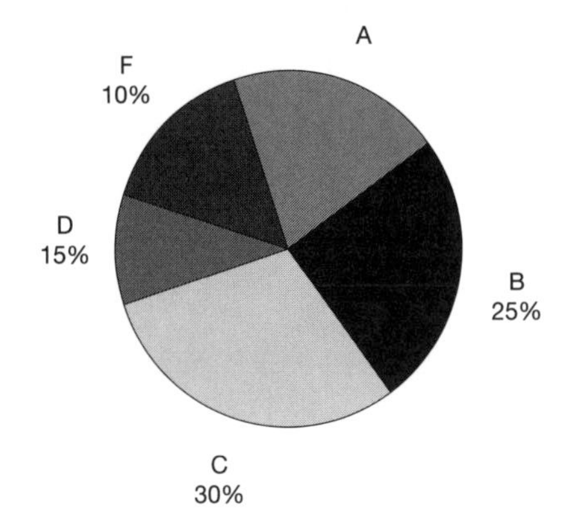

A. 10

B. 15

C. 20

D. 25

7. Which of the following steps would be correct in solving the equation $4x + 9 = 31$?

A. $4x = 31$

B. $x + 9 = 31 - 4$

C. $4x + 9 - 9 = 31 + 9$

D. $4x + 9 - 9 = 31 - 9$

8. A car salesperson sold 8 cars last month for a total of $160,000 in sales. The salesperson gets 3% commission on the first $100,000 and 4% commission on all sales over that amount. How much money did the salesperson earn?

A. $4,800

B. $5,400

C. $6,400

D. $7,200

9. The stem-and-leaf plot below shows the test scores for a history class. How many students scored in the 70's?

Stem	Leaf
9	3 5 6
8	0 2 4 8 8 9
7	0 1 2 2 5 7 8 9
6	2 6
5	5 7 7
4	9

A. 1

B. 2

C. 6

D. 8

---

10. The playing cards below are used in a game for two players. What is the probability that a player draws a card with a star on a first drawing?

A. $\frac{2}{5}$

B. $\frac{3}{10}$

C. $\frac{1}{5}$

D. $\frac{1}{10}$

11. In the diagram below, find the angle X.

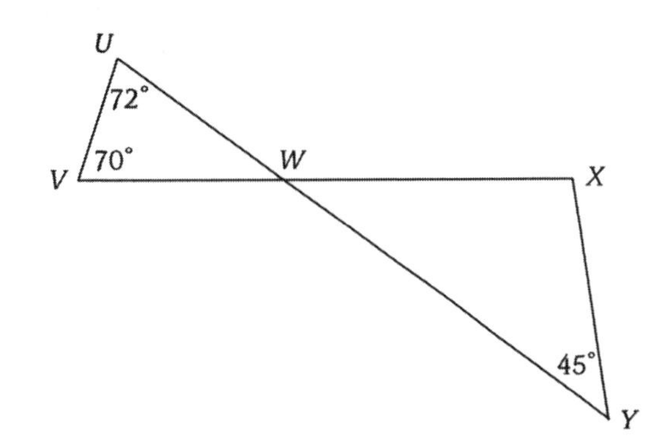

A. 45°

B. 72°

C. 81°

D. 97°

---

12. A student playing basketball records the following statistics:

Linda's Basketball Statistics

Field Goals Attempted	Field Goals Made	Points from Field Goals	Free Throws Attempted	Free Throws Made	Points from Free Throws	Total Points
12	8	16	6	4	4	20

What is the student's ratio of field goals made to field goals attempted?

A. $\frac{1}{2}$

B. $\frac{2}{3}$

C. $\frac{3}{4}$

D. $\frac{3}{2}$

13. Which of the following best describes the pattern 2, 4, 8,…

A. n, n^2, n^3

B. $n, 2n, 3n$

C. $n + 1, n + 2, n + 3$

D. $n, \frac{n}{2}, \frac{n}{3}$

14. How many yards are equal to 180 inches?

A. 4

B. 5

C. 6

D. 15

15. Evaluate $4X + 3Y$, when $X = 5$ and $Y = -3$.

A. 7

B. 11

C. 29

D. 35

16. A deliveryman delivers clean linens to a restaurant every 12 days. Another deliveryman delivers beverages to the restaurant every 15 days. If both men delivered today, how many more days until they will again deliver on the same day?

A. 30

B. 60

C. 90

D. 180

17. If you were doing a survey at a school, what is the best way to find a representative sample of students?

A. Choose a classroom of 30 students.

B. Ask the first 30 students that volunteer.

C. Put all of the students' names in into a hat and then randomly choose 30.

D. Ask the first 30 students that you see in front of the library.

18. Find the area of a rectangle having sides of 8 feet and 7 feet.

A. 15 square feet

B. 49 square feet

C. 56 square feet

D. 64 square feet

19. The table below, posted at the returns desk of a library, displays the fines, *f*, related to the number of days, *d*, a book is overdue. Select the choice that expresses the relationship specified by the table.

d	1	2	3	4	5
f	$0.05	$0.10	$0.15	$0.20	$0.25

A. $f = 0.05d$

B. $f = d + 0.05$

C. $f = 2d + 0.05$

D. $f = 0.05(d + 2)$

---

20. In a game of chance, a wheel is divided into equal sectors that are colored as shown below. If the wheel is spun at random, what is the chance that the arrow indicator will point to a blue sector when the wheel comes to rest?

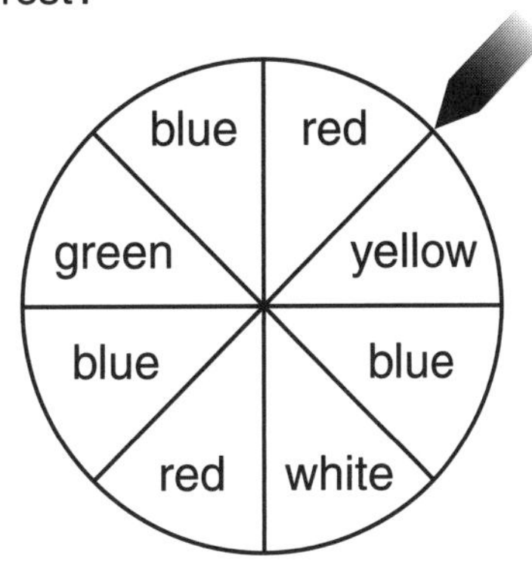

A. 1 out of 3

B. 2 out of 8

C. 3 out of 5

D. 3 out of 8

---

21. If *A* is equal to 2, what does the following expression equal?

$A^3 \times A^2 =$

A. 4

B. 8

C. 16

D. 32

22. A 24-foot telephone pole is held in place by a 30-foot cable, as shown in the figure. Find the distance from the bottom of the pole to the cable's anchor.

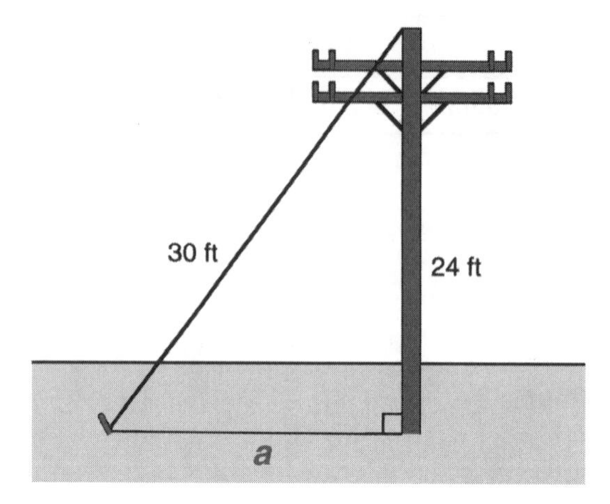

A. 54 ft

B. 27 ft

C. 18 ft

D. 12 ft

---

23. A chair priced at $120.00 costs $25 more than twice the cost of a less expensive chair made by a different manufacturer. Find the cost of the less expensive chair.

A. $190.00

B. $107.50

C. $72.50

D. $47.50

24. A family decides to take a vacation, so they pack their suitcases and set out driving. They leave their home on Saturday at 8:00 p.m. and reach their destination on Monday at 9:00 p.m. How many hours do they travel?

A. 37

B. 49

C. 54

D. 63

25. A gardener has been hired to build a flowerbed, which has an area of 36 square feet.

A. Using whole numbers, list all possible sets of dimensions (width and length) for the flowerbed.

B. How many square inches are there in each of the flowerbeds?

For full credit, you **must** do the following:

1. Show OR describe each step of your work, even if you did it in your head or used a calculator.

AND

2. Write an explanation stating the mathematical reason(s) **why** you chose each of your steps.

IF YOU NEED MORE SPACE, PLEASE USE THE NEXT PAGE

Section 2

26. Which of the following equations shows 240 written as the product of prime factors?

A. $2 \times 2 \times 60$

B. $2 \times 2 \times 2 \times 2 \times 3 \times 5$

C. $2 \times 2 \times 2 \times 2 \times 15$

D. $2 \times 2 \times 2 \times 3 \times 3 \times 3$

---

27. The two figures shown below are congruent. All of the angles equal 90 degrees. What is the length of side L in figure 2?

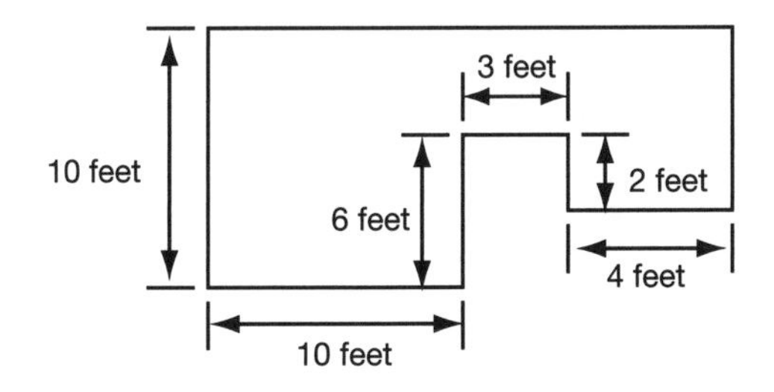

Figure 1

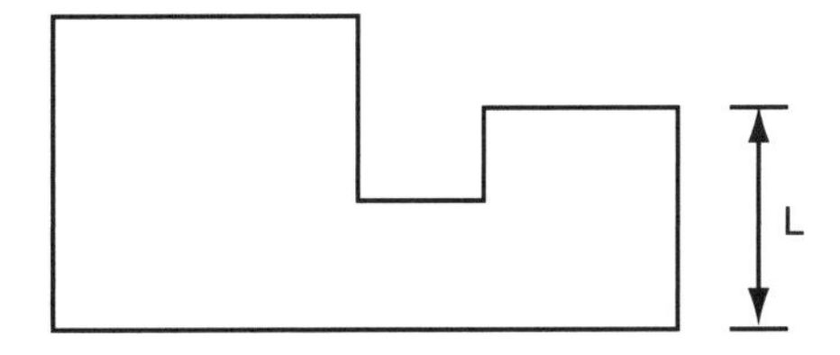

Figure 2

A. 3 feet

B. 4 feet

C. 6 feet

D. 8 feet

---

28. Which of the following equations is equal to $Z = 3X + 3Y$?

A. $Z = 3(X + Y)$

B. $Z = 3(2X + 2Y)$

C. $Z = 3(X + 2Y)$

D. $Z = XY(3 + 3)$

29. If the perimeter of a square is 4 feet, what will the perimeter be if the length of each side is tripled?

A. 8 feet

B. 12 feet

C. 24 feet

D. 36 feet

30. Given triangle ABC, if angle A was made larger and AB and AC stay the same length, what would happen to CB?

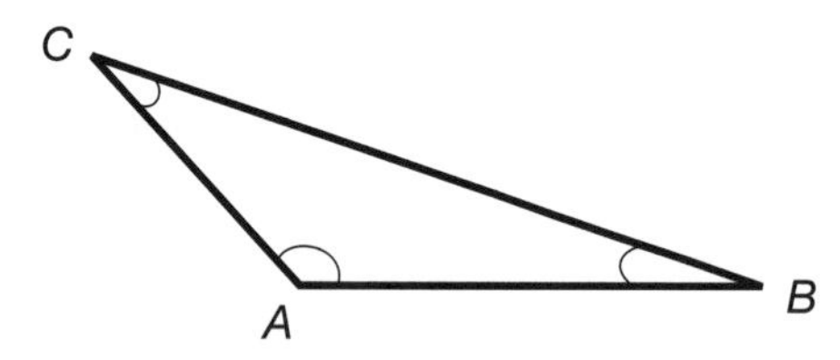

A. CB gets smaller

B. CB gets larger

C. CB stays the same

D. None of the above

31. In the sequence below, find the two end terms.

1, 3, 7, 15, 31, ___, ___.

A. 53, 117

B. 63, 127

C. 73, 137

D. 83, 147

32. Find the value of x from the equation

$5x + 2 = 9x - 4$

A. $\frac{3}{2}$

B. $\frac{2}{3}$

C. $\frac{-2}{3}$

D. $\frac{-3}{2}$

33. $0.685 \div 25 =$

A. 0.0274

B. 0.274

C. 2.74

D. 27.74

34. What is 6% of 78?

A. 0.0468

B. 0.468

C. 4.68

D. 46.8

35. $\frac{15}{3} + 4^3 - \sqrt{81} =$

A. 15

B. 24

C. 38

D. 60

36. What is the length of x in the triangle below?

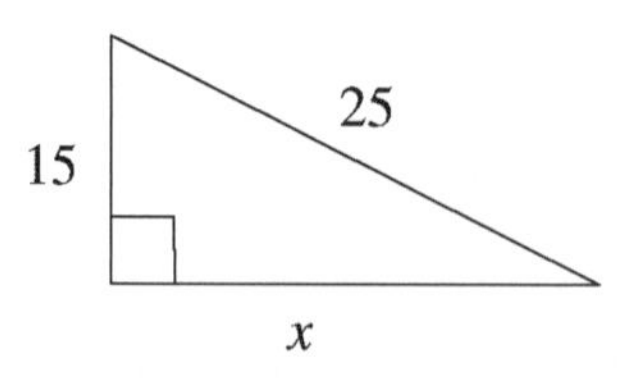

A. 18

B. 20

C. 22

D. 24

37. Which of the following is an irrational number?

A. $\frac{2}{5}$

B. –0.45

C. $\sqrt{9}$

D. $\sqrt{13}$

38. The equation of a line is $y = 3x + 2$. Which of the ordered pairs (x, y) below makes this equation true?

A. (2, 6)

B. (3, 15)

C. (4, 14)

D. (5, 19)

39. The scatter plot below has what type of correlation?

A. Positive

B. Negative

C. No correlation

D. Mixed correlation

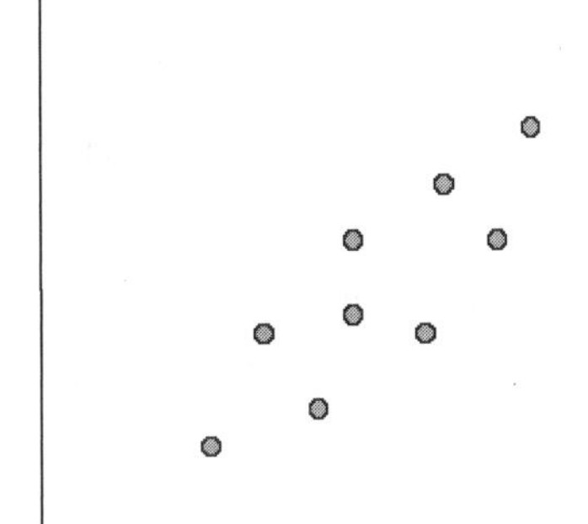

40. What is the equation of the line shown below?

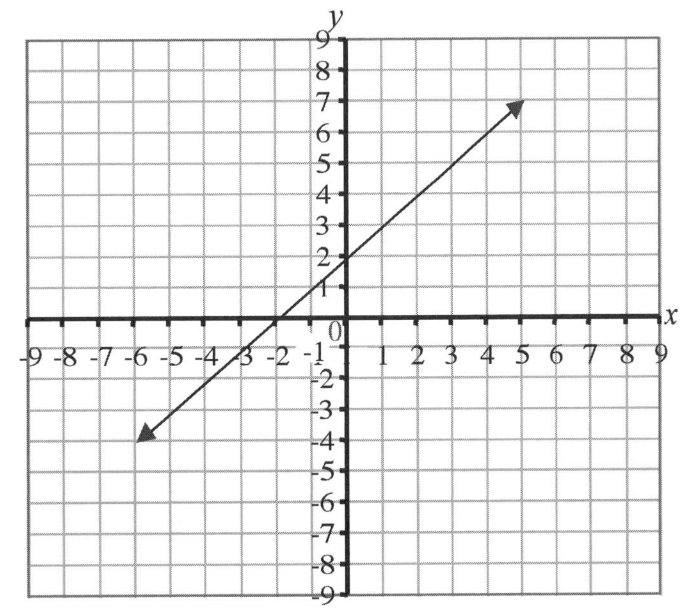

A. $y = 2x + 1$

B. $y = 2x - 1$

C. $y = x - 2$

D. $y = x + 2$

41. The angles of a triangle sum to which number?

A. 45

B. 90

C. 120

D. 180

42. An architect is designing a parking lot for a shopping center. The plan is to build the parking lot 300 feet wide and 450 feet long. A scaled drawing is made in which 1 inch represents 75 feet. What are the dimensions of the scaled drawing?

A. 4 inches × 6 inches

B. 6 inches × 8 inches

C. 8 inches × 10 inches

D. 10 inches × 12 inches

43. A group of 4 students has an average age of 12. When the teacher joins the group, the average age is 18. How old is the teacher?

A. 24

B. 36

C. 42

D. 48

44. Which of the following numbers represents 7.2×10^3 written in standard notation?

A. 0.0072

B. 0.072

C. 720

D. 7,200

45. Given $y = x^2 - x + 5$, which of the following values of x would make y the largest?

A. –4

B. –1

C. 1

D. 4

46. Using the data in the table below, predict the total number of votes that Williams would receive from 550 voters at a school election.

Sample Voting

Williams	64 votes
Becker	36 votes
total	**100 votes**

A. 276

B. 352

C. 380

D. 418

47. If x represents an even integer, which of the following represents the next three consecutive even integers?

A. $x, x + 1, x + 2$

B. $x + 2, x + 4, x + 6$

C. $x + 1, x + 2, x + 3$

D. $2x, 3x, 4x$

48. The figure below shows two parallel lines cut by a transversal. Which of the following are alternate interior angles?

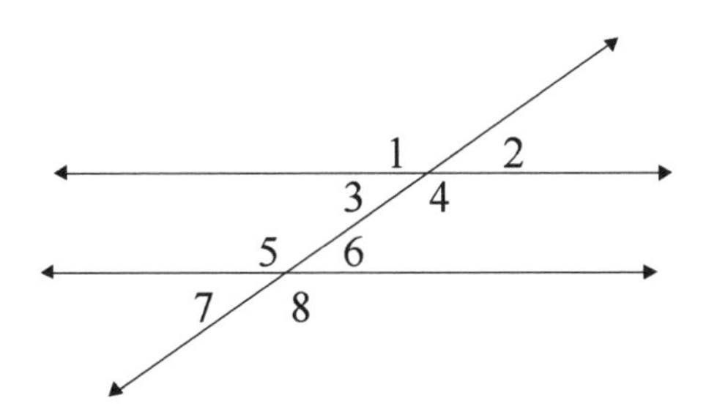

A. angle 1 and angle 2

B. angle 5 and angle 3

C. angle 3 and angle 6

D. angle 7 and angle 8

49. Find the measurement, in degrees, of the largest angle in this triangle.

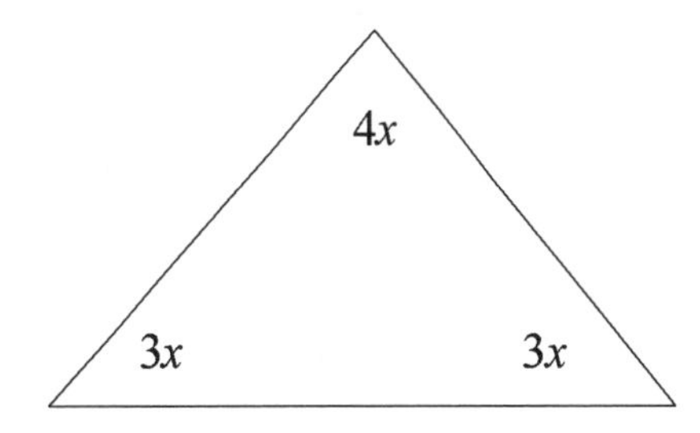

A. 42°

B. 58°

C. 65°

D. 72°

50. One telephone company (Company A) charges its customers a monthly base fee of $20, plus 3 cents a minute. Another telephone company (Company B) charges its customers a monthly base fee of $40, plus 2 cents a minute for total calls over 800 minutes. If a person talks for 950 minutes during the month, which telephone company would cost less **and** how much less would it cost?

For full credit, you **must** do the following:

1. Show OR describe each step of your work, even if you did it in your head or used a calculator.

AND

2. Write an explanation stating the mathematical reason(s) **why** you chose each of your steps.

IF YOU NEED MORE SPACE, PLEASE USE THE NEXT PAGE

Section 3

51. A fruit punch recipe calls for 2 quarts of pineapple juice and serves 8 people. How many quarts of pineapple juice will be needed to serve 20 people?

A. 3

B. 4

C. 5

D. 6

52. The stem-and-leaf plot below presents Mary's History test scores. What is the median score?

Mary's History Test Scores

Stem	Leaf
6	1
7	4 7
8	2
9	2 3 6 8 9

A. 82

B. 92

C. 93

D. 96

53. The scatter plot below shows the relationship between grade level and hours of reading each week. Which of the following statements describes this relationship?

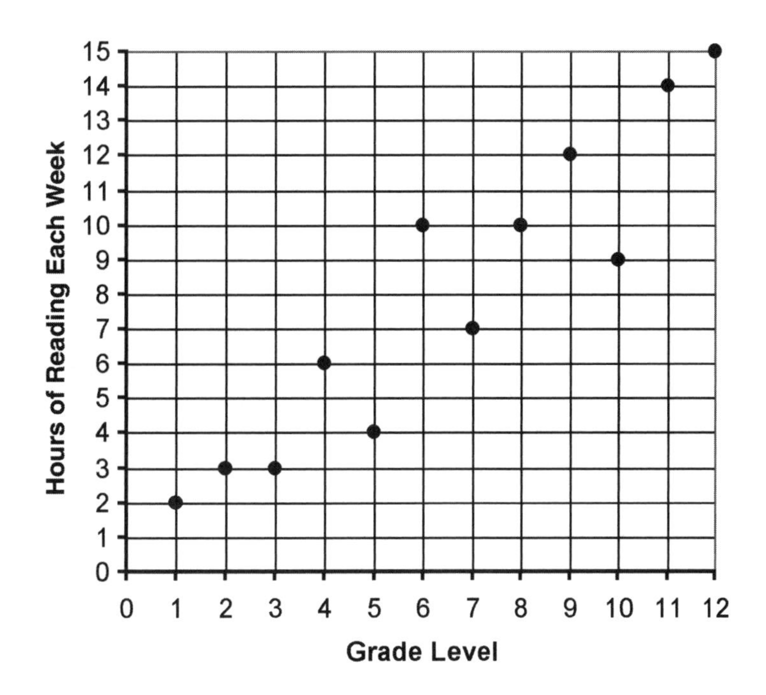

A. As grade level goes down, the number of reading hours goes up.

B. As grade level goes up, the number of reading hours goes down.

C. As grade level goes up, the number of reading hours goes up.

D. Grade level and reading hours are unrelated.

54. If angle *A* equals 37 degrees, what is the measurement of angle *B*?

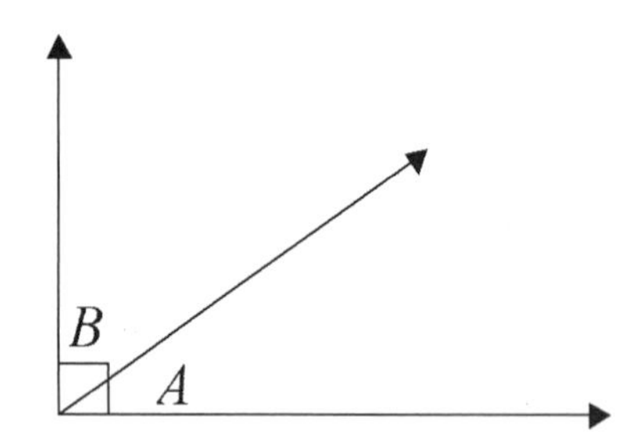

A. 45 degrees

B. 53 degrees

C. 70 degrees

D. 143 degrees

55. What is the value of the following expression?

$5^2 - 32 + 10^4$

A. 993

B. 1,057

C. 9,993

D. 10,057

56. The cost of cable TV service rose from $25.00 to $33.00 per month. What percent increase is this?

A. 12%

B. 17%

C. 28%

D. 32%

57. For the set of data shown below, which measurement (mean, median, mode) has the least value?

21, 30, 26, 31, 13,
21, 10, 21, 26, 31

A. mean

B. median

C. mode

D. all are equal

58. What is the equation of the line shown below?

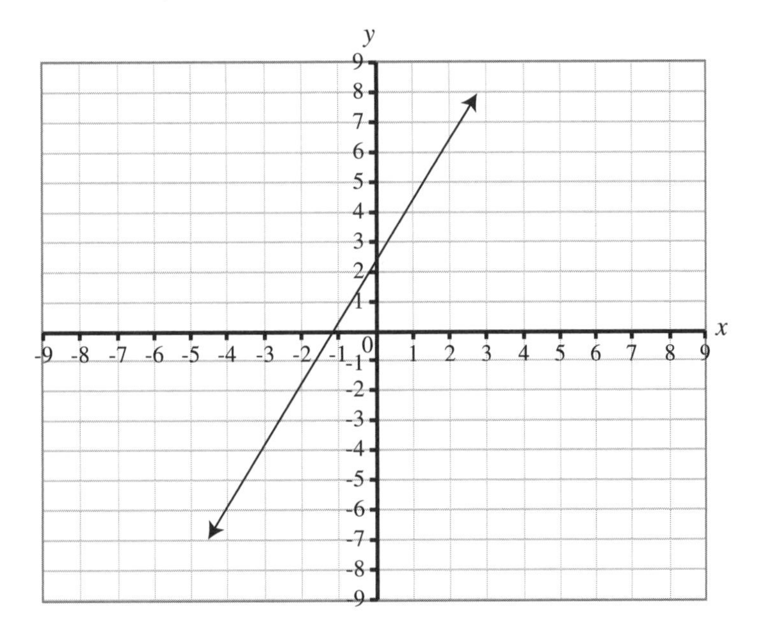

A. $y = 2x + 2$

B. $y = 2x - 2$

C. $y = (1/2)x - 2$

D. $y = (1/2)x + 2$

---

59. Janice charges $10-per-hour for house cleaning, plus $7 to walk the dog. What formula would be used to calculate the total dollar cost C if she worked x hours and walked the dog once?

A. $C = 10x + 7$

B. $C = 7x + 10$

C. $C = x(7 + 10)$

D. $C = 10 + 7 + x$

60. The figure below shows two parallel lines cut by a transversal. If angle 8 is 120 degrees, what is the measurement of angle 1 in degrees?

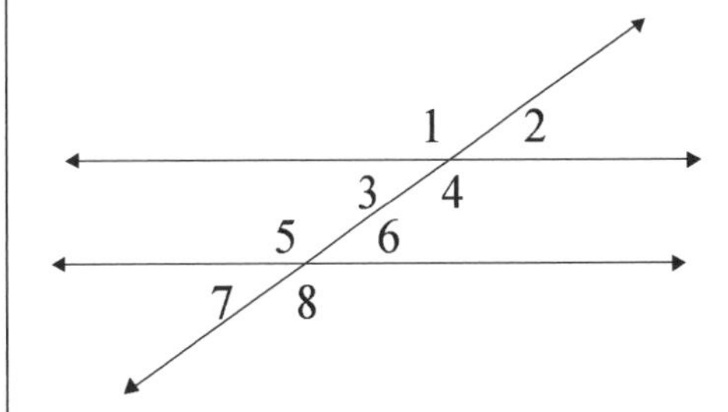

A. 80°

B. 90°

C. 100°

D. 120°

61. List the numbers shown below from greatest to least.

$\frac{1}{6}, \sqrt{6}, 6, 0.6$

A. $6, \sqrt{6}, 0.6, \frac{1}{6}$

B. $0.6, \frac{1}{6}, 6, \sqrt{6}$

C. $\frac{1}{6}, 6, \sqrt{6}, 0.6$

D. $\sqrt{6}, 6, 0.6, \frac{1}{6}$

62. The product of an odd number and an odd number is _________.

A. always odd

B. always even

C. sometimes odd and sometimes even

D. always a prime number

---

63. Given two spin wheels (shown below), which may be spun once and the results added together, determine all the sets that are possible.

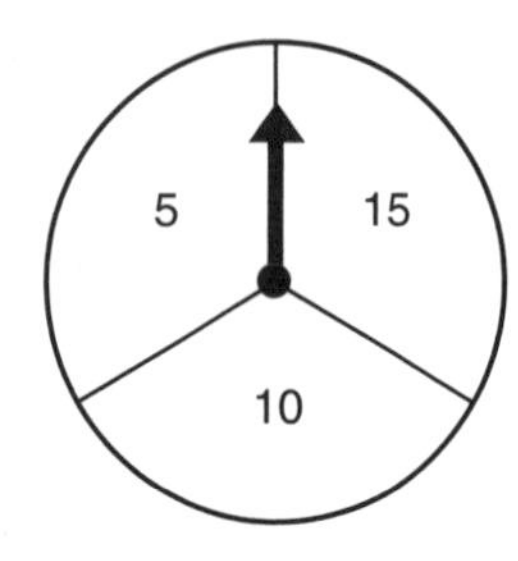

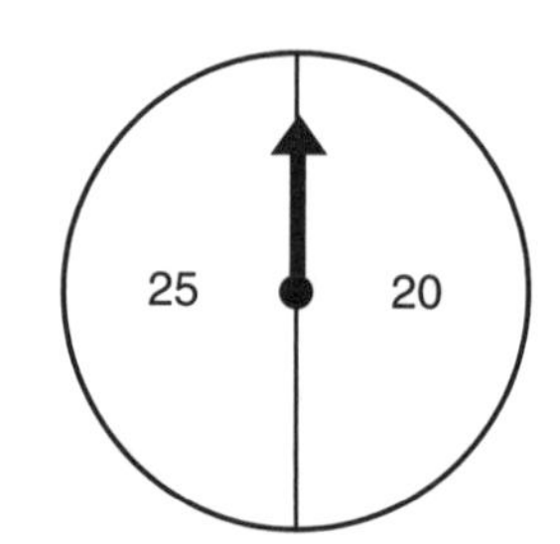

A. 25, 30, 35, 40

B. 15, 20, 25, 30, 35, 40

C. 20, 25, 30, 35, 40, 45

D. 10, 15, 20, 25, 30, 35, 40, 45, 50

64. In the diagram below, if the distance between city A and city B is 160 miles, and the distance between city B and city C is 120 miles, find the distance between cities A and C.

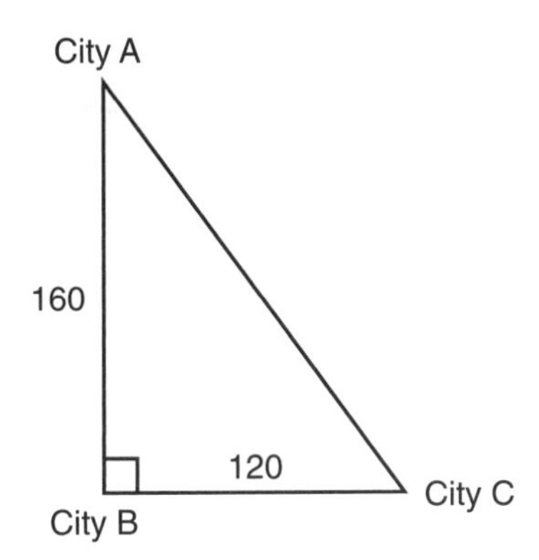

A. 80 miles

B. 140 miles

C. 200 miles

D. 280 miles

65. Which two angles are complementary?

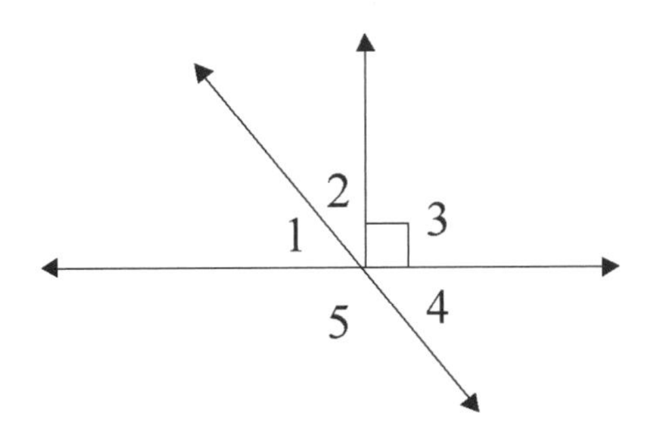

A. ∠ 5 and ∠ 4

B. ∠ 1 and ∠ 2

C. ∠ 3 and ∠ 4

D. ∠ 5 and ∠ 3

66. In which set of data is the mean, median, and mode equal to 6?

A. { 6, 4, 5, 5, 9, 6, 5 }

B. { 6, 4, 5, 6, 10, 6, 5 }

C. { 4, 5, 3, 5, 10, 9, 6 }

D. { 2, 4, 4, 12, 6, 6, 6 }

67. Evaluate 5^5.

A. 25

B. 125

C. 625

D. 3125

68. In the triangle below, if angle *A* is 40° and angle *B* is 32°, what is angle *C*?

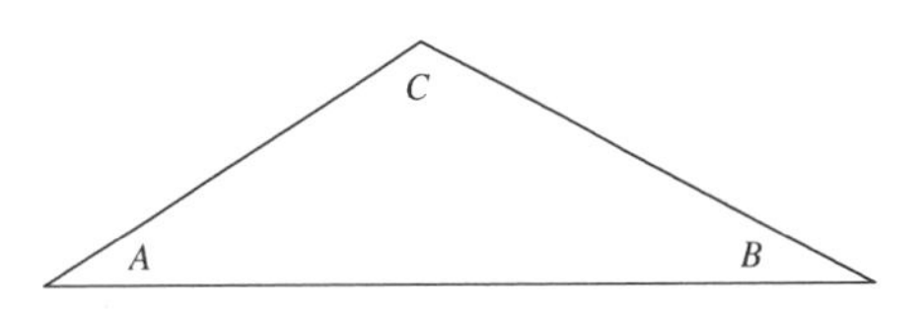

A. 18°

B. 32°

C. 108°

D. 142°

69. A brother and sister have \$21 between them, though the brother has \$9 less than his sister. Which of the following equations represents this relationship, if x represents the sister's money.

A. $(x + 9) + x = 21$

B. $(x - 9) + x = 21$

C. $x - 9 = 21 + x$

D. $x = 21 + x - 9$

70. Sammy caught 7 fish in 2 hours. At this rate, how long will it take him to catch 28 fish?

A. 4 hours

B. 6 hours

C. 8 hours

D. 16 hours

71. A runner along a track starts slowly and increases his speed gradually to the finish. Which graph below illustrates this process?

A.
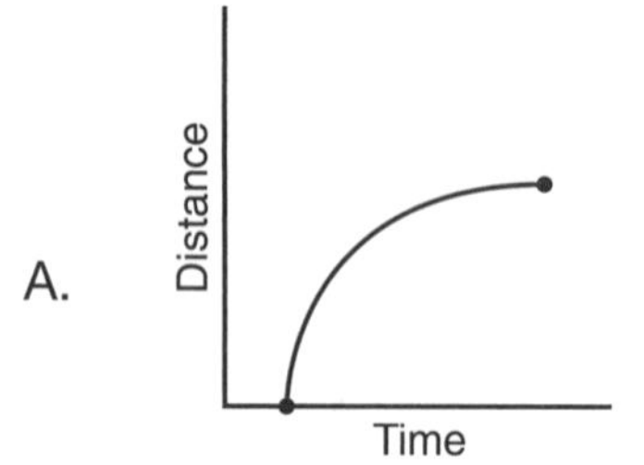

B.
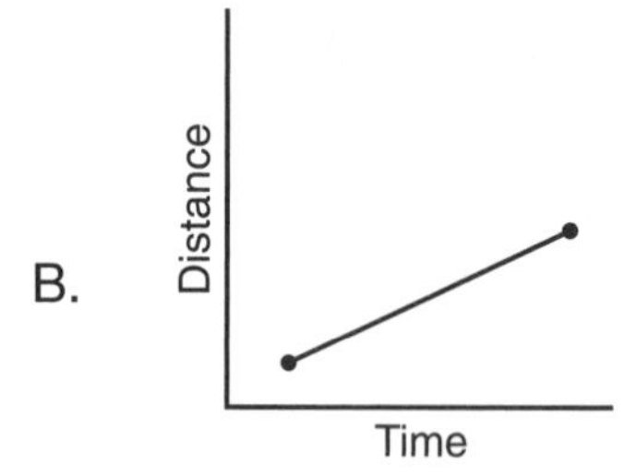

C.
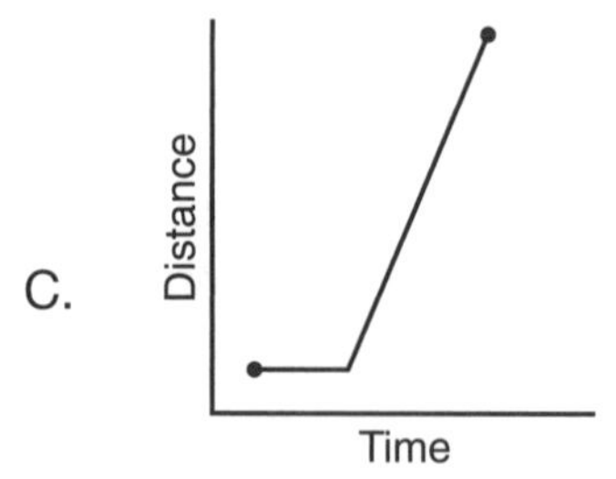

D.
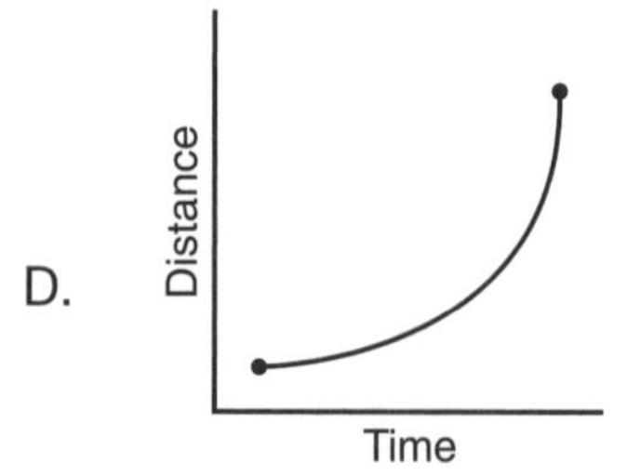

72. What is the radius of this circle?

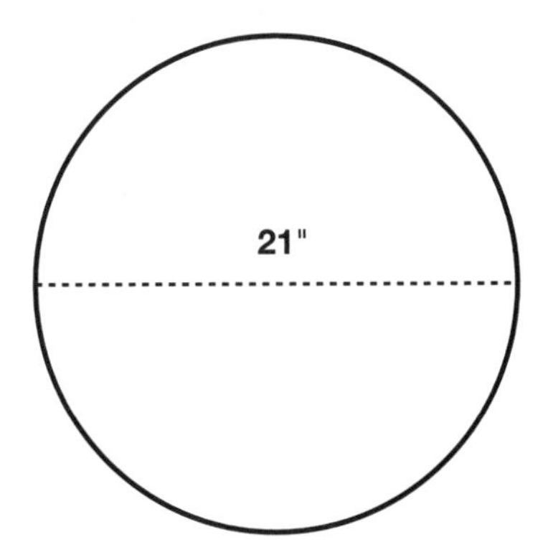

A. 5.25"

B. 42"

C. 21"

D. 10.5"

73. The distance between Earth and Mars is about 4.8732×10^7 miles when expressed in scientific notation. Express this distance in standard form.

A. 4,873,200 miles

B. 48,732,000 miles

C. 487,320,000 miles

D. 4,873,200,000 miles

74. For which equation is the coordinate (6, 4) true?

A. $y + x = 9$

B. $y - x = 2$

C. $x - y = 2$

D. $2x + y = 14$

75. A nursery landscape architect is designing a lawn for a homeowner. The lawn area is to be 27 feet by 30 feet. What is the area of the lawn in square feet? If sod grass comes in one-square-yard pieces, how many pieces of sod will this lawn need?

For full credit, you **must** do the following:

1. Show OR describe each step of your work, even if you did it in your head or used a calculator.

AND

2. Write an explanation stating the mathematical reason(s) **why** you chose each of your steps.

IF YOU NEED MORE SPACE, PLEASE USE THE NEXT PAGE

ANSWER KEY

Section 1		Section 2		Section 3	
1.	(B)	26.	(B)	51.	(C)
2.	(D)	27.	(C)	52.	(B)
3.	(D)	28.	(A)	53.	(C)
4.	(A)	29.	(B)	54.	(B)
5.	(C)	30.	(B)	55.	(C)
6.	(C)	31.	(B)	56.	(D)
7.	(D)	32.	(A)	57.	(C)
8.	(B)	33.	(A)	58.	(A)
9.	(D)	34.	(C)	59.	(A)
10.	(A)	35.	(D)	60.	(D)
11.	(D)	36.	(B)	61.	(A)
12.	(B)	37.	(D)	62.	(A)
13.	(A)	38.	(C)	63.	(A)
14.	(B)	39.	(A)	64.	(C)
15.	(B)	40.	(D)	65.	(B)
16.	(B)	41.	(D)	66.	(B)
17.	(C)	42.	(A)	67.	(D)
18.	(C)	43.	(C)	68.	(C)
19.	(A)	44.	(D)	69.	(B)
20.	(D)	45.	(A)	70.	(C)
21.	(D)	46.	(B)	71.	(D)
22.	(C)	47.	(B)	72.	(D)
23.	(D)	48.	(C)	73.	(B)
24.	(B)	49.	(D)	74.	(C)
25.	See page 375	50.	See page 378	75.	See page 380

DETAILED EXPLANATIONS OF ANSWERS

Section 1

1. **(B)** $492 \times 847 = 416{,}724$.

2. **(D)** $17\frac{3}{5} - 8\frac{3}{5} = 9$. Three-fifths minus three-fifths equals 0. Then, $17 - 8 = 9$.

3. **(D)** $4 \div 0.16 = 25$

4. **(A)** $14 - 4\frac{5}{8} = 13\frac{8}{8} - 4\frac{5}{8} = 9\frac{3}{8}$

5. **(C)** $5{,}400 \div 72 = 75$ minutes = 1 hour 15 minutes.

6. **(C)** In a pie chart, the percentages must add up to 100%. In order to find out the percentage of A's, add up the percentages of all other grades and then subtract from 100%. 25% + 30% + 15% + 10% = 80%. 100% – 80% = 20%.

7. **(D)** In order to solve an equation, you add or subtract the same amount from both sides of an equation. The idea is to get the variable, in this case x, by itself. Given the equation $4x + 9 = 31$, the first step is to subtract 9 from both sides. Therefore, $4x + 9 - 9 = 31 - 9$ is correct.

8. **(B)** The salesman had $160,000 in sales. For the first $100,000 the salesman earns $100,000 × 0.03 = $3,000. For sales over $100,000, which is $160,000 – $100,000 = $60,000, the salesman earns $60,000 × 0.04 = $2,400. Adding the amounts together, the salesman earned $3,000 + $2,400 = $5,400 in total.

9. **(D)** In this stem-and-leaf plot, the stem represents the tens place and the leaf represents the number in the ones place. According to this plot, the scores in the 70's are: 70, 71, 72, 72, 75, 77, 78, and 79. Therefore, 8 students scored in the 70's.

10. **(A)** There are a total of 10 cards, of which 4 cards have stars. The probability of drawing a card with a star, therefore, is $\frac{4}{10} = \frac{2}{5}$.

11. **(D)** The triangle *UVW* has angles that must add up to 180°. Therefore, the missing angle at *W* is 180 – 72 – 70 = 38°. This is the same angle between *WX* and *WY*. Since triangle *WYX* must also have angles totaling 180°, angle *X* = 180 – 38 – 45 = 97°.

12. **(B)** Referring to the table, Field Goals Made is 8, and Field Goals Attempted is 12. The ratio becomes, therefore, $\frac{8}{12}$, which can be reduced to $\frac{2}{3}$.

13. **(A)** The pattern 2, 4, 8,... is best described by the equation

$n^1, n^2, n^3...$ where $n = 2$

14. **(B)** There are 36 inches in one yard. 180 ÷ 36 = 5 yards.

15. **(B)** Given the equation $4X + 3Y$, substitute 5 for X and –3 for Y.

$4(5) + 3(-3) = 20 - 9 = 11$

16. **(B)** This question wants you to find the smallest number that both 12 and 15 can divide into evenly. One way to solve this problem is to multiply each number times 1, then 2, then 3, and so on until a common number is found.

12, 24, 36, 48, 60, 72 15, 30, 45, 60, 75, 90

The number 60 is the smallest number that both 12 and 15 can be divided into evenly.

17. **(C)** A random sample is the best way to get a representative sample from a population. Therefore, putting all of the students' names into a hat and then randomly choosing a sampling is the right answer.

18. **(C)** The area of a rectangle equals width times length. 8 × 7 = 56 square feet.

19. **(A)** Examining the table of how the fines are imposed with respect to the number of overdue days, we see that a fine of $0.05 is imposed for each day a book is overdue. Therefore, multiplying $0.05 by the number of overdue days gives the amount of the fine to be paid, i.e., $f = 0.05d$.

20. **(D)** The wheel has 8 sectors. Three of these 8 sectors are colored blue. The probability that the indicator will point to a blue sector when the wheel stops is $\frac{3}{8}$, or 3 out of 8.

21. **(D)** Given $A^3 \times A^2$, and $A = 2$, then $2^3 \times 2^2 = 8 \times 4 = 32$.

22. **(C)** The pole and the cable form a right triangle with the ground, as shown. The problem involves finding one side of a right triangle when the other two sides are known. Therefore, the Pythagorean theorem is applicable:

$$24^2 + a^2 = 30^2;\ 576 + a^2 = 900$$

$$a^2 = 900 - 576 = 324$$

$$a = 18$$

23. **(D)** Let the cost of the less expensive chair be represented by x. Then, twice the cost of this chair is $2x$. Following the wording of the question,

$$25 + 2x = 120$$

from which $x = \frac{120 - 25}{2} = 47.50$.

24. **(B)** Add the number of hours traveled each day, beginning and ending at midnight. On Saturday, they traveled for 4 hours. On Sunday, they traveled for 24 hours. On Monday, they traveled for 21 hours. $4 + 24 + 21 = 49$ hours.

25. A. The possible dimensions are: 6×6, 4×9, 3×12, 2×18, and 1×36.

B. There are 144 square inches in one square foot because 12×12 equals 144. The number of square inches in 36 square feet equals $36 \times 144 = 5{,}184$ square inches.

Section 2

26. **(B)** A prime number is a number that can only be divided by one and itself.

$$2 \times 2 \times 2 \times 2 \times 3 \times 5 = 240$$

27. **(C)** First, find the side of figure 1 that is congruent to side L on figure 2. Then, calculate the length of side L by starting with the measurement of the side opposite it, which is 10. Next, working towards the unknown side, subtract 6 from 10, which equals 4. Lastly, add 2 to the 4, which equals 6. The length of side L is 6 feet.

28. **(A)** The equation $Z = 3X + 3Y$ is equal to $Z = 3(X + Y)$ because the number 3 is factored out from both X and Y.

29. **(B)** If the length of each side of a square were tripled, the perimeter would be tripled. In this case, $3 \times 4 = 12$ feet.

30. **(B)** If angle A was made larger and AB and AC stayed the same length, CB would get larger.

31. **(B)** Look at the first three terms to see what they have in common. In the second term, 3 can be obtained by multiplying by 3, but that would not fit for the third term. The second term can also be obtained by multiplying the first term by 2 and adding one. This procedure would also be applicable to the third term and the other terms thereafter. Therefore, applying this procedure to the term 31 gives 63, and multiplying 63 by 2 and adding 1 results in 127.

32. **(A)** Collecting the terms expressed in x on one side of the equation and the numerical amounts on the other, we obtain

$$-9x + 5x = -4 - 2$$

$$-4x = -6$$

$$x = \frac{-6}{-4} = \frac{3}{2}$$

33. **(A)** $0.685 \div 25 = 0.0274$

34. **(C)** 6% of 78 is 4.68

35. **(D)** $\frac{15}{3} + 4^3 - \sqrt{81} = 5 + 64 - 9 = 60$

36. **(B)** Use the formula $A^2 + B^2 = C^2$.

$$15^2 + B^2 = 25^2$$

$$225 + B^2 = 625$$

$$B^2 = 400$$

$$B = \sqrt{400}$$

$$B = 20$$

37. **(D)** An irrational number is a number that cannot be represented as a fraction. It is non-ending and non-repeating. The square root of 13 is an irrational number.

38. **(C)** The coordinate (4, 14) makes the equation $y = 3x + 2$ true because $14 = 3(4) + 2$.

39. **(A)** If you drew an imaginary line through the data points, the slope of the line would be such that as the value of one measurement increases, the value of the other measurement also increases. This is known as a positive correlation.

40. **(D)** The line has a slope of 1 and a y-intercept of 2. A slope of 1 means that for each 1-unit change in x, there is a 1-unit change in y. A y-intercept of 2 means that the line crosses the y-axis at $y = 2$. Therefore, using the slope-intercept form of the equation of a line—$y = \mathrm{m}x + \mathrm{b}$, where m is the slope and b is the y-intercept—the line has the equation $y = x + 2$.

41. **(D)** Added together, the angles of a triangle equal 180 degrees.

42. **(A)** We know that a length of 75 is represented by 1 inch in the scaled drawing. The width of the parking lot will be 300 feet. Because 75 divides evenly into 300 four times, we know that the scale representation will be 4 inches wide. Using the same reasoning—450 is divided by 75 six times—the length is 6 inches. Therefore, the scaled drawing should be 4 inches by 6 inches.

43. **(C)** Since the average age of the 4 boys is 12, the total years alive is $4 \times 12 = 48$. When the teacher joins the group, the average age of the 5 people is 18, so the total years alive is $5 \times 18 = 90$. To find the teacher's age, subtract:

$$90 - 48 = 42 \text{ years old}$$

44. **(D)** $7.2 \times 10^3 = 7.2 \times 1000 = 7{,}200$

45. **(A)** By substituting the possible answers into the equation, we find that y is largest when –4 is substituted for x.

$$y = x^2 - x + 5$$

$$y = (-4)^2 - (-4) + 5$$

$$y = 16 + 4 + 5 = 25$$

46. **(B)** First, find the proportion of the sample votes for Williams. $64 \div 100 = 0.64$. Next, multiply the 550 voters by this proportion. $550 \times 0.64 = 352$ votes.

47. **(B)** If x represents an even integer, the terms $x + 2$, $x + 4$, $x + 6$ represent the next consecutive even integers.

48. **(C)** Alternate interior angles are angles that lie between the parallel lines but are on opposite sides of the transversal that cuts the parallel lines. Angle 3 and angle 6 are alternate interior angles.

49. **(D)** The angles of a triangle added together equal 180 degrees. Therefore, $4x + 3x + 3x = 180$, and $10x = 180$. $x = 18$. If x equals 18, then the largest angle is equal to $4(18) = 72$ degrees.

50. Company A has a $20.00 monthly fee, plus a charge of $0.03 \times 950 = \$28.50$ for the minutes. Company A costs $\$20.00 + \$28.50 = \$48.50$.

Company B has a $40.00 monthly fee, plus 2 cents a minute for all minutes over 800. $950 - 800 = 150$. $150 \times 0.02 = \$3.00$. Company B costs $\$40.00 + \$3.00 = \$43.00$.

Company B costs less. The difference in cost is $\$48.50 - \$43.00 = \$5.50$.

Section 3

51. **(C)** First, find out how much pineapple juice is needed for each person. $2 \div 8 = 0.25$ quarts per person. Next, multiply this amount by 20 people.

$0.25 \times 20 = 5$ quarts

52. **(B)** In this stem-and-leaf plot, the stems represent the tens place and the leaves represent the ones place. This plot represents the numbers 61, 74, 77, 82, 92, 93, 96, 98, and 99. The median score is the score in the middle of this ranked list. Since there are 9 scores, the score in the middle is the fifth score. The median score is 92.

53. **(C)** Draw an imaginary line through the scatter-plot. The slope of the line is positive. A positive slope implies that as one measurement increases the other measurement also increases. This plot shows that as grade level goes up, the number of reading hours goes up.

54. **(B)** Angles A and B are complementary. Thus, they must add to 90 degrees.

$90 - 37 = 53$ degrees

55. **(C)** $5^2 - 32 + 10^4 = 25 - 32 + 10{,}000 = -7 + 10{,}000 = 9{,}993$

56. **(D)** First, find the dollar increase by subtracting the old cost from the new. $\$33 - \$25 = \$8$. Then, find the percent increase by dividing the dollar increase by the old cost. $8 \div 25 = 0.32 = 32\%$.

57. **(C)** The mean is the sum of the scores divided by the number of scores. $230 \div 10 = 23$. The median score is found by ordering the set of scores by rank, and finding the middle score. In this case, 23.5 is the median score. The mode is the most frequently occurring score. The mode score is 21. The mode has the least value.

58. **(A)** The line has a slope of 2 and a y-intercept of 2. A slope of 2 means that for each 1-unit change in x, there is a 2 unit change in y. A y-intercept of 2 means the line crosses the y-axis at $y = 2$. Therefore, using the slope-intercept form of the equation of a line—$y = mx + b$, where m is the slope and b is the y-intercept—the line has the equation $y = 2x + 2$.

59. **(A)** Janice gets paid $10 an hour for cleaning plus $7 to walk the dog. The equation to calculate cost C should multiply the dollars per hour times the number of hours worked x, then add the amount earned for walking the dog once. $C = 10x + 7$ is the answer.

60. **(D)** Angles 1 and 8 are alternate exterior angles. Alternate exterior angles are congruent. Since the measurement of angle 8 is 120 degrees, the measurement of angle 1 is also 120 degrees.

61. **(A)** The sequence 6, $\sqrt{6}$, 0.6, $\frac{1}{6}$ lists the numbers from greatest to least. One way to solve this problem is to convert the square root of 6 into a decimal form and to do this to the fraction 1/6 as well. Then compare all four numbers in decimal form.

62. **(A)** The product of an odd number and an odd number is always odd.

63. **(A)** If the right wheel points to the field 25, and the left wheel points along the line (zero) as shown in the drawing, then that set is 25. If the left wheel were to point instead to the field 5, the set would be 30. If the left wheel were to point to the field 10, the set would be 35. If the left wheel were to point to field 15, the set would be 40.

If, now, the right wheel were to point to the field 20, the set 20 can result, and this would fit choice (B). But 15 in this choice cannot be obtained. Similarly, sets 45 and 50 in choices (C) and (D) cannot be obtained. Therefore, choice (A) is the only possible answer.

64. **(C)** As seen from the diagram, the distance AC is the hypotenuse of a right triangle. We solve by using the Pythagorean Theorem.

$$160^2 + 120^2 = (AC)^2$$

$$25{,}600 + 14{,}400 = (AC)^2$$

$$40{,}000 = (AC)^2$$

$$200 = AC$$

65. **(B)** Complementary angles are two angles that add together to equal 90 degrees. The figure shows that angle 3 is a right angle, which means that it is equal to 90 degrees. Therefore, angle 1 and angle 2 added together must also be equal to 90 degrees. Angles 1 and 2 are complementary.

66. **(B)** The data set { 6, 4, 5, 6, 10, 6, 5 } has a mean, median, and mode equal to 6.

67. **(D)** $5^5 = 5 \times 5 \times 5 \times 5 \times 5 = 3{,}125$

68. **(C)** First, add the measurements of the first two angles. 40 + 32 = 72 degrees. Next, subtract 72 from 180 degrees (because there are 180 degrees in a triangle). 180 – 72 = 108 degrees.

69. **(B)** Let x = sister's money. Then, the brother's money can be represented by $(x - 9)$. The sum of the two quantities, $x + (x - 9)$, is equal to 21. Expressed as an equation,

$$(x - 9) + x = 21$$

70. **(C)** First, calculate the number of fish caught per hour. 7 ÷ 2 = 3.5 fish-per-hour. Next, divide the number of fish needed by the fish-per-hour rate. 28 ÷ 3.5 = 8 hours.

71. **(D)** The condition of starting slowly and increasing the speed gradually implies a smooth curve. Graph A does not meet the condition because it illustrates decreasing speed, not increasing speed. Graph B does not meet the condition because it illustrates a constant speed. Graph C does not meet the condition because it illustrates a combination of two constant rates of speed. Graph D meets the required condition because it illustrates an increasing rate of speed.

72. **(D)** The radius of a circle is the distance from the center of a circle to its edge. The distance all the way across a circle is called the diameter. The radius is equal to 1/2 of the diameter. In this circle, the diameter is 21 inches, so the radius is equal to 10.5 inches.

73. **(B)** The decimal point has to be moved seven places to the right, as indicated by the term 107. Therefore, we need to add three zeros after 4.8732 to result in 48,732,000 miles.

74. **(C)** The coordinate (6, 4) is true for the equation $x - y = 2$, because $6 - 4 = 2$.

75. The are 27 × 30 = 810 square feet. There are 9 square feet in one square yard, so there are 90 pieces of sod grass (810 ÷ 9 = 90).

ANSWER SHEET

Record your answers by filling in the corresponding ovals next to the question number.

Section 1	Section 2	Section 3
1. (A) (B) (C) (D)	26. (A) (B) (C) (D)	51. (A) (B) (C) (D)
2. (A) (B) (C) (D)	27. (A) (B) (C) (D)	52. (A) (B) (C) (D)
3. (A) (B) (C) (D)	28. (A) (B) (C) (D)	53. (A) (B) (C) (D)
4. (A) (B) (C) (D)	29. (A) (B) (C) (D)	54. (A) (B) (C) (D)
5. (A) (B) (C) (D)	30. (A) (B) (C) (D)	55. (A) (B) (C) (D)
6. (A) (B) (C) (D)	31. (A) (B) (C) (D)	56. (A) (B) (C) (D)
7. (A) (B) (C) (D)	32. (A) (B) (C) (D)	57. (A) (B) (C) (D)
8. (A) (B) (C) (D)	33. (A) (B) (C) (D)	58. (A) (B) (C) (D)
9. (A) (B) (C) (D)	34. (A) (B) (C) (D)	59. (A) (B) (C) (D)
10. (A) (B) (C) (D)	35. (A) (B) (C) (D)	60. (A) (B) (C) (D)
11. (A) (B) (C) (D)	36. (A) (B) (C) (D)	61. (A) (B) (C) (D)
12. (A) (B) (C) (D)	37. (A) (B) (C) (D)	62. (A) (B) (C) (D)
13. (A) (B) (C) (D)	38. (A) (B) (C) (D)	63. (A) (B) (C) (D)
14. (A) (B) (C) (D)	39. (A) (B) (C) (D)	64. (A) (B) (C) (D)
15. (A) (B) (C) (D)	40. (A) (B) (C) (D)	65. (A) (B) (C) (D)
16. (A) (B) (C) (D)	41. (A) (B) (C) (D)	66. (A) (B) (C) (D)
17. (A) (B) (C) (D)	42. (A) (B) (C) (D)	67. (A) (B) (C) (D)
18. (A) (B) (C) (D)	43. (A) (B) (C) (D)	68. (A) (B) (C) (D)
19. (A) (B) (C) (D)	44. (A) (B) (C) (D)	69. (A) (B) (C) (D)
20. (A) (B) (C) (D)	45. (A) (B) (C) (D)	70. (A) (B) (C) (D)
21. (A) (B) (C) (D)	46. (A) (B) (C) (D)	71. (A) (B) (C) (D)
22. (A) (B) (C) (D)	47. (A) (B) (C) (D)	72. (A) (B) (C) (D)
23. (A) (B) (C) (D)	48. (A) (B) (C) (D)	73. (A) (B) (C) (D)
24. (A) (B) (C) (D)	49. (A) (B) (C) (D)	74. (A) (B) (C) (D)
25. ____________	50. ____________	75. ____________

Index

INDEX

A

B

C

D

E

F

Q

R

S

T

V

W

X

Y

Z

WORKSHEET

WORKSHEET